Chemistry Research and Applications

nova science publishers
www.novapublishers.com

Chemistry Research and Applications

Introduction to Multidisciplinary Science in an Artificial-Intelligence Age: The Matter in our Universe, Biological Cells, and Plate Tectonics
Luc Ikelle
2023. ISBN: 978-1-68507-992-5 (Hardcover)
2023. ISBN: 979-8-88697-498-0 (eBook)

Introduction to Multidisciplinary Science in an Artificial-Intelligence Age: Chemical, Nuclear, and Thermonuclear Reactions, and Oxygenic and Anoxygenic Photosyntheses
Luc Ikelle
2023. ISBN: 979-8-88697-048-7 (Hardcover)
2023. ISBN: 979-8-88697-499-7 (eBook)

What to Know about Lanthanum
Catherine C. Bradley (Editor)
2023. ISBN: 979-8-88697-615-1 (Softcover)
2023. ISBN: 979-8-88697-623-6 (eBook)

The Future of Biorefineries
Waldemar Nyström (Editor)
2023. ISBN: 979-8-88697-524-6 (Hardcover)
2023. ISBN: 979-8-88697-528-4 (eBook)

The Essential Guide to Alkaloids
Deepak Kumar Semwal, PhD (Editor)
2023. ISBN: 979-8-88697-456-0 (Hardcover)
2023. ISBN: 979-8-88697-526-0 (eBook)

Properties and Uses of Antimony
David J. Jenkins (Editor)
2022. ISBN: 979-8-88697-081-4 (Softcover)
2022. ISBN: 979-8-88697-088-3 (eBook)

More information about this series can be found at
https://novapublishers.com/product-category/series/chemistry-research-and-applications/

Thomas W. Gerdes
Editor

A Review of Hydrazine and Its Applications

nova science publishers
www.novapublishers.com

NOTICE TO THE READER

Additional color graphics may be available in the e-book version of this book.

Library of Congress Cataloging-in-Publication Data

Names: Gerdes, Thomas W., editor.
Title: A review of hydrazine and its applications / Thomas W. Gerdes, editor.
Description: New York : Nova Science Publishers, [2023] | Series: Chemistry research and applications | Includes bibliographical references and index. |
Identifiers: LCCN 2023007235 (print) | LCCN 2023007236 (ebook) | ISBN 9798886976717 (hardcover) | ISBN 9798886976786 (adobe pdf)
Subjects: LCSH: Hydrazine. | Hydrazines. | Bioactive compounds--Therapeutic use.
Classification: LCC TP248.H85 R48 2023 (print) | LCC TP248.H85 (ebook) | DDC 660/.042--dc23/eng/20230302
LC record available at https://lccn.loc.gov/2023007235
LC ebook record available at https://lccn.loc.gov/2023007236

Published by Nova Science Publishers, Inc. † New York

Contents

Preface

This book contains three chapters reviewing hydrazine and its applications. Chapter One overviews derivatives of hydrazine as massive sources of pharmaceutically bioactive drugs. Chapter Two looks to hydrazine itself as a source of pharmaceutically bioactive drugs. Chapter Three explores the use of carbon based nanomaterials for the electrochemical determination of hydrazine.

Chapter 1 - Heterocyclic compounds derived from hydrazine (pyrazole, pyridazine, etc.) are vital in drug design and pharmaceutical research. Many structurally diverse heterocyclic derivatives (pyrazole, pyridazine, and other miscellaneous heterocyclic compounds containing N-N linkage) possess interesting therapeutical properties, including sedative, antidepressant, non-steroidal anti-inflammatory drugs, anticancer, uricosuric drugs, steroids, anorectic, alcohol dehydrogenase inhibitory, antipsychotic, antibiotic, anticoagulant, antibacterial, antifungal, antihistaminic, vasodilator, and many more to cite. More than 65 drugs consisting of heterocycles derived from hydrazine cater to the needs of many patients suffering from numerous diseases and disorders. In addition to the established convenient chemistry, medicinal chemists get enticed to claim these pharmaceutically attractive moieties in their potential bioactive compounds. This marching of drugs from this heterocyclic category containing *N-N* linkage towards the century mark would be achieved sooner as many newer bioactive molecules claim to have these moieties in their structure. This review provides a comprehensive analysis of drugs consisting of heterocyclic nucleuses consisting of *N-N* linkage.

Chapter 2 - Hydrazine-containing drugs are of immense importance in drug design and pharmaceutical research. Many structurally diverse hydrazine analogs such as isoniazid, nitrofurazone, nifurtimox, procarbazine, selinexor, hydralazine, rimonabant, benserazide, phenelzine, isocarboxazid, carbidopa, cadralazine, and clopamide show a landscape of fascinating therapeutic properties including antitubercular, antimicrobial, anticancer, vasodilator, antidepressant, anti-obesity, anti-parkinsonian and many more to mention. Due to the importance of the hydrazine moiety, many scientists have synthesized compounds comprising *N–N* bond and evaluated their therapeutic activity.

It would be interesting to note that hydrazine is a highly reactive nucleophilic reagent, which reacts with different functional groups, including ketones, aldehydes, carboxylic acids, acyl halides, alkyl halides/other electrophilic alkylating reagents and many other to ultimately generate an array of functional groups such as hydrazones, hydrazines, acyl hydrazides, semicarbazides, etc. Additionally, hydrazines is a very convenient reducing agent. The combination of suitable, convenient, and reactive organic reagents as well as the presence of it in many pharmaceutically active drugs, enable it to be widely used moiety in the drug design and development processes. The authors have reviewed its role in the pharmaceutical industry in the development of various drugs.

Chapter 3 - Nanomaterials have been widely explored as an amenable sensing platform to further boost the electrochemical kinetics with diminution of overpotential. Besides, carbon-based nanomaterials have gained substantial amenities of being stable, economical, chemically inert which will nurture significant properties like relatively wide potential window with diminished background current made them as an appropriate electrocatalyst for sensing applications. In addition, it has a promising conductivity with higher surface area and easy to functionalize. Herein, unique configuration of carbonaceous materials and their composites are explored for assaying the detection of hydrazine (HZ) owing to the possibility of quantitative measurement with high precision, sensitivity and reproducibility, which are necessary for accurate detection in water and environmental samples. In this context, the authors mainly focused on overview of carbon-based nanomaterials for the detection of toxic HZ in a worthy manner. Additionally, carbon-based materials with surface modifications were summarized along with the inclusion of carbon nanotubes, carbon quantum dots, graphene, ordered mesoporous carbon (OMC), metal organic framework (MOF) derived carbon, carbon nanofibers and carbon black and their optimized sensing parameters are outlined. This chapter mainly discusses the advantages and limitations of existing electrode alterations and also providing a futuristic idea to improve the detection strategy for HZ assay and covers the in-depth mechanism of hydrazine oxidation in water and other environmental samples.

Chapter 1

Hydrazine: A Source of Pharmaceutically Bioactive Drugs

Shubhangi S. Savale[1]
Vijayendran K. K. Praneeth[2]
Ganesh U. Chaturbhuj[1,*]
and Pravin R. Bhansali[3,†]

[1] Department of Pharmaceutical Sciences and Technology,
Institute of Chemical Technology, Mumbai, Maharashtra, India
[2] Department of Chemistry, School of Advanced Sciences,
Kalasalingam Academy of Research and Education (Deemed to be University),
Krishnankoil, Tamilnadu, India
[3] Department of Chemistry, Faculty of Science and Technology,
Alliance University, Bengaluru, Karnataka, India

Abstract

Hydrazine-containing drugs are of immense importance in drug design and pharmaceutical research. Many structurally diverse hydrazine analogs such as isoniazid, nitrofurazone, nifurtimox, procarbazine, selinexor, hydralazine, rimonabant, benserazide, phenelzine, isocarboxazid, carbidopa, cadralazine, and clopamide show a landscape of fascinating therapeutic properties including antitubercular, antimicrobial, anticancer, vasodilator, antidepressant, antiobesity, anti-parkinsonian and many more to mention. Due to the importance of the hydrazine moiety, many scientists have synthesized compounds comprising *N–N* bond and evaluated their therapeutic activity.

* Corresponding Author's Email: gu.chaturbhuj@ictmumbai.edu.in.
† Corresponding Author's Email: pravin.bhansali@alliance.edu.in.

In: A Review of Hydrazine and Its Applications
Editor: Thomas W. Gerdes
ISBN: 979-8-88697-671-7

It would be interesting to note that hydrazine is a highly reactive nucleophilic reagent, which reacts with different functional groups, including ketones, aldehydes, carboxylic acids, acyl halides, alkyl halides/other electrophilic alkylating reagents and many other to ultimately generate an array of functional groups such as hydrazones, hydrazines, acyl hydrazides, semicarbazides, etc. Additionally, hydrazines is a very convenient reducing agent. The combination of suitable, convenient, and reactive organic reagents as well as the presence of it in many pharmaceutically active drugs, enable it to be widely used moiety in the drug design and development processes. We have reviewed its role in the pharmaceutical industry in the development of various drugs.

Introduction

Hydrazine is a frequently used chemical in general in the chemistry arena and especially heavily utilized in the pharmaceutical industry to synthesize various drugs and analogs because of its usefulness, high reactive nucleophilicity, and reduction properties. As mentioned above, it is also used in Wolff-Kishner reduction along with a base such as KOH and alcoholic solvent, including ethylene glycol. It readily reacts with carbonyl derivatives, aldehydes, and ketones to generate hydrazones. Because of its nucleophilicity, *N*-alkylation of it can be done very neatly. Upon reaction with cyanides, it gets converted to amide hydrazides (acyl hydrazides). Because of the availability of pair of nucleophiles, it reacts with difunctional electrophilic reagents to make a variety of heterocyclic compounds, including pyrazoles, pyridazines, and many more. This is also utilized in the last cleavage step in the Gabriel phthalimide primary amine synthetic method. Hydrazine is a very handy reducing agent as it generates simple side products such as nitrogen gas and water [1].

As the pharmaceutical industry is rich with drugs containing hydrazine core, this chapter provides a comprehensive review of drugs comprising substituted hydrazine moiety featuring *N–N* bond. The medicinal importance, pharmacological actions, efficacy, adverse effects, structural features, metabolism studies, and synthetic protocols reported to date for these drugs, such as isoniazid, nitrofurazone, nifurtimox, procarbazine, selinexor, hydralazine, rimonabant, benserazide, phenelzine, isocarboxazid, carbidopa, cadralazine, and clopamide are compiled, analysed, and reported here, which would benefit the scientific community around the globe for the development of various hydrazine-based novel bioactive drug molecules.

Hydrazine-Containing Drugs

Table 1. Hydrazine-containing drugs

Sr. No.	Drug	Class	Structure
1	Isoniazid	Anti-tubercular	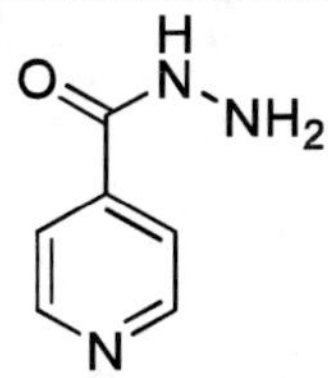
2	Nitrofurazone	Anti-microbial	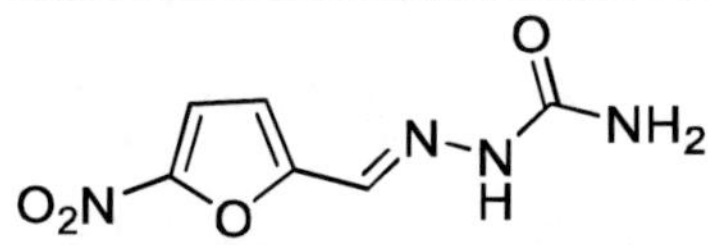
3	Nifurtimox	Treatment of Chagas disease	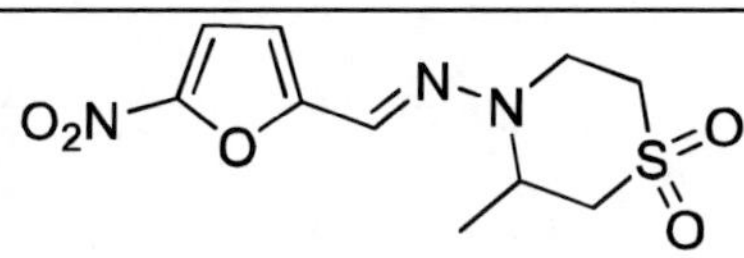
4	Procarbazine	Anti-cancer	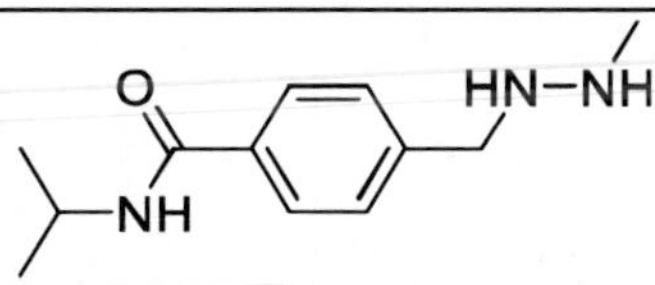
5	Selinexor	Anti-cancer	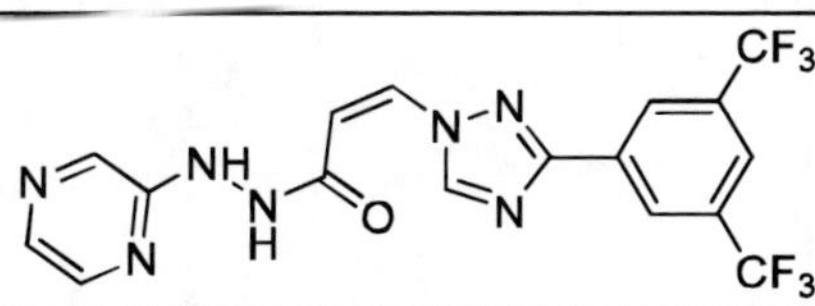
6	Rimonabant	Anorectic	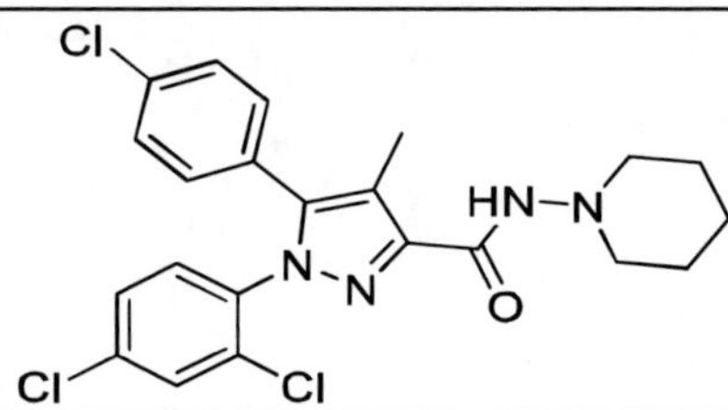

Table 1. (Continued)

Sr. No.	Drug	Class	Structure
7	Phenelzine	MAOI (anti-depressant)	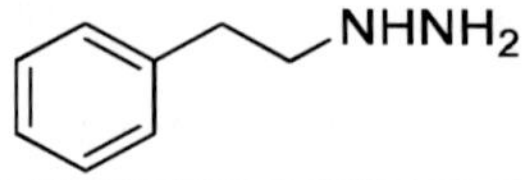
8	Isocarboxazid	MAOI (anti-depressant)	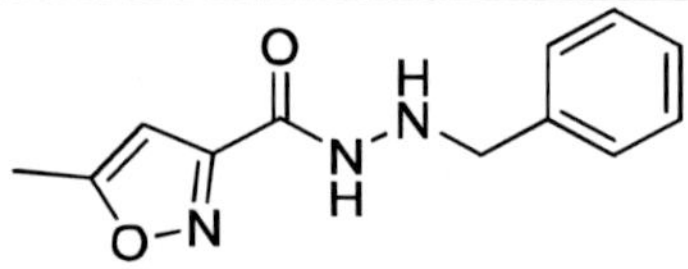
9	Carbidopa	Decarboxylase inhibitor (Parkinson's disease)	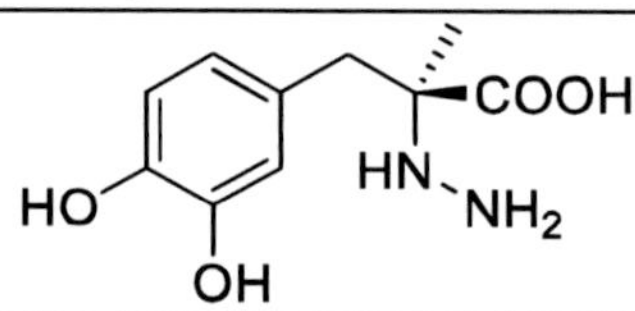
10	Benserazide	Decarboxylase inhibitor (Parkinson's disease)	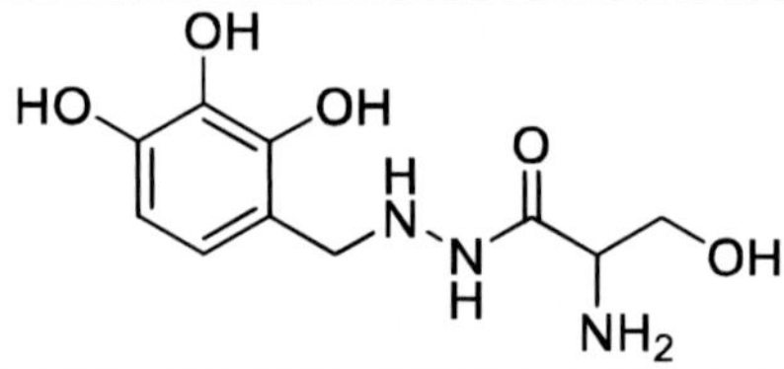
11	Hydralazine	Vasodilator	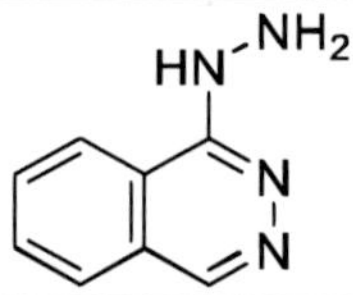
12	Cadralazine	Vasodilator	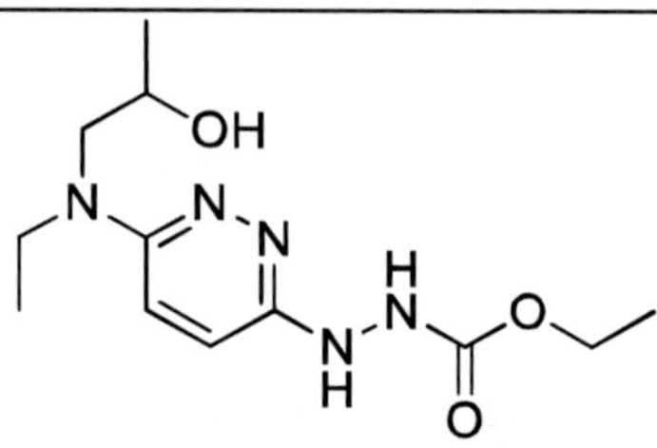
13	Clopamide	Diuretic	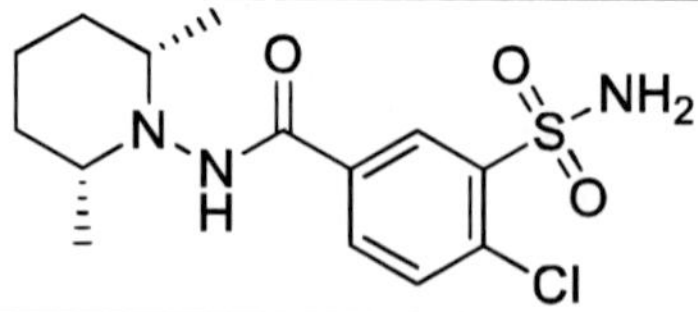

Isoniazid

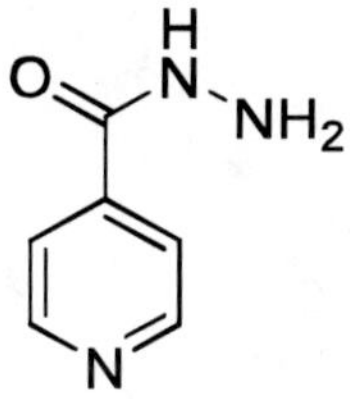

1

Isoniazid (**1**), isonicotinohydrazide is one of the prominent, pretty old first-line drug in the management of tuberculosis. Tuberculosis is caused by Mycobacterium *tuberculosis* bacteria and perturbs the entire world, which is reflected by the higher number of cases infected with it each year.

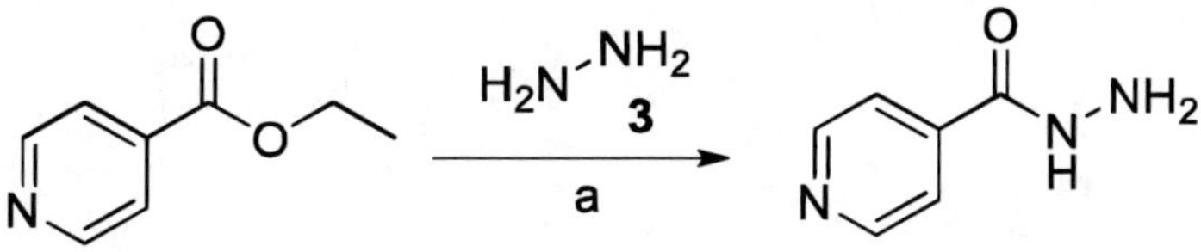

2 **1**

Reagents and conditions: a) Methanol

Scheme 1. Synthesis of isoniazid (**1**).

Isoniazid is synthesized (Scheme 1) by reacting ethyl ester of isonicotinic acid **2** with hydrazine **3** in methanol to yield isoniazid (**1**) [2, 3].

Metabolism of isoniazid (Scheme 2) majorly occurs by *N*-acetyltransferase 2 (NAT 2) to N-acetyl hydrazine **6**, which is hydrolyzed by amidase enzyme to acetyl hydrazine **7**. CYP2E1 enzyme is also involved in isoniazid metabolism converting acetyl hydrazine to a reactive metabolite which experiences further phase II metabolism by glutathione conjugation. Amidases can act on isoniazid to cleave amide bond to generate isonicotinic acid (**8**)and hydrazine **3,** which can undergo acetylation by NAT 2 to acetyl hydrazine **7**. Minor metabolites include ring oxidation amide metabolite **4** and isoniazid-N-oxide **5** [4].

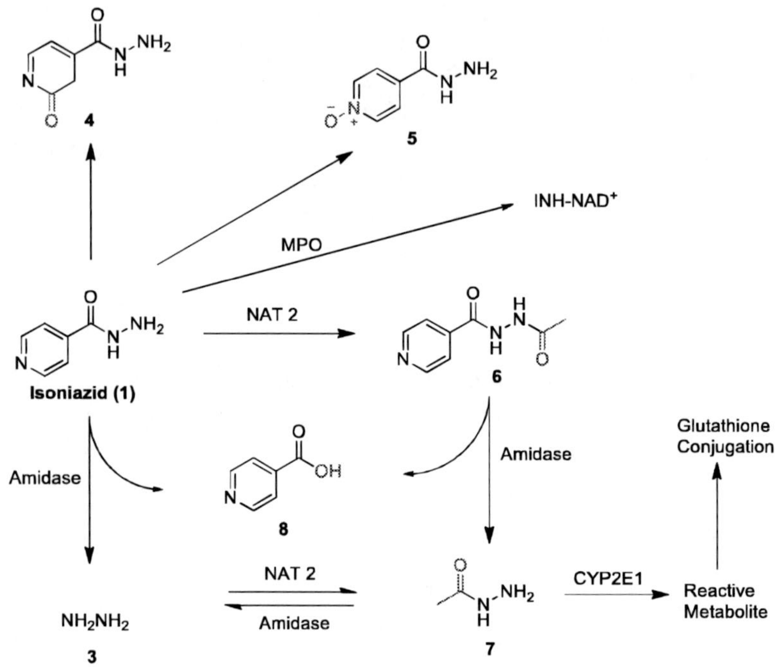

Scheme 2. Metabolic pathway of isoniazid (**1**), MPO: myeloperoxidase; NAT2: N-acetyltransferase 2; INH: isoniazid; NAD: nicotinamide adenine dinucleotide.

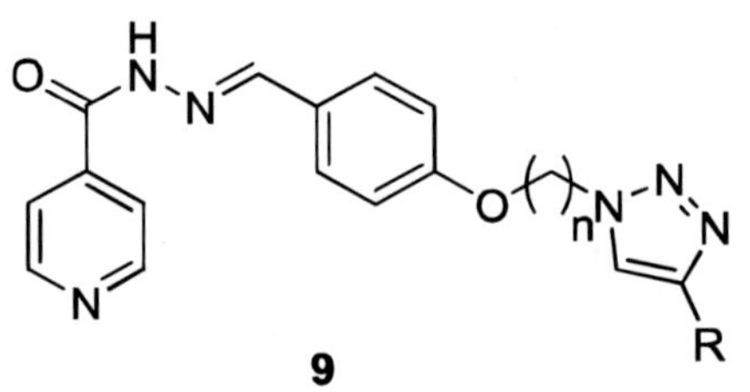

Figure 1. Prominent triazolylalkoxy banzaldehyde analog of isoniazid.

Kumar *et al.* synthesized various isoniazid analogs (Figure 1) to perform structure-activity relationship (SAR) studies. They have synthesized analogs by reacting isoniazid with several 4-(substituted triazolylalkoxy)benzaldehyde to prepare an array of imines. These synthesized analogs, when tested in an *in-vitro* assay against Mycobacterium *tuberculosis* H37Rv, displayed potent activity. However, when tested *in-vivo* using a mouse model challenged with Mycobacterium *tuberculosis* H37Rv strain, all of these analogs turned out to be less active. This may be a result of a solubility issue [5].

Santos *et al.* synthesized acyl hydrazones (Figure 2) using isoniazid and various substituted aldehydes, including substituted benzaldehydes, and heterocyclic aldehydes. To ascertain anti-Alzheimer's activity of these 16 analogs, acetylcholinesterase and myeloperoxidase inhibition properties were determined. Additionally, antioxidant properties were also evaluated against the reference standard of quercetin and isoniazid. Isoniazid acted as a reference standard for myeloperoxidase inhibition, while rivastigmine was used as a reference standard for acetylcholinesterase inhibition studies. Most of the synthesized compounds were less active compared to the reference standards. However, chromonyl substituted analog was almost equipotent to all three reference standards. Also, hydroquinone analog showed almost equipotency to the reference standard for myeloperoxidase inhibition and antioxidant property [6].

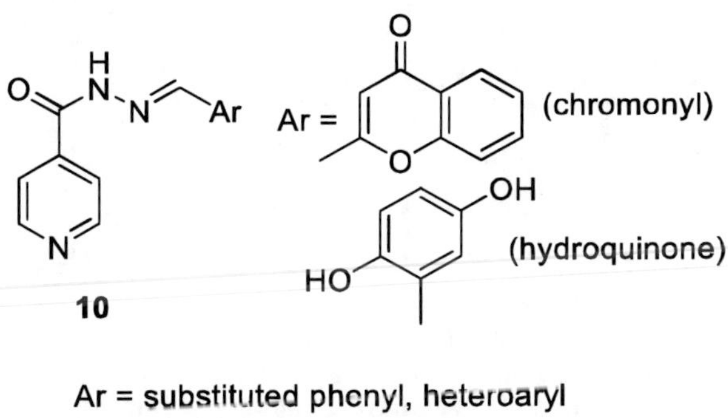

Figure 2. Prominent chromonyl and hydroquinone analog of isoniazid.

In another repurposing project, Mohanram *et al.* synthesized isoniazid analogs (Figure 3) for anti-inflammatory and anthelmintic properties. Using fluorite as a catalyst, isoniazid was reacted with substituted benzaldehyde and 1-naphthol or 8-hydroxyquinoline to synthesize a variety of analogs. In an anti-inflammatory study, the percent inhibitions of carrageenin-induced paw edema of Wistar albino rats were determined using diclofenac as a reference standard. Some of these analogs exhibited better activity than diclofenac. Pheretima *posthuma*, Indian earthworm species, was utilized to assess the anthelmintic activity. Compounds and reference albendazole were tested to varying doses of 25, 50, 100, and 150 mg/ml. Few analogs produced the desired anthelmintic activity even at the lowest dose [7].

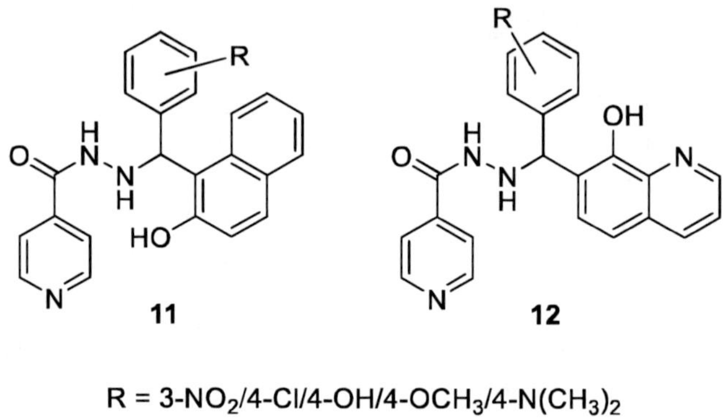

Figure 3. Prominent chromonyl and hydroquinone analog of isoniazid.

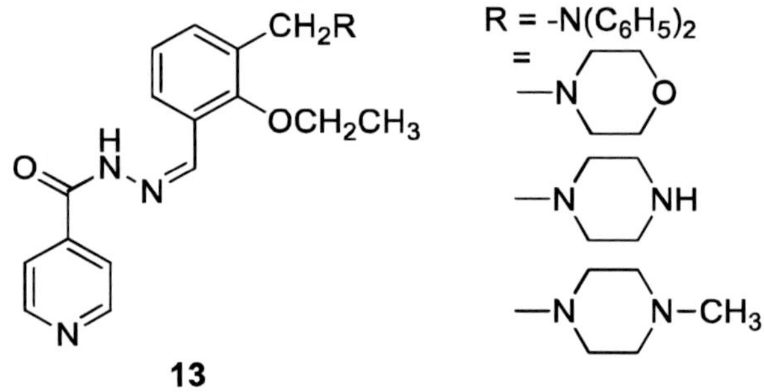

Figure 4. Prominent 2-ethoxybenzaldehyde analog of isoniazid.

In another interesting work, Malhotra *et al.* synthesized a few analogs (Figure 4) using isoniazid, 2-ethoxybenzaldehyde, which will form an imine, in situ was reacted with formaldehyde and secondary amine to carry out Mannich reaction to synthesize various analogs of it. Synthesized compounds were tested for antimicrobial and antifungal activities using amoxycillin and nystatin as reference standards. Zone of inhibition studies was carried out against Gram-positive bacterias: *Staphylococcus aureus*, and *Bacillus subtilis*; Gram-negative bacterias: *Pseudomonas aeruginosa*, *Escherichia coli,* and fungal strains of *Candida albicans*, *Candida gabrata.* MIC (minimum inhibitory concentrations) and MFC (minimum fungicidal concentrations) were also assessed. Many analogs displayed better activity than the reference standards [8].

Castelo-Branco *et al.* reported a very extensive design, synthesis, and antimycobacterial activity evaluation of analogs of acyl hydrazides (Figure 5). These acyl hydrazides were divided into two series: one consisting of isoniazid moiety and the other devoid of it. Many analogs from these series were active against susceptible mycobacterial strain (H37Rv) and their potency was even higher than the first-choice drugs in tuberculosis management. Few analogs showed better activity than isoniazid for treating resistant strains of the microbe. The most relieving feature of this research includes one analog that is less hepatotoxic than the isoniazid. As isoniazid is part of the primary treatment of tuberculosis, which needs to be administered for at least 6 months depending on the infection severity, and at the same time isoniazid is hepatotoxic, liver damage could be a big concern while on the therapy [9].

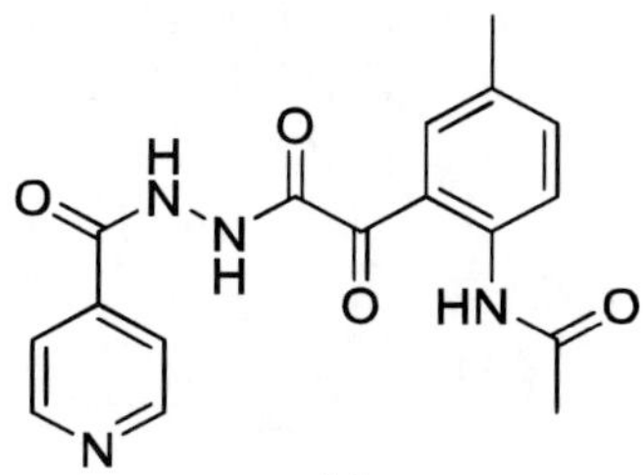

14

Figure 5. Prominent acyl hydrazide analog of isoniazid.

In a quest to design antitubercular drugs, Matei *et al.* synthesized novel analogs of isoniazid (Figure 6) by reacting isoniazid with thioamide to generate novel analogs of isoniazid. However, synthesized compounds were less active than the parent compound. Synthesized compounds were also tested for antibacterial activity against a few Gram-positive strains of *Enterococcus faecalis*, *Streptococcus hominis*, *Staphylococcus aureus,* and Gram-negative strains of *Acinetobacter baumanii*, *Klebsiella pneumoniae*, *Escherichia coli*, *Morganella morganni*, *Proteus mirabilis*, *Pseudomonas aeruginosa,* and *Citrobacter koseri*. All analogs were more active than isoniazid. However, when tested for cytotoxicity evaluation, these compounds showed prominent cytotoxicity, which demands extensive structure activity relationship (SAR) study to enhance desired activity and minimize side effects [10].

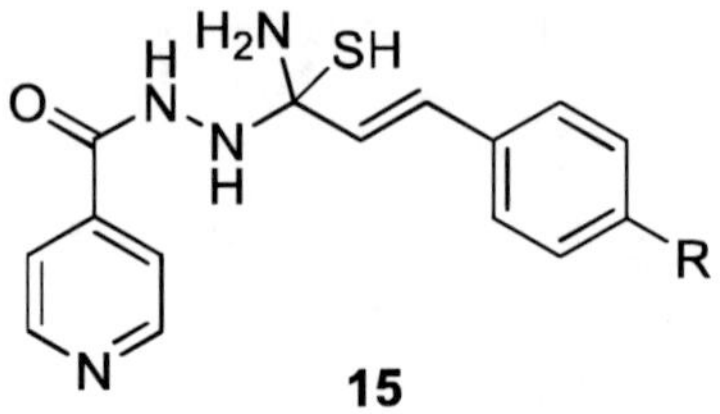

15

R = H/Cl/CH_3/OCH_3/C_2H_5

Fig. 6

Figure 6. Prominent thioamide analog of isoniazid.

Volynets *et al.* synthesized various imines by reacting isoniazid with aldehydes or ketones, mainly heterocyclic aldehydes and ketones. These compounds, when tested against Mycobacterium tuberculosis H37Rv strain under aerobic conditions, almost all of these exhibited single digits or sub-micromolar MIC values except one. However, when tested against isoniazid-resistant strain SRI 1369 (INH-R) all were inactive except one with a MIC value of 0.14 μM. This same analog was also effective against rifampin-resistant and ofloxacin-resistant strains with MIC values of 0.13 and 0.14, respectively. As this one analog (Figure 7) exhibited potent activity against all three drug-resistant strains, has good pharmacokinetic properties and lesser hepatotoxicity than isoniazid, it demands further exploration of in-depth drug design and development studies [11].

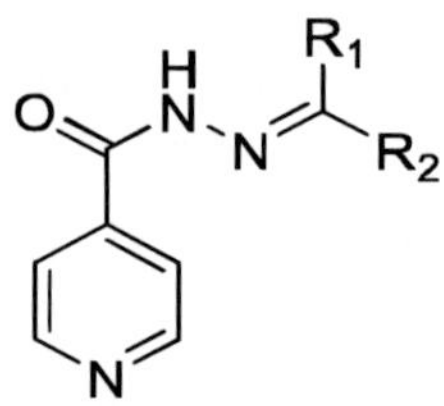

16

Most potent when R_2 = H, R_1 = 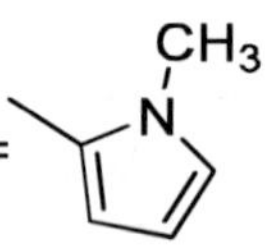

Figure 7. Prominent thioamide analog of isoniazid.

In another fascinating drug repurposing study by Rizvi *et al.*, as pyridine is common in isoniazid molecule, which is an anti-tubercular drug, similarly it is shared in a few urease inhibitors such as pantoprazole, lansoprazole, omeprazole which acts as anti-ulcer agents and also found in few non-steroidal anti-inflammatory drugs (NSAIDs) for instance in piroxicam and etoricoxib. Considering these, thiosemicarbazides were synthesized by retaining the entire backbone of isoniazid. The two sides of thiosemicarbazide are 1) the isonicotinylic part and 2) various substituted aromatic/cycloalkyl nuclei. Among the synthesized compounds with cyclohexyl analog (Figure 8) exhibited the greatest activity on both fronts; that is, it inhibited urease with IC_{50} value of 22.4 ±1.83 μM and an anti-inflammatory effect at a 12.3±1.2 μg/mL. Other compounds were either active on both fronts or only at one target [12].

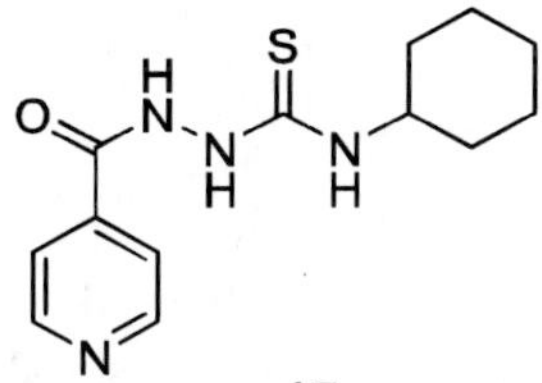

17

Figure 8. Prominent thiosemicarbazides analog of isoniazid.

Kaur *et al.* synthesized substituted pyrimidine conjugates of isoniazid (Figure 9). However, none of the synthesized analogs were as active as the parent drug, isoniazid, against the H37Rv strains of *Mycobacterium tuberculosis* [13].

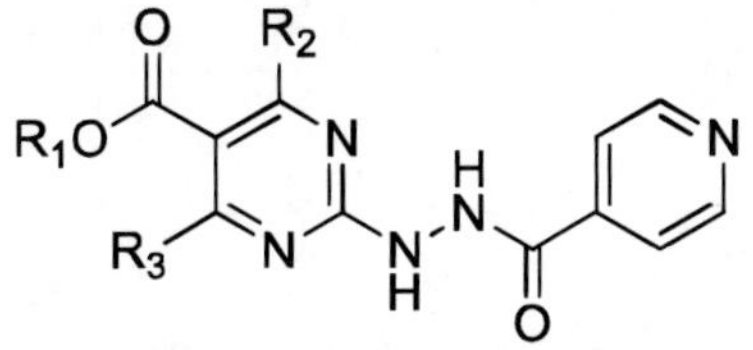

18

R_1 = alkyl,
R_2 = alkyl/ phenyl/substituted phenyl
R_3 = Me/Ph

Figure 9. Prominent thiosemicarbazides analog of isoniazid.

In 2017, Hu *et al.* reviewed isoniazid analogs for antitubercular activity [14]. Another review article appeared in 2019 by Vilchèze *et al.*, which described isoniazid's story, including random discovery, mechanism of action finding, resistance development, locating genes related to resistance, persistence, and all in great detail [15]. Readers are highly encouraged to read these reviews for further interest and deliberations.

Nitrofurazone

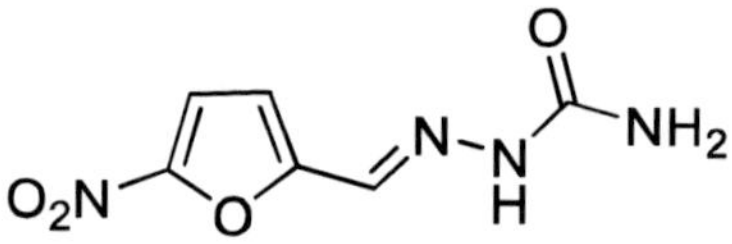

19

Nitrofurazone **(19)**, (*E*)-2-((5-nitrofuran-2-yl)methylene)hydrazine-1-carboxamide is one of the nitrofuran antibacterial compounds. These are mainly used for veterinary treatment of gastrointestinal infections and mixed with animal food [16].

Synthesis of nitrofurazone as in Scheme 3; furfural **20** was reacted with a nitrating mixture and acetic anhydride to form intermediate **21**, which was reacted with semicarbazide hydrochloride **22** in the ethanol-water mixture to form nitrofurazone **19** [17].

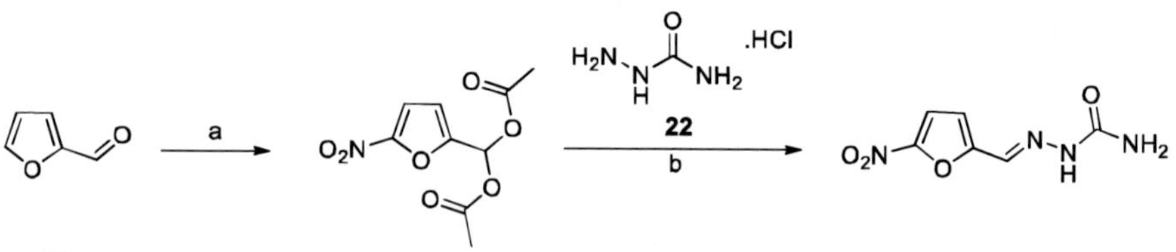

20 **21** **19**

Reagents and conditions: a) i) HNO_3, H_2SO_4, ii) $(CH_3CO)_2O$; b) water,ethanol

Scheme 3. Synthesis of nitrofurazone (**19**).

Nitrofurazone is metabolized in the gastrointestinal tract by acid hydrolysis to produce semicarbazide **22**. Also, nitro-reduction produces different metabolites (**23, 24**) which further get converted to cyano metabolites (**25** and **26**) [18] (Scheme 4).

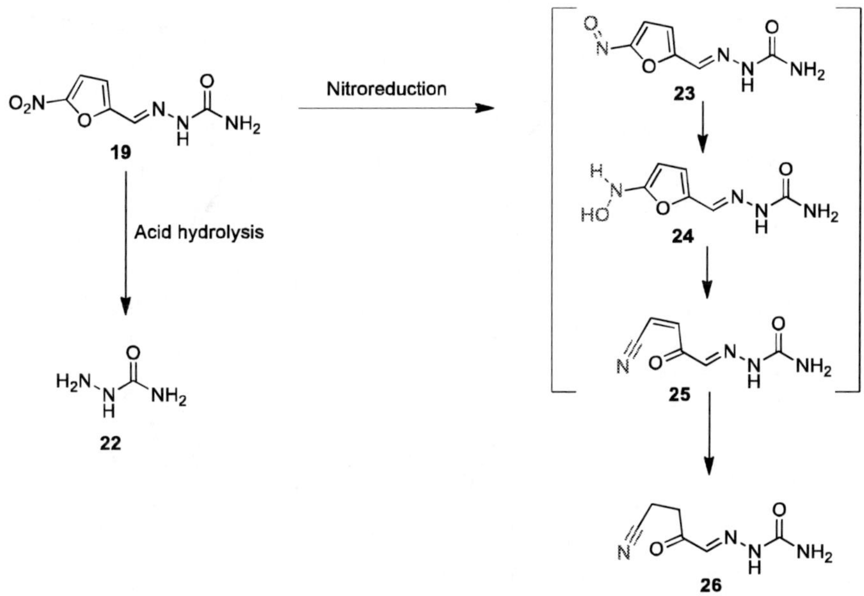

Scheme 4. Metabolic pathway of nitrofurazone (**19**).

Afghahi *et al.* synthesized various analogs by replacing terminal NH_2 with substituted CH_2, which leads to acyl hydrazones compared to semicarbazide. Synthesized compounds were tested for antibacterial activity against *B. subtilis*, *K. pneumoniae*, *S. aureus*, *S. lutea*, *B. cereus* and compared with standard nitrofurazone by zone inhibition studies. Although many of those synthesized compounds retained the antibacterial activity, all of those were less potent compared to the parent compound, nitrofurazone. Synthesized compounds were also screened for antifungal activity against *C. albicans*, *P. notatum*, *A. niger* by comparing them with those of griseofulvin at 10, 25, and 50 µg/ml concentrations. However, all compounds, including griseofulvin activity, were trivial at 10 and 25 µg/ml concentrations. Few compounds, including griseofulvin, possessed antifungal activity against either all or a few species at 50 µg/ml concentration [19].

Four nitrofurans are structural congeners and were available in the market for their antibacterial properties, e.g., furazolidone, nitrofurantoin, nitrofurazone, and furaltadone for animals. In addition to the mutagenicity and genotoxicity of furazolidone, nitrofurazone and other nitrofuran analogs, considering the shorter half-life (generally 1 hour), these are extensively metabolized. Because of their toxicity issues, parent drugs/metabolites must be eliminated from the body. To detect the metabolite, high-performance liquid chromatography-tandem mass spectrometry (HPLC-tandem MS) Leitner *et al.* developed a very robust technique from animal tissues [20].

Chung *et al.* synthesized an analog of nitrofurazone as *N*-hydroxy-methylnitrofurazone (Figure 10) and tested it against *Trypanosoma cruzi,* which infects humans with Chagas disease. It is interesting to note that the researchers synthesized this analog as an intermediate for the assembly of another prodrug for targeting *Trypanosoma cruzi.* However, this hydroxymethyl nitrofurazone was found to be a potent compound for Chagas disease in *in-vitro* testing. These anti-Chagas drugs are believed to inhibit trypanothione reductase upon metabolic activation [21, 22].

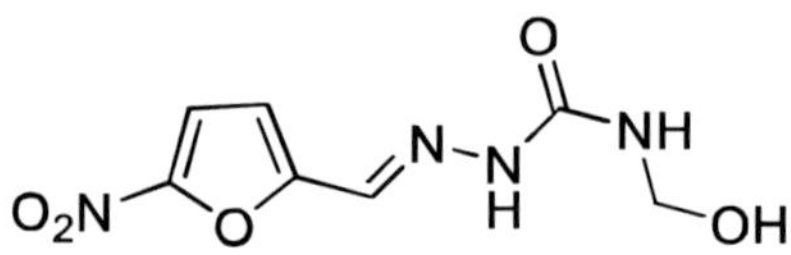

27

Figure 10. Prominent N-hydroxymethyl analog of nitrofurazone.

Hydrazide-hydrazones of nitrofurazone (Figure 11) were synthesized by Popiołek *et al.* These 21 analogs were tested for antimicrobial activity *in-vitro* against Gram-positive bacteria such as *Staphylococcus aureus*, *Staphylococcus epidermidis*, *Micrococcus luteus*, *Bacillus cereus*, *Bacillus subtilis*; Gram-negative bacteria including *Klebsiella pneumoniae*, *Escherichia coli*, *Bordetella bronchiseptica*, *Pseudomonas aeruginosa*, *Proteus mirabilis*, *Salmonella typhimurium* and antifungal activity against yeasts such as *Candida albicans*, *Candida parapsilosis.* Reference control compounds utilized for antibacterial activity were nitrofurantoin, ciprofloxacin, and cefuroxime, while fluconazole acted as a reference compound for antifungal activity. Many of these synthesized analogs showed better antibacterial properties than nitrofurantoin, ciprofloxacin, and cefuroxime. However, the antifungal activities of these compounds were negligible as compared to fluconazole [23].

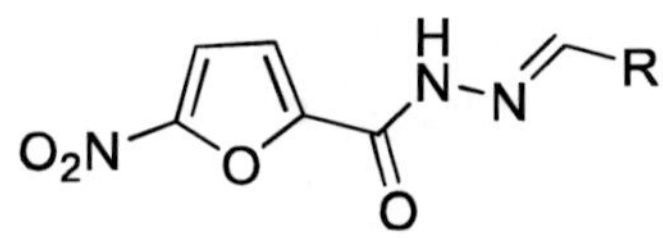

28

Figure 11. Prominent hydrazide-hydrazones analog of nitrofurazone.

Le *et al.* discovered an enzyme, AhpF, which activates 5-nitrofurans such as nitrofurantoin, nitrofurazone, and furazolidone. It would be interesting to design molecules with a higher affinity for the AhpF enzyme and determine its effect on antimicrobial activity [24].

Quorum signalling plays a role in *Pseudomonas aeruginosa* infectivity as it controls many genes, including those required for biofilm component formation, generation of metabolic toxins, and development of resistance. In a repurposing study by Baldelli *et al.*, researchers found nitrofurazone's usefulness in inhibiting the quorum sensing system in *Pseudomonas aeruginosa.* At a lower dose than required for antibacterial activity, nitrofurazone minimized biofilm formation as well as a virulence factor, pyocyanin. Prominently, it does not interfere with other antibacterial compounds' activity [25].

In one study by Krátký *et al.*, semicarbazone of nitrofurazone was replaced by aminoguanidine and tested for antimicrobial, antifungal, and anticholinergic activities. Minimum inhibitory concentrations were determined against *Staphylococcus aureus*; methicillin-resistant *Staphylococcus aureus* (MRSA); *Staphylococcus epidermidis* (SE); *Enterococcus faecalis* (EF); *Escherichia coli (EC); Klebsiella pneumoniae* (KP); *Acinetobacter baumannii* (ACI); *Pseudomonas aeruginosa* (PA). Few analogs (Figure 12) exhibited good activity, e.g., molecule **29** had activity in the range of 7.8-62.5 μM against most of the microorganisms except *Acinetobacter baumannii* (ACI) and *Pseudomonas aeruginosa* (PA). Piperacillin sodium was used as a reference standard. Most of these analogs have poor antifungal activity when tested against *Candida albicans* (CA); *Candida krusei* (CK); *Candida parapsilosis* (CP); *Candida tropicalis* (CT); *Trichophyton interdigitale* (TI) by using fluconazole as a reference standard. However, most analogs exhibited potent inhibition of acetylcholine esterases (AChE) and butyrylcholine esterases (BuChE) using rivastigmine and galantamine as reference standards. Molecule **30** (with IC_{50} =1.69 ± 0.17) was almost more potent than galantamine (IC_{50} = 2.77 ± 0.15 μM for butyrylcholine esterase inhibition) with its IC_{50} values. Few analogs had better selectivity for AChE

(AChE/BuChE inhibition to almost 0.05), and few had a ratio of 16.66 for AChE/BuChE inhibition, clearly showing selectivity for BuChE [26].

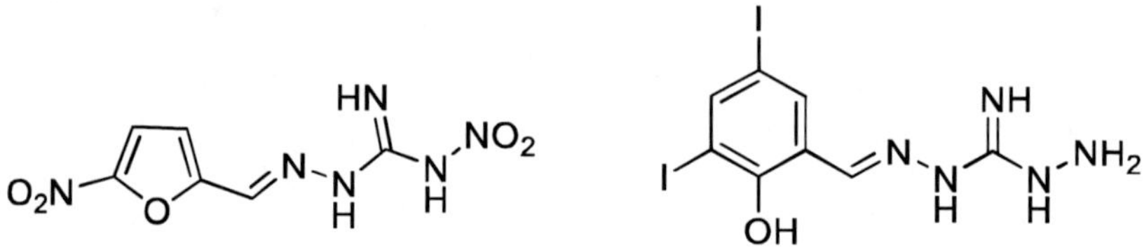

29 **30**

Figure 12. Prominent hydrazide-hydrazones analog of nitrofurazone.

Wakil *et al.* synthesized novel nitrofurazone (Figure 13), *N*-(α-acrylonitrylcarbonyl)-5-nitrofurfural hydrazone, and determined anti-microbial, antifungal, and anticancer activities. Synthesized compounds were tested for antimicrobial properties against *S. aureus*, *E. fecalis*, MRSA (Gram-positive strains), *E. coli*, *P. aeruginosa*, *K. pneumoniae* (Gram negative strains), and *C. albicans* for antifungal activity. However, almost all analogs did not improve antimicrobial and antifungal activity compared to nitrofurazone. Few of those were less active than the parent compound, nitrofurazone. However, for anticancer properties, many analogs depicted better potency as compared to the parent compound, nitrofurazone, and or standard reference anticancer drug doxorubicin against colon (CaCo-2), breast (MDA-MB 231) and liver (HepG-2) cancers using MTT assay [27].

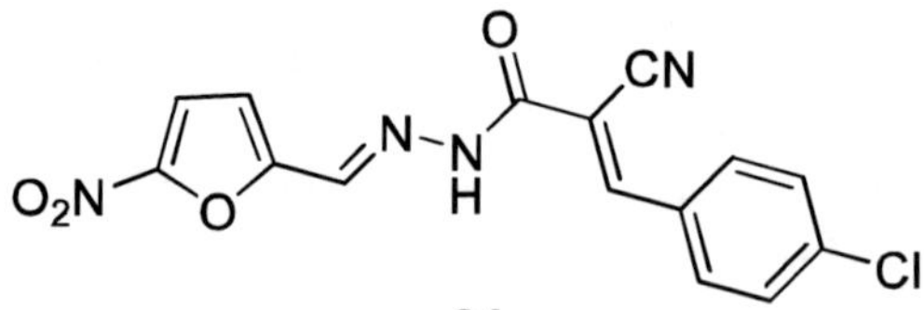

31

Figure 13. Prominent N-(α-acrylonitrylcarbonyl)-5-nitrofurfural hydrazone analog of nitrofurazone.

Wahid *et al.* synthesized nitrofurazones and tested these for inhibition studies against α-glucosidase enzyme. α-Glucosidase enzyme is involved in glycogen-to-glucose conversion. This enzyme inhibition could be used to treat diabetes mellitus and other ailments such as hepatitis, obesity, cancer, and hyperlipoproteinuria. All synthesized analogs had better activity than the parent compound nitrofurazone; however, few analogs (Figure 14) were more potent than the acarbose reference standard [28].

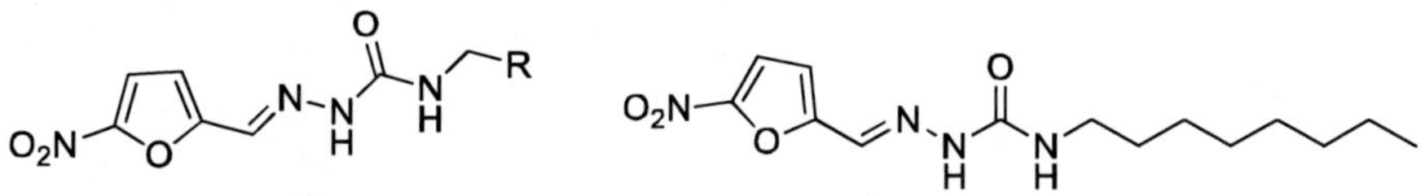

32 33

IC_{50} of analog for alpha-glucosidase = 0.54 ± 0.30 μM
IC_{50} of nitrofurazone for alpha-glucosidase = 37.5 ± 2.87 μM
IC_{50} of acarbose for alpha-glucosidase = 2.05 ± 0.41 μM

Figure 14. Prominent analog of nitrofurazone.

Nifurtimox

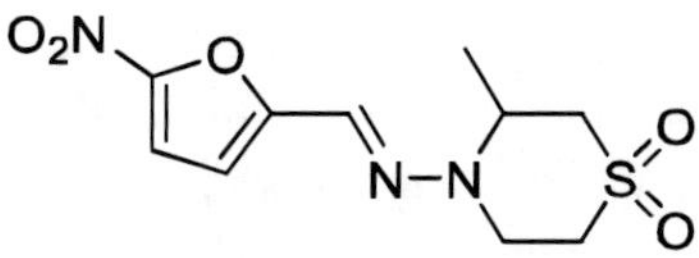

34

Nifurtimox (**34**), (*E*)-3-methyl-4-(((5-nitrofuran-2-yl)methylene)amino)thiomorpholine 1,1-dioxide is one of the leading drugs in treating Chagas disease. World Health Organization (WHO) has categorized Chagas disease (American Trypanosomiasis) as a neglected disease. The only other treatment option for Chagas disease includes benznidazole. Hall *et al.* reported that type I nitroreductases encoded by the parasite *Trypanosoma brucei* are required to activate nifurtimox to generate cytotoxic nitrile metabolites that possess activity against *Trypanosoma brucei.* All these developments indicate that nifurtimox is a prodrug [29, 30].

The synthesis of nifurtimox is given in Scheme 5. Propylene oxide (**35**) is treated with 2-mercaptoethanol (**36**) to produce 2-(hydroxyethyl)-(2-hydroxypropyl)sulfide (**37**), which cyclizes with the help of $KHSO_4$ to 2-methyl-1,4-oxathiane (**38**). Oxidation of compound **38** using hydrogen peroxide gives 2-methyl-1,4-oxathian-4,4-dioxide (**39**), which, when reacted with hydrazine, transforms to 4-amino-3-methyltetrahydro-1,4-thiazin-1,1-dioxide **40**, Reacting this with 5-nitrofurfural **41** gives nifurtimox **34** [31].

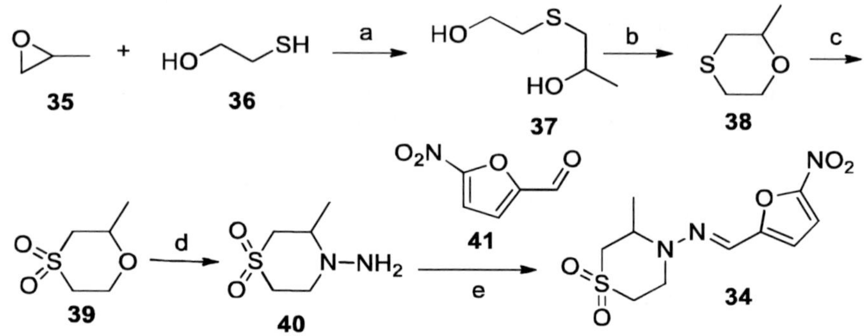

Reagents and conditions: a) NaOEt, ethanol, reflux; b) $KHSO_4$; c) H_2O_2; d) NH_2NH_2, aq. NaOH, reflux; e) ethanol, reflux

Scheme 5. Synthesis of nifurtimox (**34**).

Nifurtimox, a prodrug, is metabolized (Scheme 6) by cytochrome P450 reductase and nitroreductases 2 (NTR2) to amine **44** through nitroso intermediate **42** initially and hydroxylamine intermediate **43** later. Hydroxylamine intermediate upon dehydration, furan ring opens to generate unsaturated nitrile **45,** which is cytotoxic. Nitrile metabolite inhibits both parasite and mammalian cell growth. This is the basis for nifurtimox's mechanism [32].

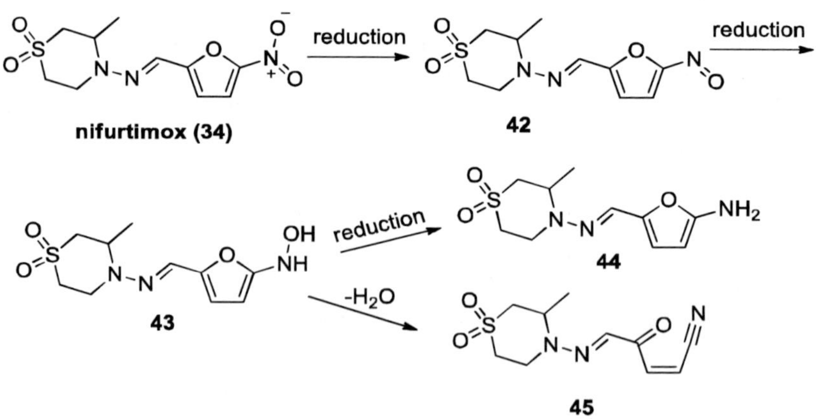

Scheme 6. Metabolic pathway of nifurtimox (**34**).

Procarbazine

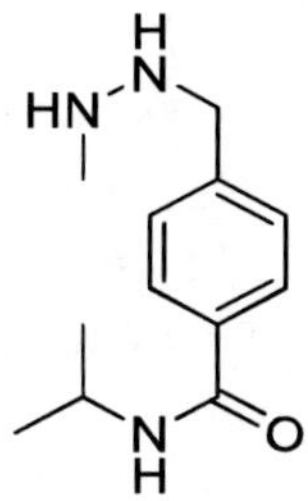

46

Procarbazine (**46**), *N*-isopropyl-4-((2-methylhydrazinyl)methyl)benzamide is an alkylating agent, used as an anticancer agent mainly in Hodgkin's disease but also useful in other non-Hodgkin's cancers such as bronchogenic carcinoma, multiple myeloma, malignant melanoma, polycythaemia vera. It is also reported to have an immunosuppressant activity [33]. The enhancement of antihypertensive and sympathomimetic properties occurs when procarbazine is administered to a patient taking tyramine-rich food. Depression of the central nervous system directly or indirectly by inhibiting hepatic microsomal mixed function oxidase(s) is the outcome after amplifying barbiturate's effect by procarbazine [34].

Procarbazine is synthesized by *p*-methylbenzoic acid **47** reacted with $SOCl_2$ to form *p*-methylbenzoyl chloride **48**, reacted with excess isopropyl amine resulted in N-isopropyl-4-methylbenzamide **49**. 4-Methyl of amide **49** is oxidized to 4-formyl intermediate **50**, converted into procarbazine **46** by reductive amination with methylhydrazine using $NaCNBH_3$ (Scheme 7) [35].

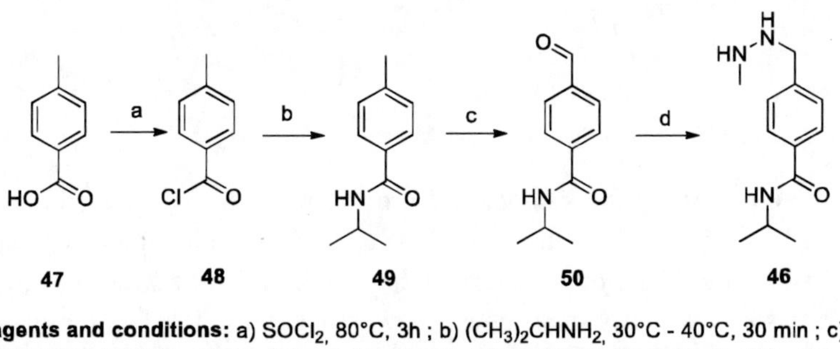

Reagents and conditions: a) $SOCl_2$, 80°C, 3h ; b) $(CH_3)_2CHNH_2$, 30°C - 40°C, 30 min ; c) $Ce(NH_4)_2(NO_3)_6/HNO_3$, 100°C, 24h ; d) $CH_3NHNH_2.HCl/NaCNBH_3/Et_3N$, EtOH, 0°C - RT, overnight

Scheme 7. Synthesis of procarbazine (**46**).

Metabolism studies provide useful information. It not only offers biotransformation data that could be utilized for drug development processes but also deliver information regarding bioactivity/side effects associated with the metabolites. As procarbazine was introduced in 1969, several metabolism studies have been reported. Interestingly enough, Swaffar *et al.* in 1989 reported that one of its metabolites, methyl azoxyprocarbazine with IC_{50} value of 0.15 mM is 10 times more active than the parent drug procarbazine with IC_{50} value of 1.5 mM in a soft-agar clonogenic assay. The metabolism takes place in two steps. Initially, procarbazine undergoes oxidative dehydrogenation to azocarbazine and later gets transformed into an isomeric mixture of methylazoxyprocarbazine and benzyl azoxyprocarbazine, as shown in Scheme 8 [36].

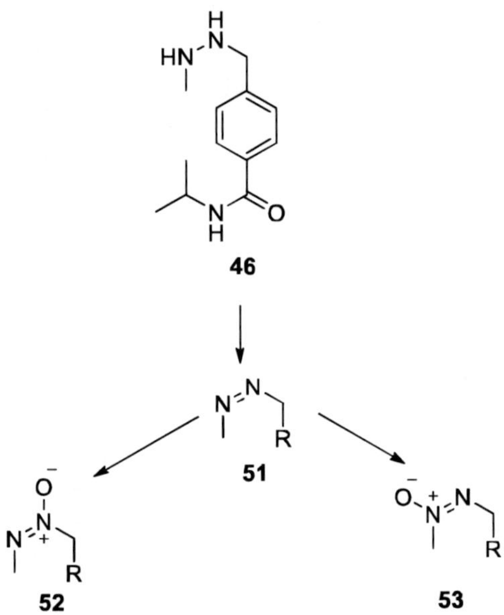

Scheme 8. Metabolic pathway of procarbazine (**46**).

Although most of the synthesized procarbazine analogs were either less potent or lost activity [37, 38] one of the SAR/metabolic studies has shown that benzylic H's oxidation would help the metabolite to be genotoxic [39]. In a separate study by Grossman *et al.*, it was observed that the simultaneous administration of enzyme-inducing anti-epileptic drugs along with procarbazine does not affect the pharmacokinetics of procarbazine to a great extent [40].

Selinexor

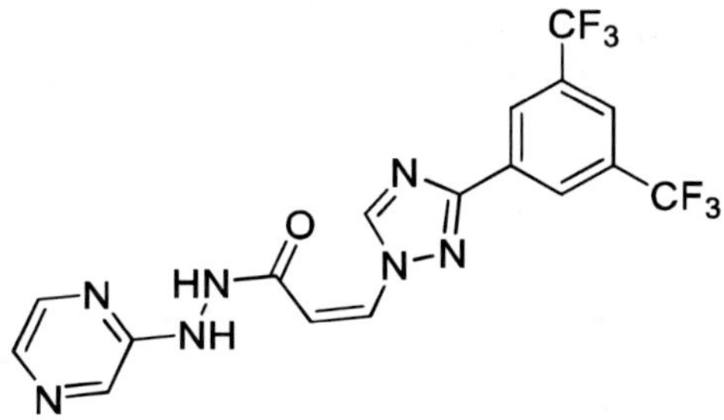

54

Selinexor (**54**), (*Z*)-3-(3-(3,5-bis(trifluoromethyl)phenyl)-1*H*-1,2,4-triazol-1-yl)-*N'*-(pyrazin-2-yl)acrylohydrazide is an accelerated approved anticancer drug by US FDA for the first time in July 2019 for the treatment of multiple myeloma [41, 42, 43]. Later on, it was approved for treating adults with relapsed or refractory diffuse large B-cell lymphoma (DLBCL). It has serious side effects such as hyponatremia, thrombocytopenia, neutropenia, leukopenia, and anemia [44].

Selinexor is also being tested in non-small cell lung cancer (NSCLC) characterized by modest prognosis and inadequate treatment choices. Selinexor would be a good choice there as it inhibits exportin-1 (XPO-1), a nuclear export protein, and has shown a highly promising treatment option in the A549 mouse xenograft model by reducing the mean tumor volume significantly compared to the positive cisplatin control [45]. Although overexpression of XPO-1 generally occurs in most cancers, its prognosis is insignificant in many cancer types, such as glioma, acute myeloid leukemia, mantle-cell lymphoma, and pancreatic, cervical, and ovarian cancers [46].

Synthesis of Selinexor is shown in Scheme 9, condensation of acid **55** and 2-hydrazinylpyrazine **56** in the presence of T3P and DIPEA forms selinexor **54** [47]. The multi-step synthetic protocol developed by Karyopharm has been reviewed 48]. The improved synthetic protocol was published by Chen *et al.* to enhance yield and avoid the generation of unwanted *E* isomer [49]. Urine and plasma analysis of patients receiving selinexor indicates that selinexor primarily undergoes glucuronidation metabolism [50].

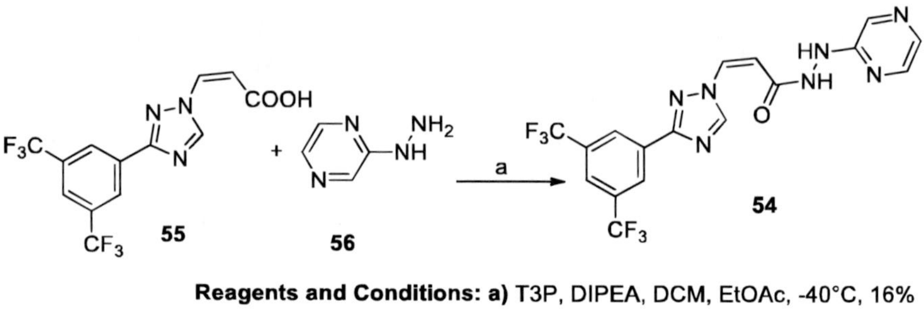

Scheme 9. Synthesis of selinexor (54).

Rimonabant

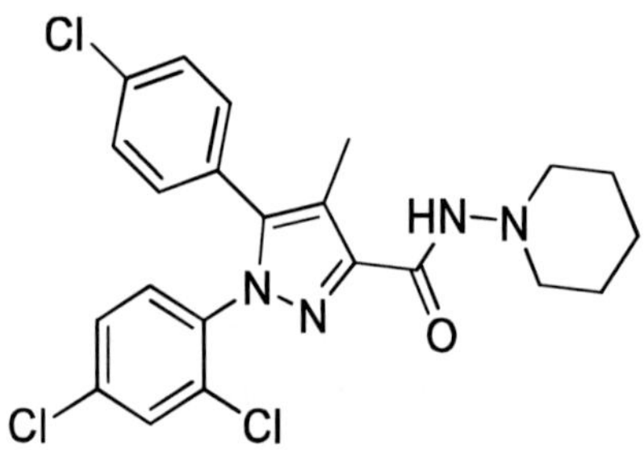

57

Rimonabant (**57**), 5-(4-chlorophenyl)-1-(2,4-dichlorophenyl)-4-methyl-*N*-(piperidin-1-yl)-1*H*-pyrazole-3-carboxamide, a pyrazole analog, acts as anti-obesity drug. It is the first drug of this class. The endocannabinoid system is found in the brain to regulate food intake and the CNS reward system. Rimonabant interacts with the endocannabinoid system via selective antagonism of cannabinoid type 1 receptor [51]. The drug was withdrawn from the market in 2009 due to its significant side effects like serious psychiatric and suicidal tendencies [52].

Despres *et al.* reported a study on rimonabant's effect on overweight patients' metabolic risk factors. It was observed that rimonabant significantly improves the profile of several metabolic risk factors in high-risk patients [53]. Russell *et al.* reported the study of rimonabant on an animal model. It shows that rimonabant effectively acts on renal failure and internal lipid over-secretion related to metabolic syndrome, and it does not affect insulin secretion and macrovascular function [54].

Synthesis of rimonabant (Scheme 10) involves the preparation of lithium salt of ester **59** which undergoes acid-mediated cyclization followed by de-esterification to yield intermediate **61**. This intermediate serves to DCC-mediated amidation to yield rimonabant (**57**) [52].

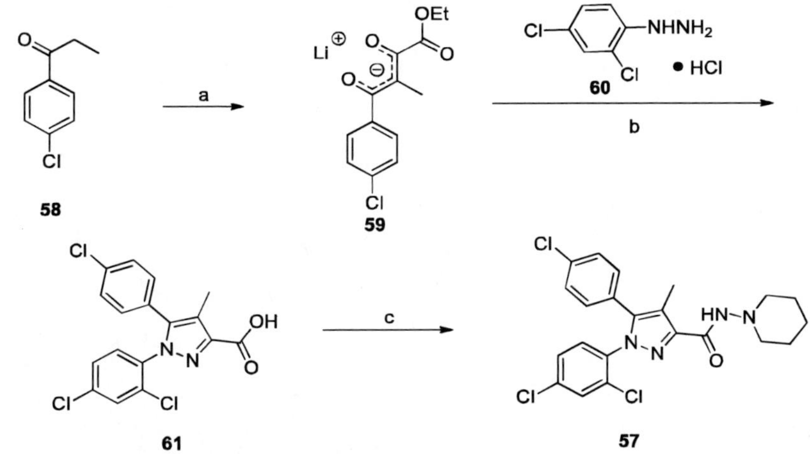

Reagents and conditions: a) LHMDS, Ethyl oxalate, methyl cyclohexane, 25°C, 6hr ; b) i) 50%, aq. H_2SO_4, ethanol, 79°C 14hr ; ii) NaOH; c) 1-aminopiperidine, DCC, HOBt, CH_2Cl_2, 25°C, 2.5h

Scheme 10. Synthesis of rimonabant (**57**).

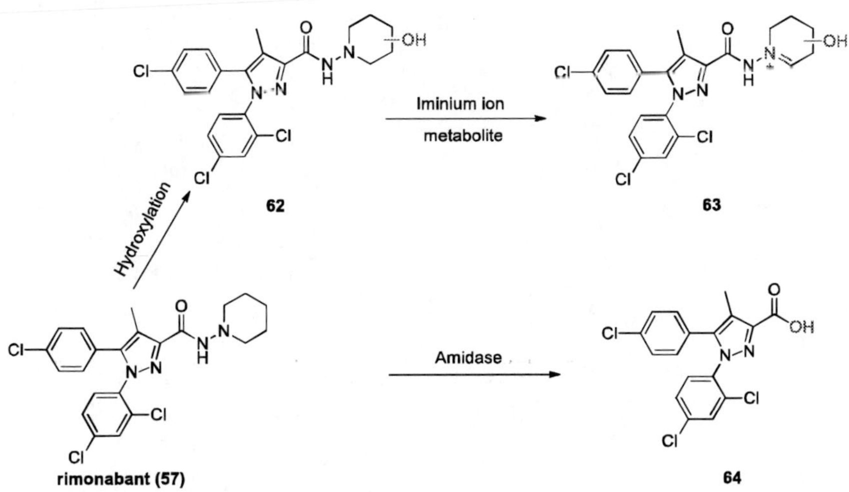

Scheme 11. Metabolic pathway of rimonabant (**57**).

CYP450 enzymes are involved in the metabolism of rimonabant in humans, as shown in Scheme 11. Hydroxylated metabolite **62** is a major metabolite which is converted to iminium ion metabolite **63**. Amidase enzyme is involved in the metabolism of the drug to carboxylic acid metabolite **64** [55].

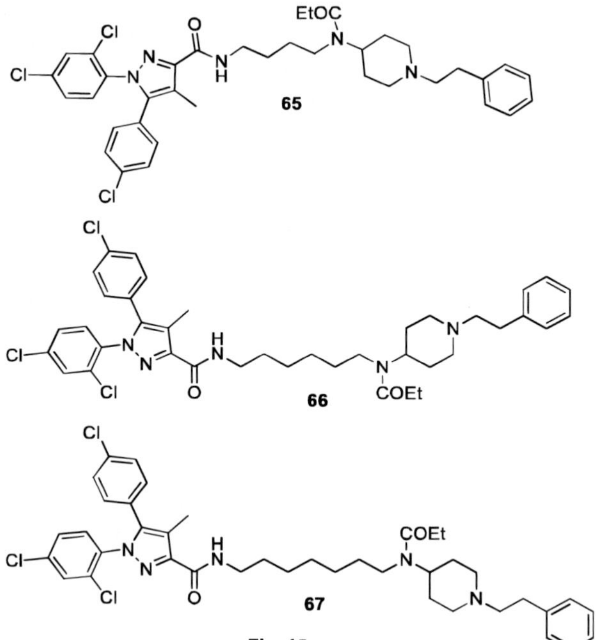

Fig. 15

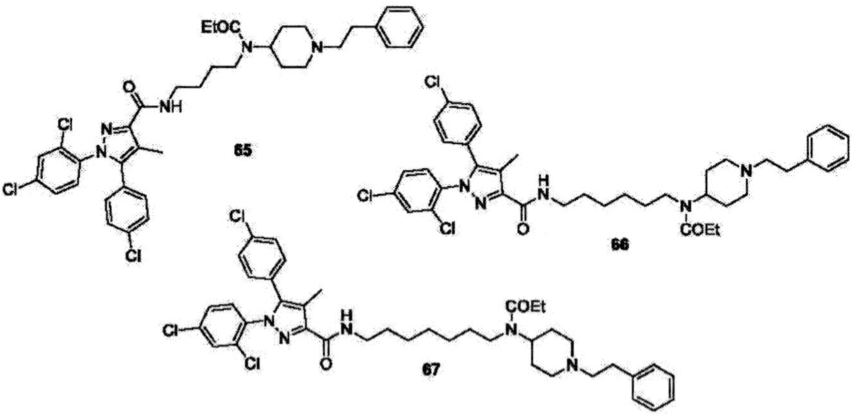

Figure 15. Prominent hybrid analog of rimonabant.

Fernández *et al.* synthesized and evaluated the hybrid analog of rimonabant (Figure 15) (selective cannabinoid type 1 receptor antagonist) and fentanyl (μ-opioid receptor antagonist). Molecules **65**, **66**, and **67** show promising activity in the assay. These three molecules show potent CB1 receptor antagonism similar to rimonabant. But only compounds **65** and **66** show μ-opioid receptor antagonism. *In-silico*, ADME properties indicate that only compound **67** is likely to penetrate through the blood brain barrier (BBB). It has been suggested that co-administration of both receptor antagonists may increase the therapeutic efficacy of the treatment [56].

Menozzi *et al.* synthesized and evaluated the rimonabant 5-substituted analogs **68** (Figure 16). Structural changes improved the activity, like substituting 2,4-dichlorophenyl at 1st position, *p*-substituted phenyl at 5th position, and carboxamide group at the 3rd position of pyrazole. This series of compounds produced potent activity [57].

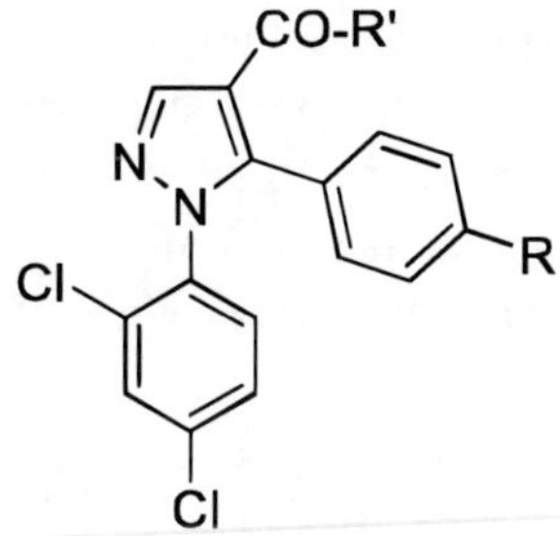

68

Figure 16. Prominent hybrid analog of rimonabant.

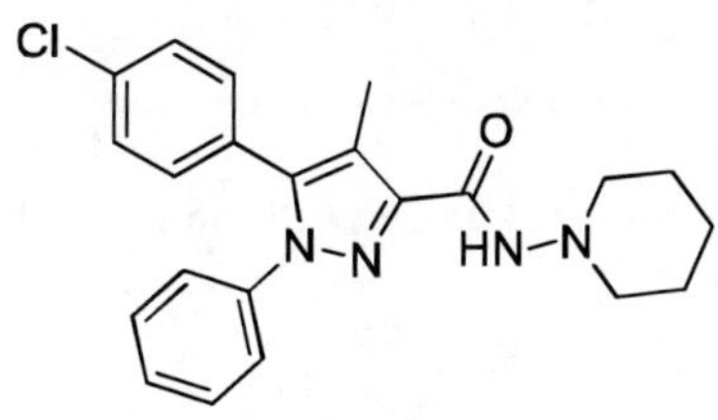

JMG-14

Gajbhiye *et al.* reported the rimonabant analogs evaluated for anti-TB activity *in-vitro* and *in-silico*. This study shows that **JMG-14** has potent anti-TB activity [58].

Phenelzine

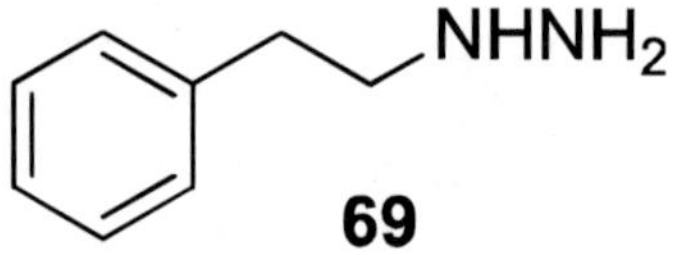

69

The antidepressant activity of phenelzine (**69**) is due to the monoamine oxidase (MAO) inhibition. However, phenelzine, 2-phenylethyl-1-hydrazine also elicits actions in other systems such as enhancement of levels of γ-aminobutyric acid (GABA) in the central nervous system (CNS) [59], GABA transaminase inhibition, depletion of free aldehydes through phenelzine sequestration. All these events lead to elevated GABA levels. Therefore, it could be utilized as a neuroprotective agent [60]. Additionally, the formation of the active metabolite, primary amine oxidase inhibition, elevation of ornithine and *N*-acetylamino acids levels in the brain would help the neuroprotection role of phenelzine [61]. Phenelzine is also undergoing clinical trials for the treatment of prostate cancer (NCT02217709) [62]. Additionally, phenelzine is being tested as an antimicrobial agent as antidepressants, including phenelzine, were known to interfere with the intestinal microbiota [63].

Phenelzine (**69**) is synthesized by reacting 2-phenylethyl chloride **70** with hydrazine **3** in ethanol (Scheme 12) [64].

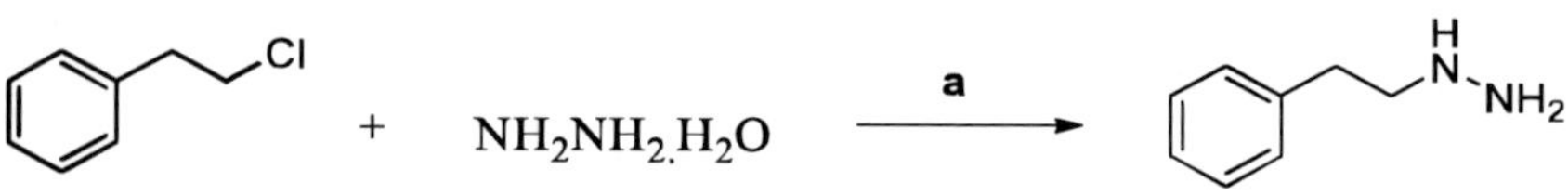

70 **3** **69**

Reagents and conditions: a) Ethanol

Scheme 12. Synthesis of phenelzine (**69**).

Lysine-specific demethylases 1 (LSD1) are implicated in many malignant tumors. LSD1 and MAO have structural homology. Thus, MAO inhibitors were utilized as a starting point in developing inhibitors for LSD1 [65].

Phenelzine's direct hydroxylation leads to the formation of *p*-hydroxyphenelzine (**72**), which may be converted to *p*-tyramine (**74**) (Scheme 13). Phenelzine is metabolized to 2-phenylethylamine (**73**), which can undergo hydroxylation to form *p*-tyramine (**74**). As both phenelzine and 2-phenylethylamine (**73**) are the substrates of monoamine oxidase (MAO), they are oxidized to phenylacetic acid (**75**). *p*-hydroxyphenyl acetic acid (**76**) is formed by MAO catalysis of *p*-hydroxyphenelzine (**72**) and *p*-tyramine (**74**) as well. Additionally, phenyl acetic acid (**75**) hydroxylation can take place to form *p*-hydroxyphenyl acetic acid (**76**) [66].

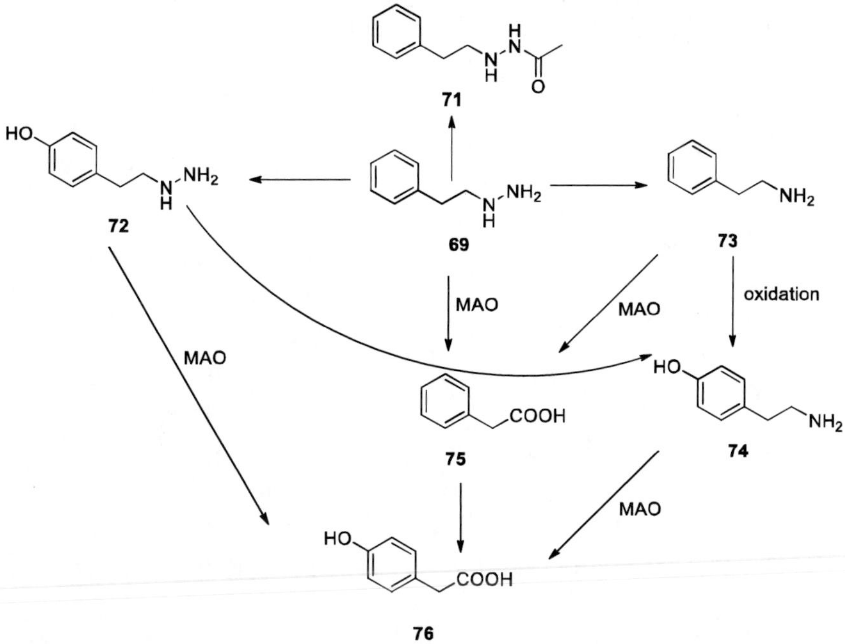

Scheme 13. Metabolic pathway of phenelzine (**69**).

Isocarboxazid

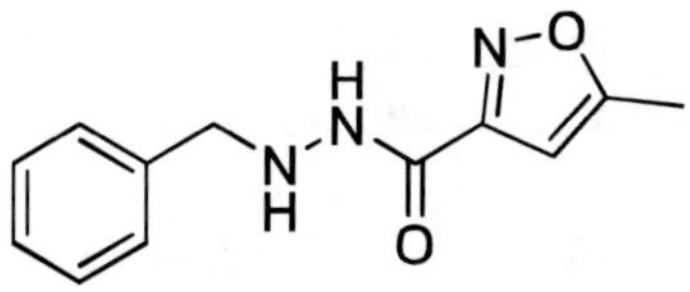

77

Isocarboxazid (77), *N'*-benzyl-5-methylisoxazole-3-carbohydrazide is a monoamine oxidase inhibitor (MAOI) useful as an antidepressant. It was observed that isocarboxazid enhances levels of noradrenaline and 5-hydroxytryptamine. However, it was experienced to heighten the 5-hydroxytryptophan-induced head twitches [67].

Synthesis of isocarboxazid (Scheme 14) starts from acetylacetone **78** get converted to 5- methyl-isoxazol-3-carboxylic acid (**79**) upon treating it with HNO_2. Esterification of this product **79** produces the ethyl ester of 5-methyl-isoxazol-3-carboxylic acid (**80**). The synthesized ester is further reacted with benzylhydrazine (**81**) to yield isocarboxazid (77) [65].

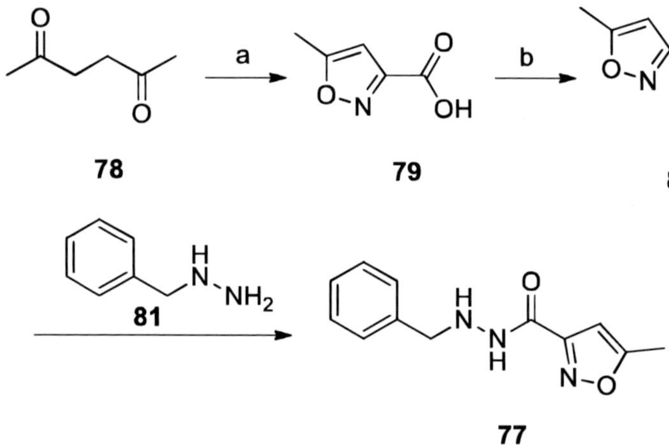

Reagents and conditions: a) HNO_2 ; b) Ethanol / H^+

Scheme 14. Synthesis of isocarboxazid (**77**).

Once administered, it is seen that isocarboxazid retains more nitrogen and continues to do so for 2-4 weeks. Upon isocarboxazid administration, protein digestion is negatively affected, and this effect existed for several weeks. A drug repurposing study showed that isocarboxazid prevents irinotecan-induced diarrhea (anticancer drug). Isocarboxazid elicits its action by inhibiting bacterial β-glucuronidase and thus can help to cope with irinotecan-induced side effects [68].

Structure-activity relationship studies were carried out by replacement of isoxazole to imidazole and a few other changes. However, none of these synthesized analogs showed significant activity as monoamine oxidase inhibitors [69]. However, in another study by Nulu *et al.,* imidazole analogs of isocarboxazid showed better hexobarbital-induced sleep prolongation properties compared to isocarboxazid-and-reversal of reserpine-induced hypothermia and ptosis [70].

Carbidopa

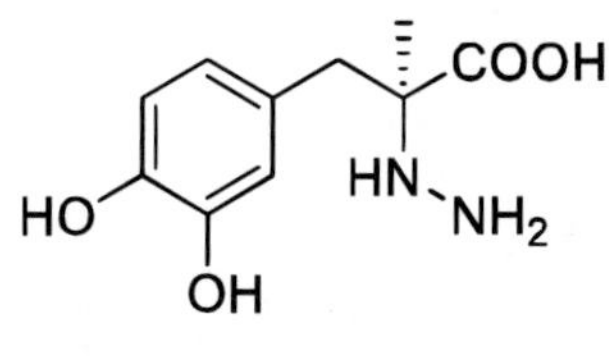

82

Carbidopa (**82**), (*S*)-3-(3,4-dihydroxyphenyl)-2-hydrazinyl-2-methylpropanoic acid, an inhibitor of the systemic decarboxylase, influences the level of levodopa in the CNS and assists in maintaining the level of it there [71]. Carbidopa, not only affects the metabolism of levodopa but also modifies by inhibiting synthesis of noradrenaline in sympathetic neurons in a study carried out by Wurtman *et al.* in a rat model. However, it does not affect catecholamine synthesis in the brain [72]. In another exciting research finding, it was realized that carbidopa was metabolized peripherally to α-methyldopa (AMD) (**83**) and *α*-methyldopamine (**84**). In fact, *α*-methyldopamine (**84**) is converted to *α*-methylnorepinephrine (**85**) by *β*-hydroxylation. Peripherally produced AMD crosses the blood-brain barrier (BBB) to elicit its effect in the brain (Figure 17) [73].

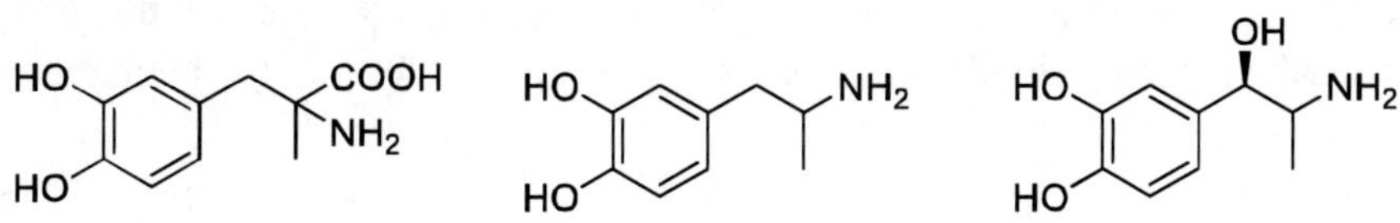

83 **84** **85**

Figure 17. Metabolites of carbidopa.

The enantioselective synthetic approach (Scheme 15) described by Pericas *et al.* for the assembly of *L*-carbidopa involved a seven-step protocol including a crucial enantioselective step catalyzed by one of the lanthanides, europium catalyst with 95% yield with excellent 98% enantiomeric excess. Although it is a 7-step synthetic strategy for constructing *L*-carbidopa, the overall yield of 50% for this protocol because of high-yielding individual steps makes it an acceptable route [74].

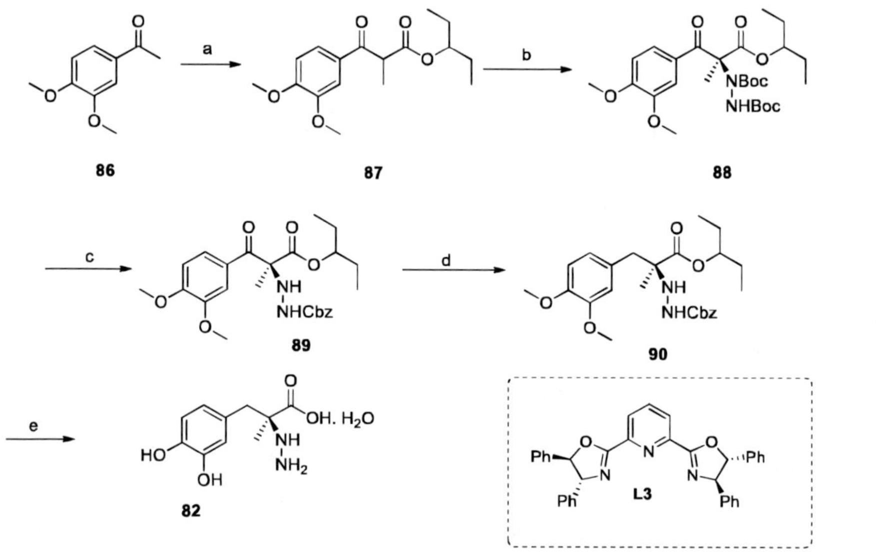

Reagents and Conditions: a) 1) dimethylcarbonate, NaH, THF, 97% ; 2) ZnO (cat.), toluene, 3-pentanol, 99% ; 3) MeI, K_2CO_3, acetone, 96% ; b) BocN=NBoc,10 mol% $Eu(OTf)_3$,15 mol % **L_3**, -20 °C, 2d, CH_3CN, 95%, 99% ee ; c) 1) TFA, 2) $NaHCO_3$, BzOCOCl, 84% ; d) TFA, $Et(Me)_2SiH$, 72% ; e) 1) BBr_3, CH_2Cl_2, -78 °C to rt, 2) MeOH, 50 °C,3) $(Me)_2NH$, pH 6, 4, ^iPrOH, 95%

Scheme 15. Synthesis of L-carbidopa (**82**).

Asano *et al.* described a 7-step protocol for assembly of *L*-carbidopa with a crucial step catalyzed by lithium 3,3'-dibromobinaphtholate **91** with 99% yield and 99% ee. Although this has similar steps as that of the earlier described protocol by Pericas *et al.*, the overall yield for this synthetic scheme is 42% [75].

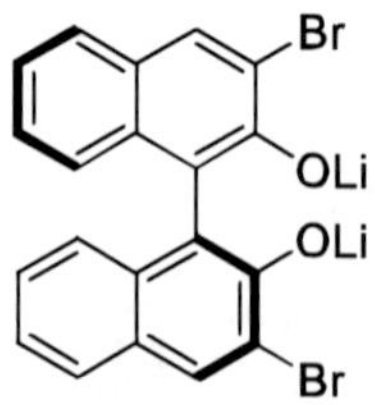

91

A metabolic study by Vickers *et al.* observed that various urine metabolites formed in humans, dogs, rats, and monkeys, and their concentration determined by gas-liquid chromatography and recovery of radioactivity are given in Scheme 16. Either single-step conversion forms these metabolites or a combination of these reactions in either sequence, such as i) dehydrazination; ii) dehydrazination, aromatic O-methylation; iii) dehydrazination, oxidative decarboxylation; iv) dehydrazination and α-hydroxylation v) dehydrazination, α-hydroxylation, aromatic O-methylation vi) dehydrazination and dehydroxylation [76].

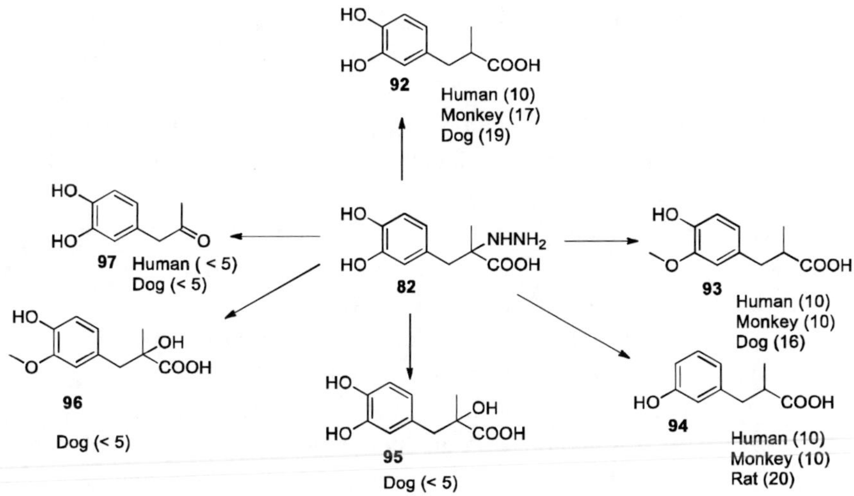

Scheme 16. Metabolic pathway of carbidopa (**82**).

In a comparative study of low-dose bromocriptine therapy versus low-dose levodopa-carbidopa combination by Hely *et al.*, it was observed that low-dose levodopa-carbidopa combination not only managed those symptoms of Parkinson's disease but also prolonged the dyskinesia appearance and emergence of end of dose failure for more than 2 years than routine doses. Comparatively, prolonging these things happened only for a year in a few patients in case of low bromocriptine doses. However, lower emergence dyskinesia appearance or dose failure onset was seen in the case of low-dose bromocriptine as compared to the low-dose levodopa-carbidopa combination [77].

Dopaminergic drugs have shown potential promise in treating restless legs (RL) and periodic leg movements in sleep (PLMS). They have demonstrated better effects than high doses of opiate analgesic propoxyphene. The levodopa/carbidopa combination was found to be effective in the management of PLMS. However, the combination potentiates the symptoms of RL and demands alternative therapy. Pergolide, an alternative dopaminergic drug, has revealed promising effects in managing RL symptoms [78].

In another alternative therapy study by Leguire *et al.*, the role of catecholamine neurotransmitters-neuromodulators drugs in the management of symptoms of amblyopia especially in improving visual sharpness was studied. A combination of levodopa-carbidopa and occlusion helps to improve visual acuity in older amblyopic children [79].

It was scrutinized that carbidopa inhibited human lung carcinoid cell line NCI-H727 (IC_{50} = 29 μM), NCI-H209 (IC_{50} = 22 μM) small cell lung cancer (SCLC) cell line, and NCI-H146 SCLC cell line (IC_{50} = 12 μM). 73% cell inhibition was observed on carbidopa exposure at 100 μM dose in the case of the SK-N-SH neuroblastoma cell line. In comparison, cell inhibition was 58% at the same dose of carbidopa in the A204 rhabdomyosarcoma cell line. However, no inhibition occurred in MCF7 breast carcinoma, NCI-H460 large cell lung cancer, and Du 145 prostate carcinoma cell line at a 100 μM concentration of carbidopa. Additive anticancer properties were seen in carcinoid cells upon combining carbidopa with etoposide, while topotecan indicated synergistic activity with carbidopa in SCLC cells [80].

In another assessment of finding newer applications of older drugs, carbidopa had a positive impact as an alternative for the treatment of familial dysautonomia (FD) to relieve patients from its symptoms such as nausea, retching, and vomiting. Conventional drugs such as clonidine and benzodiazepines suffer from sedative effects, while carbidopa is devoid of this side effect. However, this study had several limitations, such as a lack of dose fixation and non-confirmation of BBB crossing of carbidopa. Additionally, few patients had no relief from symptoms even after carbidopa administration [81].

Cansino-Torres *et al.* reported an allergic response of a 77-year-old woman patient who was diagnosed with idiopathic Parkinson disease who was on carbidopa-levodopa medication. Allergic reactions such as autonomic and dermatological signs: hypertension, conjunctival injection, excessive sweating, rhinorrhoea, and a generalized pruritic rash appeared quickly after medication, and these symptoms are unrelated to the manufacturer of the formulation. Dexamethasone, along with chloropyramine hydrochloride, managed these allergic reactions. These symptoms were specific to carbidopa, as the benserazide-levodopa combination did not produce these allergic responses. However, it was unclear if the allergy was due to active constituents or excipients used for formulations [82].

Rojo-Sebastián *et al.* found that little more than 79% of the Parkinsonian patients taking the levodopa-carbidopa combination suffered from vitamin B6 deficiency. This deficiency is due to the interaction of levodopa, especially of carbidopa, with pyridoxal pyrophosphate (vitamin B6 coenzyme). A direct corelation between carbidopa/levodopa-induced vitamin B6 deficiency and the appearance of dyskinesia was observed [83].

Benserazide

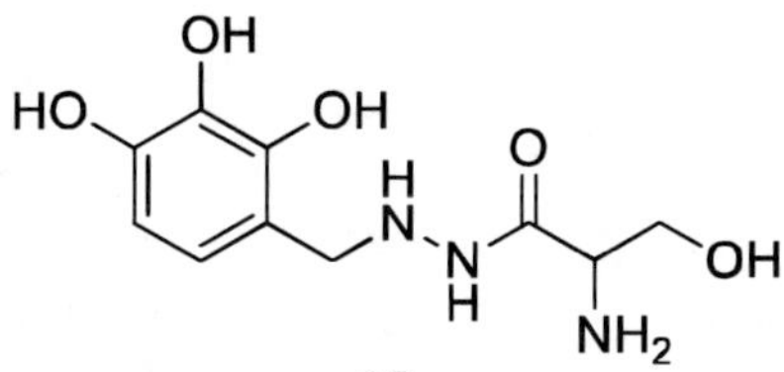

98

Benserazide (**98**), 2-amino-3-hydroxy-*N*'-(2,3,4-trihydroxybenzyl)propane-hydrazide, a dopa decarboxylase inhibitor, is combined with levodopa to treat Parkinson's disease. Dopa decarboxylase carries out extensive metabolism of levodopa peripherally which results in less amount of it crossing BBB to reach to the CNS. However, when combined with dopa decarboxylase inhibitors such as benserazide, and carbidopa, a significantly higher amount of levodopa could cross the BBB and, hence, be utilized in such combined form to achieve the maximum from the treatment. A combination of levodopa with either carbidopa or benserazide is equally effective, except frequent nausea-vomiting was associated with the carbidopa-levodopa combination as compared to the benserazide-levodopa combination [84].

In another interesting study, a combination of levodopa-benserazide was tried for restless leg syndrome, one of the major causes of insomnia. It was observed that the combination relieved restless leg syndrome symptoms but restored symptoms when the therapy was discontinued. It was also observed that dopaminergic receptor agonists reduce symptoms while dopaminergic receptor antagonists worsen them. Benserazide is a good substitute compared to codeine, a narcotic analgesic, because of the habit-forming tendency of codeine [85].

One prominent side effect of levodopa in the treatment of Parkinson's disease (PD) is that it induces dyskinesia called as levodopa-induced dyskinesia (LID). In one of the research studies by Yang *et al.*, co-workers used methyl ester of levodopa (LDME)/benserazide-loaded nanoparticles for the management of PD. It was realized that LDME/benserazide nanoparticles have better outcomes than the LDME/benserazide normal formulation in managing LID because of the controlled and longer release of levodopa-benserazide in the case of nanoparticle formulation [86].

An exciting study recognized that benserazide inhibits hexokinase-2, an enzyme in the early glycolytic pathway. In addition to its prominent role in tumor induction and maintenance, hexokinase-2 is overexpressed in malignant tumor than in normal cells. Hence, by inhibiting hexokinase-2, benserazide acts as an anticancer agent. Moreover, benserazide was responsible for decreased uptake of glucose which ultimately leads to decreased lactate and ATP level, which eventually can result in apoptosis. Also, it is accountable for the higher loss of mitochondrial membrane potential. Lysosomal formulation of benserazide improved efficacy which is liable to a reduction of doses of benserazide [87].

In melanoma cells, the M2 splice isoform of pyruvate kinase (PKM2) is over-expressed. Pyruvate kinase is one of the early-phase enzymes in the energy-generating glycolytic pathway. Benserazide selectively inhibits PKM2, which was revealed during a screening study for inhibition of PKM2 as benserazide^{-} does not bind to PKM1. As BRAFi-resistant cell lines overexpress PKM2, these cell lines have shown higher efficacy for benserazide. It was confirmed that benserazide had less efficacy in PKM2-suppressed melanoma cell lines, which suggests its specificity for PKM2 [88].

In a rare incidence, one patient with Parkinson's disease developed protracted pancolitis and ileitis upon treatment with carbidopa-levodopa for 5 months. Upon diagnosis, it was later realized that it was a irritable bowel syndrome (IBD) case. Replacement of levodopa-benserazide combination for levodopa-carbidopa combination relieved patients without IBD recurrence [89].

The synthesis of benserazide is given in Scheme 17. 2,3,4-Trihydroxybenzaldehyde **99** reacted with Serine hydrazide **100** in methanol to obtain N-(DL-seryl)-2,3,4-trihydroxybenzaldehyde hydrazone **101**. Reduction of hydrazone **101** yields benserazide (**98**) [90].

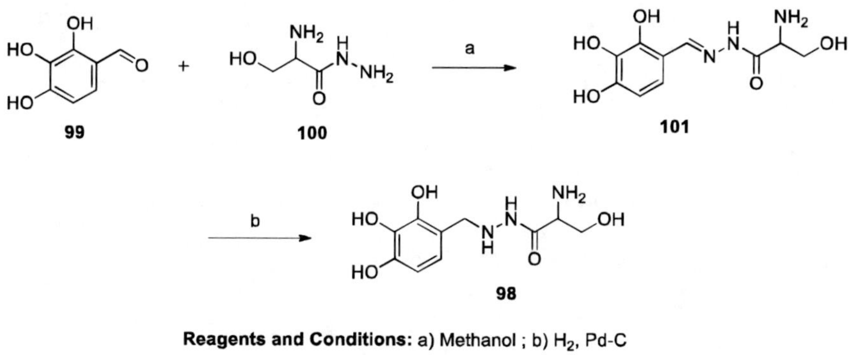

Scheme 17. Synthesis of benserazide (98).

Hydralazine

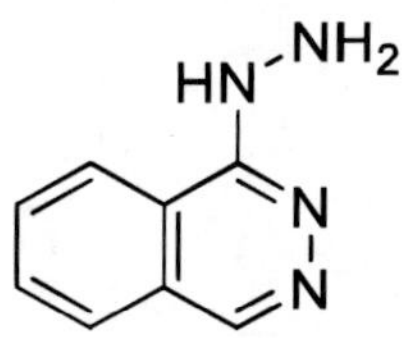

102

Hydralazine **102**, 1-hydrazinylphthalazine an interesting molecule that has a simple structure consisting of phthalazine with the presence of hydrazine at position 1, which acts as a peripheral vasodilator through relaxation of arterial smooth muscle and is utilized as an antihypertensive agent [91]. As this drug is a few decades old, lots of information is available regarding various aspects of it. Hydralazine suffers from many side effects, and the lupus-like syndrome is most prominent among them, as 4-20% of patients reported this effect.

Synthesis of Hydralazine (**Scheme 18**) is by oxidative chlorination of phthalide (**103**) with simultaneous hydrolysis of product, which results in hydroxyphthalide **104**, which upon reaction with hydrazine changes to phthalazone **105**. This undergoes a reaction with phosphorous oxychloride, forming 1-chlorophthalazine **106**, reacts with hydrazine, to yield hydralazine (**102**) [92, 93].

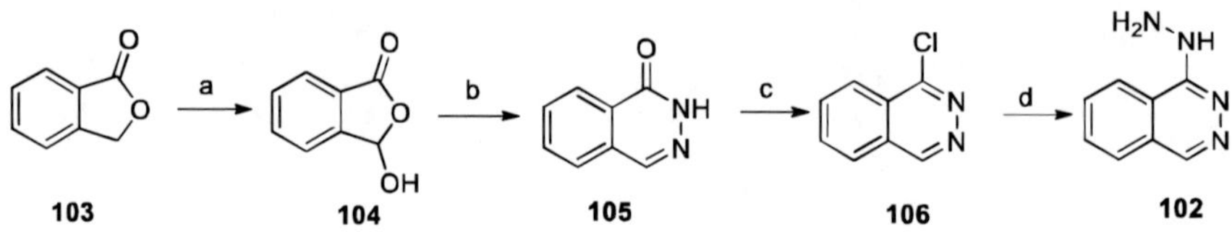

Reagents and conditions: a) Cl_2/H_2O ; b) H_2NNH_2 ; c) $POCl_3$; d) H_2NNH_2

Scheme 18. Synthesis of hydralazine (**102**).

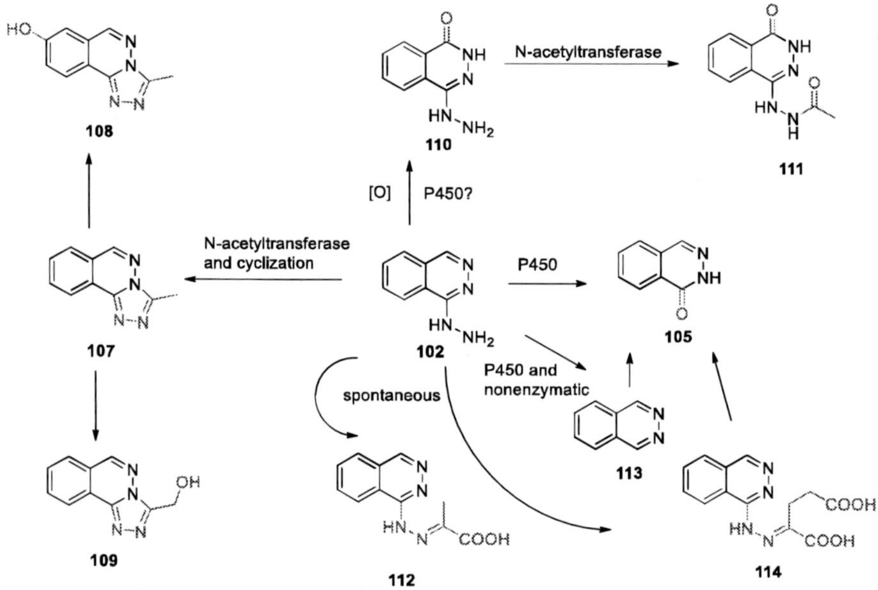

Scheme 19. Metabolic pathway of hydralazine (**102**).

The metabolic pathway of hydralazine is given in Scheme 19. Oxidation of hydralazine produces reactive species that might bind to proteins, eventually leading to lupus-like syndrome [94, 95, 96]. Although hydralazine is not a good substrate for peroxidase-catalyzed oxidation, other peroxidase substrates could enhance hydralazine metabolism. These reactive metabolites generated by peroxidase oxidation would be responsible for DNA strand damage [97]. It is interesting to know that nitrogens should be unsubstituted to maintain its potency. Hydroxyl, mercapto, amino, morpholino, or piperazinyl substitution for hydrazine leads to depletion in antihypertensive activity [98]. In repurposing studies, hydralazine has shown promise as an anticancer agent to reactivate tumor suppressor gene expression in vitro and in vivo testing with mice as a demethylating agent [99, 100]. Additionally, another repurposing study realized it has potential application in advanced heart failure treatment in black patients [101].

The factors that initiate the building of atherosclerotic plaque in the blood vessels are carbonyl stress, LDL oxidation, and oxidative stress. Another fascinating repurposing study showed that hydralazine analog bisvanillyl-hydralazone inhibited LDL oxidation, reduced oxidative stress, and prevented carbonyl stress. The effectiveness of bisvanillyl-hydralazone, in tackling atherosclerotic symptoms indicates its usefulness in atherosclerosis [102].

In one of the derivatization and repurposing studies, El-Sherif *et al.* synthesized imine by reacting hydralazine with salicylaldehyde. Synthesized imine metal complexes of Cu, Ni, and Co were prepared. These synthesized metal chelators of hydralazine imine analogs were tested for antimicrobial and antifungal activity. Both activities were moderate compared to the prototype drug, ampicillin [103].

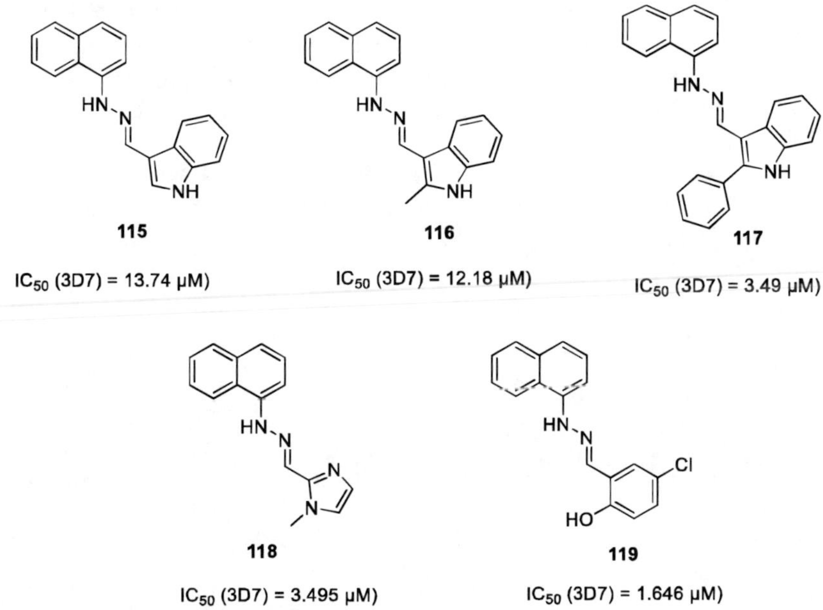

Figure 18. Prominent aromatic heterocyclic analog of hydralazine.

In an attempt of drug repurposing, Subramanian *et al.* explored the antimalarial properties of hydralazine analogs in their quest to search for different therapeutic properties associated with numerous heterocycles. Hydralazine was condensed with several aromatic and/or heteroaromatic or aliphatic aldehydes to synthesize an array of imines that were evaluated for antimalarial activity. Five of those tested compounds shown in Figure 18 exhibited activity against laboratory grade (3D7) of *Plasmodium falciparum* at the trophozoite stage with an IC_{50} value less than 20 μM [104].

Cadralazine

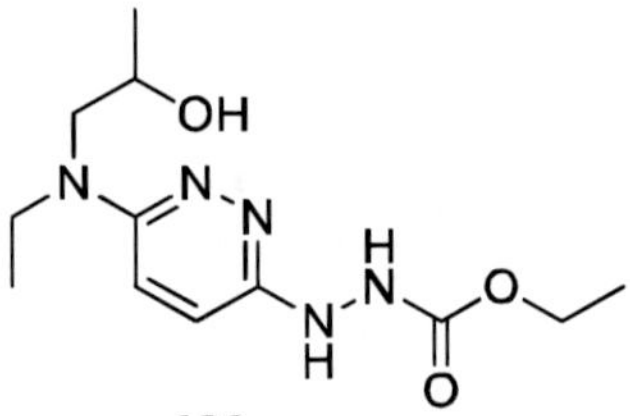

120

Cadralazine (**120**), ethyl 2-(6-(*N*-ethyl-*N*-(2-hydroxypropyl)amino)pyridazin-3-yl)hydrazine-1-carboxylate is a potent peripheral vasodilator used in hypertensive patients. Cadralazine is chemically related to hydralazine and has a smooth muscle relaxant effect [105, 106]. Voto *et al.* reported a cadralazine study for the treatment of preeclampsia [107].

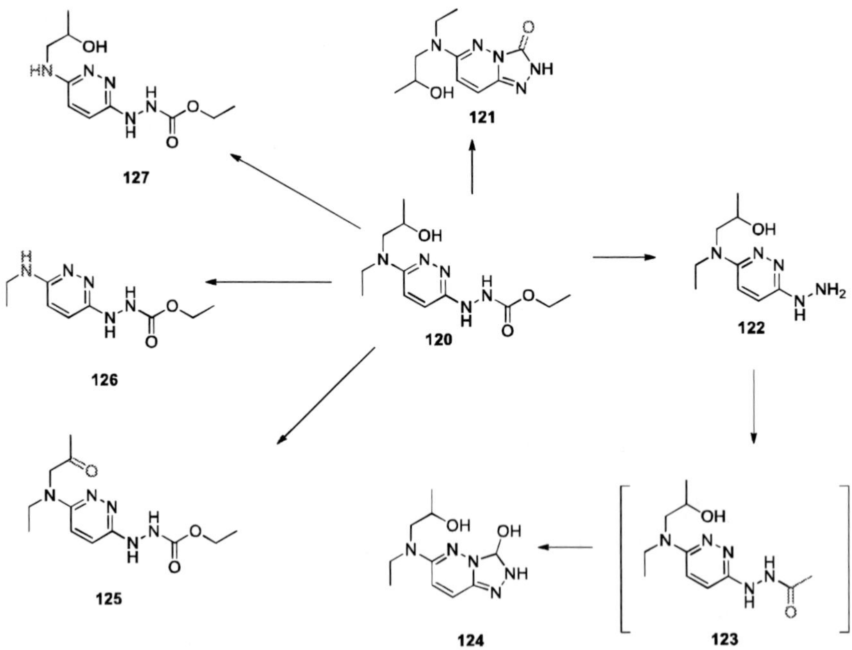

Scheme 20. Metabolic pathway of cadralazine (**120**).

Cadralazine undergoes direct cyclization to metabolites **121**. Cadralazine gets metabolized to metabolite **122** to form free hydrazine, which is further acetylated to form metabolite **123,** and ultimately, cyclization of **123** takes place to the metabolite **124**. Metabolite **122** shows a significant dose-dependent smooth muscle relaxing effect combined with potent antihypertensive action. Metabolite **125** is formed via oxidation, while metabolites **126** and **127** are formed by N-dealkylation (Scheme 20) [108, 109].

Clopamide

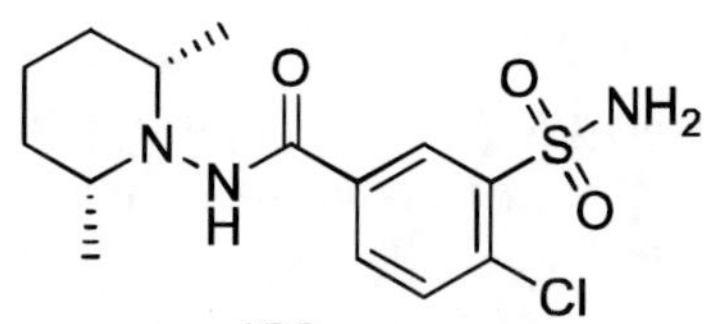

128

Clopamide (**128**), 4-chloro-*N*-((2*S*,6*R*)-2,6-dimethylpiperidin-1-yl)-3-sulfamoylbenzamide is a oral diuretic agent who has aromatic sulfonamide moiety same as of thiazide diuretics. It is widely used for the treatment of hypertension and cardiac failure. It acts on the nephron's distal convoluted tubule (DCT) by inhibiting the sodium-chloride symporter [110]. Clopamide and pindolol are used in combination to treat hypertension [111]. Gupta *et al.* reported the photodegradation study of clopamide during drug storage [112].

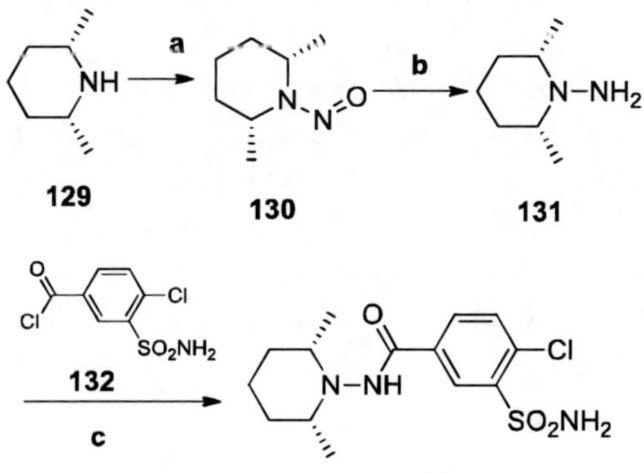

Reaction and conditions: a) H_2SO_4, $NaNO_2$,10-15 °C; b) Zn, AcOH, 0-3 °C; c) TEA, $CHCl_3$, rt, 48 h

Scheme 21. Synthesis of clopamide (**128**).

Clopamide is synthesized from 2,6-dimethylpiperidine **129** by nitration to yield *N*-nitropiperidine **130**, which upon reduction by Zn and acetic acid, yields *N*-aminopiperidine **131**. Coupling of **131** with sulfonamide **132** using triethylamine generates clopamide (**128**) (Scheme 21)[113].

Conclusion

The list of drugs containing hydrazine moiety consists of pretty old drugs such as isoniazid, nitrofurazone, nifurtimox, procarbazine, hydralazine, phenelzine, isocarboxazid, carbidopa, cadralazine and clopamide to newer ones such as selinexor, rimonabant which has applications in an array of diseases and disorders. Additionally, hydrazine is a useful reagent in synthesizing various heterocyclic (pyrazole, pyridazine, and many more) drugs that cover a large area of therapeutic utilities. Some of these drugs, such as isoniazid, still act as a primary source of drugs for the first-line treatment in tuberculosis. Although carbidopa is a very old drug, still prominently used in the treatment of Parkinson's disease.

In numerous repurpose studies, some of these drugs and their analogs are being explored for newer therapeutic applications for other diseases. These developments indicate that hydrazine is an important functional group to provide various pharmaceutical weapons utilized in several human ailments management. Also, newer drugs containing hydrazine are being launched. Many analogs having hydrazine moiety are poured daily into the research pipeline by the academic and pharmaceutical industry, signifying its importance in medicinal chemistry.

Close to a hundred drugs are already in the market, from hydrazine and hydrazine derivatives (pyrazole, pyridazine, and miscellaneous), and current research frequently synthesizes hydrazine analogs for bioactivity determination. This trend will continue to enrich drugs from the hydrazine analogs class to widen the treatment available and provide the first breakthrough for managing various diseases and disorders, including terminal and rare/orphan diseases.

References

[1] Barron, A. R., C. E. Hamilton, C. Smith "Hydrazine" *Chemistry of the Main Group Elements* (2010) 382-384.

[2] Meyer, H., J. Mally. "On hydrazine derivatives of pyridine carbonic acids." *Monatshefte Chemie Verwandte Teile Anderer Wissenschaften* 23 (1912): 393-414.
[3] Vardanyan, R., and V. Hruby. *Synthesis of Essential Drugs*. Elsevier, 200:512.
[4] Wang, P., K. Pradhan, X. Zhong, X. Ma. "Isoniazid metabolism and hepatotoxicity." *Acta Pharmaceutica Sinica B* 6, 5 (2016): 384-392.
[5] Kumar, D., G. Khare, S. Kidwai, A. K. Tyagi, R. Singh, D. S. Rawat. "Synthesis of novel 1, 2, 3-triazole derivatives of Isoniazid and their in vitro and in vivo antimycobacterial activity evaluation." *European Journal of Medicinal Chemistry* 81 (2014): 301-313.
[6] Santos, D. C., R. R. Henriques, M. A. de Abreu Lopes Junior, A. Borges Farias, T. L. do Couto Nogueira, J. V. Fernandes Quimas, N. C. Romeiro, L. L. da Silva, A. L. Ferreira de Souza. "Acylhydrazones as Isoniazid derivatives with multi-target profiles for the treatment of Alzheimer's disease: Radical scavenging, myeloperoxidase/acetylcholinesterase inhibition and biometal chelation." *Bioorganic & Medicinal Chemistry* 28, 10 (2020): 115470.
[7] Ipsita, M., J. Meshram. "Design, synthesis, and evaluation of Isoniazid derivatives acting as potent anti-inflammatory and anthelmintic agents via Betti reaction." *Medicinal Chemistry Research* 23, 2 (2014): 939-947.
[8] Malhotra, M., S. Sharma, A. Deep. "Synthesis, characterization and antimicrobial evaluation of novel derivatives of Isoniazid." *Medicinal Chemistry Research* 21, 7 (2012): 1237-1244.
[9] Castelo-Branco, F. S., E. C. de Lima, J. L. de Oliveira Domingos, A. C. Pinto, M. C. S. Lourenço, K. M. Gomes, M. M. Costa-Lima, C. A. Fortes Aiub, I. Felzenszwalb, T. E. M M Costa, C. Penido, M. G. Henriques, N. Boechat "New hydrazides derivatives of Isoniazid against Mycobacterium tuberculosis: Higher potency and lower hepatocytotoxicity."*European Journal of Medicinal Chemistry* 146 (2018): 529-540.
[10] Matei, L., C. Bleotu, I. Baciu, C. Draghici, P. Ionita, A. Paun, M. C. Chifiriuc, A. Sbarcea, I. Zarafu. "Synthesis and bioevaluation of some new Isoniazid derivatives." *Bioorganic & Medicinal Chemistry* 21, 17 (2013): 5355-5361.
[11] Volynets, G. P., M. A. Tukalo, V. G. Bdzhola, N. M. Derkach, M. I. Gumeniuk, S. S. Tarnavskiy, S. M. Yarmoluk. "Novel Isoniazid derivative as promising antituberculosis agent." *Future Microbiology* 15, 10 (2020): 869-879.
[12] Rizvi, F., M. Khan, A. Jabeen, H. Siddiqui, M. I. Choudhary. "Studies on Isoniazid derivatives through a medicinal chemistry approach for the identification of new inhibitors of urease and inflammatory markers." *Scientific Reports* 9, 1 (2019): 6738-6742.
[13] Kaur, H., L. Singh, K. Chibale, and K. Singh. "Structure elaboration of Isoniazid: Synthesis, *in silico* molecular docking and antimycobacterial activity of Isoniazid–pyrimidine conjugates." *Molecular Diversity* 24, 4 (2020): 949-955.
[14] Hu, Y. Q., S. Zhang, F. Zhao, C. Gao, L-S Feng, Zao-Sheng Lv, Z. Xu, X. Wu. "Isoniazid derivatives and their anti-tubercular activity." *European Journal of Medicinal Chemistry* 133 (2017): 255-267.

[15] Vilchèze, C., W. R. Jacobs. "The Isoniazid paradigm of killing, resistance, and persistence in Mycobacterium tuberculosis." *Journal of Molecular Biology* 431, 18 (2019): 3450-3461.

[16] Leitner, A., P. Zöllner, W. Lindner. "Determination of the metabolites of nitrofuran antibiotics in animal tissue by high-performance liquid chromatography–tandem mass spectrometry." *Journal of Chromatography A* 939, 1-2 (2001): 49-58.

[17] Tehrani, H., M. H. A. Zarghi, S. Fathali. "A modified method for the synthesis of nitrofurazone." *Iranian Journal of Pharmaceutical Research* 2 (2010): 67-69.

[18] European Food Safety Authority (EFSA), M. O'Keeffe, A. Christodoulidou, C. Nebbia. "Presence of nitrofurans and their metabolites in gelatine." *EFSA Journal* 19, 10 (2021): e06881.

[19] Afghahi, F., S. Yazdany, I. Lalezari. "Synthesis and antibacterial and antifungal activities of 5-nitro-2-furfurylidene polyhalophenoxyacethydrazides VIII." *Journal of Pharmaceutical Sciences* 64, 5 (1975): 858-859.

[20] Leitner, A., P. Zöllner, W. Lindner. "Determination of the metabolites of nitrofuran antibiotics in animal tissue by high-performance liquid chromatography–tandem mass spectrometry." *Journal of Chromatography A* 939, 1-2 (2001): 49-58.

[21] Chung, M-C., R. V. Carvalho Güido, T. F. Martinelli, M. F. Gonçalves, M. C. Polli, K. C. Alves Botelho, E. A. Varanda, W. Colli, M. T. M. Miranda, E. I. Ferreira. "Synthesis and in vitro evaluation of potential antichagasic hydroxymethylnitrofurazone (NFOH-121): a new nitrofurazone prodrug." *Bioorganic & Medicinal Chemistry* 11, 22 (2003): 4779-4783.

[22] Trossini, G. H. G., A. Malvezzi, A. T.-do Amaral, C. O. Rangel-Yagui, M. A. Izidoro, M. Helena S. Cezari, L. Juliano, C. Man Chin, C. MS Menezes, E. Igne Ferreira. "Cruzain inhibition by hydroxymethylnitrofurazone and nitrofurazone: investigation of a new target in Trypanosoma cruzi." *Journal of Enzyme Inhibition and Medicinal Chemistry* 25, 1 (2010): 62-67.

[23] Popiołek, L., B. Rysz, A. Biernasiuk, M. Wujec. "Synthesis of promising antimicrobial agents: hydrazide-hydrazones of 5-nitrofuran-2-carboxylic acid." *Chemical Biology & Drug Design* 95, 2 (2020): 260-269.

[24] Le, V. V. H., I. G. Davies, C. D. Moon, D. Wheeler, P. J. Biggs, J. Rakonjac. "Novel 5-nitrofuran-activating reductase in Escherichia coli." *Antimicrobial Agents and Chemotherapy* 63, 11 (2019): e00868-19.

[25] Baldelli, V., F. D'Angelo, V. Pavoncello, E. V. Fiscarelli, P. Visca, G. Rampioni, L. Leoni. "Identification of FDA-approved antivirulence drugs targeting the Pseudomonas aeruginosa quorum sensing effector protein PqsE." *Virulence* 11, 1 (2020): 652-668.

[26] Krátký, M., Š. Štěpánková, K. Konečná, K. Svrčková, J. Maixnerová, M. Švarcová, O. Janďourek, F. Trejtnar, J. Vinšová. "Novel Aminoguanidine Hydrazone Analogs: From Potential Antimicrobial Agents to Potent Cholinesterase Inhibitors." *Pharmaceuticals* 14, 12 (2021): 1229.

[27] El-Wakil, M. H., M. A. Meheissen, M. M. Abu-Serie. "Nitrofurazone repurposing towards design and synthesis of novel apoptotic-dependent anticancer and antimicrobial agents: Biological evaluation, kinetic studies and molecular modeling." *Bioorganic Chemistry* 113 (2021): 104971.

[28] Wahid, S., S. Jahangir, M. A. Versiani, K. M. Khan, Y. Y. Sung, J. Iqbal, A. Wadood, A. U. Rehman, A. M. Uzair, I. A. Khan, M. Taha, S. Perveen "Biology-oriented drug synthesis of nitrofurazone derivatives: Their α-glucosidase inhibitory activity and molecular docking studies." *Arabian Journal of Chemistry* 15, 6 (2022): 103806.

[29] Hall, B. S., C. Bot, S. R. Wilkinson. "Nifurtimox activation by trypanosomal type I nitroreductases generates cytotoxic nitrile metabolites." *Journal of Biological Chemistry* 286, 15 (2011): 13088-13095.

[30] Scarim, C. B., D. H. Jornada, R. C. Chelucci, L de Almeida, J. L. Dos Santos, and M. C. Chung. "Current advances in drug discovery for Chagas disease." *European Journal of Medicinal Chemistry* 155 (2018): 824-838.

[31] Vardanyan, R., V. Hruby. *Synthesis of Essential Drugs*. Elsevier, 200:512: 578.

[32] Rice, A. M., Y. Long, S. B. King. "Nitroaromatic antibiotics as nitrogen oxide sources." *Biomolecules* 11, 2 (2021): 267.

[33] Spivack, S. D., "Drugs five years later: Procarbazine." *Annals of Internal Medicine* 81, 6 (1974): 795-800.

[34] Lee, I. P., R. L. Dixon. "Mutagenicity, carcinogenicity and teratogenicity of procarbazine." *Mutation Research/Reviews in Genetic Toxicology* 55, 1 (1978): 1-14.

[35] Wang, R., C. Zhang, C. Zheng, H. Li, X. Xie, Y. Jin, Z. Liu, H. Chen. "Introduction of Z-GP scaffold into procarbazine reduces spermatoxicity and myelosuppression." *Bioorganic Chemistry* 83 (2019): 461-467.

[36] Swaffar, D. S., D. Shomin, M. G. Horstman, J-Yan Jaw, B. D. Thrall, G. G. Meadows, W. G. Harker, G. S. Yost. "Methylazoxyprocarbazine, the active metabolite responsible for the anticancer activity of procarbazine against L1210 leukemia." *Cancer Research* 49, 9 (1989): 2442-2447.

[37] Beisler, J. A., G. W. Peng, and J. S. Driscoll. "Potential antitumor agents: procarbazine analogs and other methylhydrazine derivatives." *Journal of Pharmaceutical Sciences* 66, 6 (1977): 849-852.

[38] Weinstock, L. T., L. T., and C. C. Cheng. "A structural modification study of procarbazine." *Journal of Medicinal Chemistry* 22, 5 (1979): 594-597.

[39] Holme, J. A., E. J. Søderlund, G. Brunborg, J. K. Hongslo, B. Trygg, S Nelson. "Comparative genotoxicities of procarbazine and two deuterated analogs in mammalian cells in vitro and in vivo." *Mutagenesis* 4, 5 (1989): 355-360.

[40] Grossman, S. A., K. A. Carson, T. T. Batchelor, G. Lesser, T. Mikkelsen, J. B. Alavi, S. Phuphanich, T. Hammour, J. D. Fisher, J. G. Supko. "The effect of enzyme-inducing antiseizure drugs on the pharmacokinetics and tolerability of procarbazine hydrochloride." *Clinical Cancer Research* 12, 17 (2006): 5174-5181.

[41] https://www.fda.gov/news-events/press-announcements/fda-approves-new-treatment-refractory-multiple-myeloma, updated on 07/03/2019, visited on September 04, 2022.

[42] Argueta, C., T. Kashyap, B. Klebanov, T. J. Unger, C. Guo, S. Harrington, E. Baloglu, M. Lee, W. Senapedis, S. Shacham, Y. Landesman "Selinexor synergizes with dexamethasone to repress mTORC1 signaling and induce multiple myeloma cell death." *Oncotarget* 9, 39 (2018): 25529.

[43] Syed, Y. Y. "Selinexor: first global approval." *Drugs* 79, 13 (2019): 1485-1494.
[44] https://www.fda.gov/news-events/press-announcements/fda-approves-new-treatment-refractory-multiple-myeloma, accesed on 13 Sep 2022.
[45] Crochiere, M., W. Senapedis, T. Kashyap, T. Rashal, D. McCauley, M. Kauffman, S. Shacham, Y. Landesman. "The selective inhibitor of nuclear export compound, selinexor, inhibits NF-κb and induces anti-non-small cell lung cancer activity regardless of p53 status." *Int J Cancer Res Mol Mech* 2, 2 (2016), 2381-3318.
[46] Razak, A., R. Abdul, M. Mau-Soerensen, N. Y. Gabrail, J. F. Gerecitano, A. F. Shields, T. J. Unger, J. R. Saint-Martin "First-in-class, first-in-human phase I study of selinexor, a selective inhibitor of nuclear export, in patients with advanced solid tumors." *Journal of Clinical Oncology* 34, no. 34 (2016): 4142.
[47] Yuan, S., B. Yu, H. Liu. "New drug approvals for 2019: Synthesis and clinical applications." *European Journal of Medicinal Chemistry* 205 (2020): 112667.
[48] Yin, Z., W. Hu, W. Zhang, H. Konno, H. Moriwaki, K. Izawa, J. Han, V. A. Soloshonok. "Tailor-made amino acid-derived pharmaceuticals approved by the FDA in 2019." *Amino Acids* 52, 9 (2020): 1227-1261.
[49] Chen, X, L Xu L, W. Liu. Biological Technology Co Ltd, "*Novel synthesis method of Selinexor active pharmaceutical ingredient*." CN106831731, 8 Nov. 2019.
[50] Razak, A., R. Abdul, M. Mau-Soerensen, N. Y. Gabrail, J. F. Gerecitano, A. F. Shields, T. J. Unger, J. R. Saint-Martin "First-in-class, first-in-human phase I study of selinexor, a selective inhibitor of nuclear export, in patients with advanced solid tumors." *Journal of Clinical Oncology* 34, 34 (2016): 4142.
[51] Kotagiri, V. K., S. Suthrapu, J. M. Reddy, C. P. Rao, V. Bollugoddu, A. Bhattacharya, R. Bandichhor. "An improved synthesis of rimonabant: anti-obesity drug." *Organic Process Research & Development* 11, 5 (2007): 910-912.
[52] Kotagiri, V. K., S. Suthrapu, J. M. Reddy, C. P. Rao, V. Bollugoddu, A. Bhattacharya, R. Bandichhor. "An improved synthesis of rimonabant: anti-obesity drug." *Organic Process Research & Development* 11, 5 (2007): 910-912.
[53] Després, J. P., A. Golay, L. Sjöström. "Effects of rimonabant on metabolic risk factors in overweight patients with dyslipidemia." *New England Journal of Medicine* 353, 20 (2005): 2121-2134.
[54] Russell, J. C., S. E. Kelly, A. Diane, Y. Wang, R. Mangat, S. Novak, D. F. Vine, S. D. Proctor. "Rimonabant-mediated changes in intestinal lipid metabolism and improved renal vascular dysfunction in the JCR: LA-cp rat model of prediabetic metabolic syndrome." *American Journal of Physiology-Gastrointestinal and Liver Physiology* 299, 2 (2010): G507-G516.
[55] Bergström, M. A., E. M. Isin, N. Castagnoli, C. E. Milne. "Bioactivation pathways of the cannabinoid receptor 1 antagonist rimonabant." *Drug Metabolism and Disposition* 39, 10 (2011): 1823-1832.
[56] Fernández-Fernández, C., L. F. Callado, R. Girón, E Sánchez, A. M. Erdozain, J. Antonio López-Moreno, P. Morales, F. R. de Fonseca, J. Fernández-Ruiz, P. Goya, J. J. Meana, M. Isabel Martín, N. Jagerovic "Combining rimonabant and fentanyl in a single entity: preparation and pharmacological results." *Drug Design, Development and Therapy* 8 (2014): 263–277.

[57] Menozzi, G., P. Fossa, E. Cichero, A. Spallarossa, A. Ranise, L. Mosti. "Rational design, synthesis and biological evaluation of new 1, 5-diarylpyrazole derivatives as CB1 receptor antagonists, structurally related to rimonabant." *European Journal of Medicinal Chemistry* 43, 12 (2008): 2627-2638.

[58] Gajbhiye, J. M., N. A. More, M. D. Patil, R. Ummanni, S. S. Kotapalli, P. Yogeeswari, D. Sriram, V. H. Masand. "Discovery of Rimonabant and its potential analogs as anti-TB drug candidates." *Medicinal Chemistry Research* 24, 7 (2015): 2960-2971.

[59] McManus, D. J., G. B. Baker, I. L. Martin, A. J. Greenshaw, K. F. McKenna. "Effects of the antidepressant/antipanic drug phenelzine on GABA concentrations and GABA-transaminase activity in rat brain." *Biochemical Pharmacology* 43, 11 (1992): 2486-2489.

[60] MacKenzie, E. M., M-S Song, S. M. Dursun, S. Tomlinson, K. G. Todd, G. B. Baker. "Phenelzine: An old drug that may hold clues to the development of new neuroprotective agents." *Klinik Psikofarmakoloji Bülteni-Bulletin of Clinical Psychopharmacology* 20, 2 (2010): 179-186.

[61] Matveychuk, D., E. M. MacKenzie, D. Kumpula, M-S Song, A. Holt, S. Kar, K. G. Todd, P. L. Wood, Glen B. Baker. "Overview of the neuroprotective effects of the MAO-inhibiting antidepressant phenelzine." *Cellular and Molecular Neurobiology* 42, 1 (2022): 225-242.

[62] Gaur, S., M. E. Gross, C. Liao, B. Qian, and J. C. Shih. "Effect of Monoamine oxidase A (MAOA) inhibitors on androgen-sensitive and castration-resistant prostate cancer cells." *The Prostate* 79, 6 (2019): 667-677.

[63] Ait Chait, Y., W. Mottawea, T. A. Tompkins, R. Hammami. "Unravelling the antimicrobial action of antidepressants on gut commesal microbes." *Scientific Reports* 10, 1 (2020): 17878-17891.

[64] Vardanyan, R., V. Hruby. *Synthesis of Essential Drugs*. Elsevier, 2006: 111.

[65] Mehndiratta, S., J-P Liou. "Histone lysine specific demethylase 1 inhibitors." *RSC Medicinal Chemistry* 11, 9 (2020): 969-981.

[66] Baker, G. B., L. J. Urichuk, K. F. McKenna, S. H. Kennedy. "Metabolism of monoamine oxidase inhibitors." *Cellular and Molecular Neurobiology* 19, 3 (1999): 411-426.

[67] Nakamura, M., H. Fukushima, S. Kitagawa. "Effects of amitriptyline and isocarboxazid on 5-hydroxytryptophan induced head twitches in mice." *Psycho-pharmacology* 48, 1 (1976): 101-104.

[68] Ahmad. S., M. A. Hughes, L. Yeh, and J. E. Scott. "Potential repurposing of known drugs as potent bacterial β-glucuronidase inhibitors." *Journal of Biomolecular Screening* 17, 7 (2012): 957-965.

[69] Hadizadeh, F., R. Ghodsi. "Synthesis of novel *N*-substituted imidazolecarboxylic acid hydrazides as monoamine oxidase inhibitors." *Il Farmac* 60, 3 (2005): 237-240.

[70] Nulu, J. R., J. Nematollahi. "Imidazolecarbohydrazides. III. Chemistry and biological evaluation." *Journal of Medicinal Chemistry* 12, 5 (1969): 804-806.

[71] Mars, H., "Levodopa, carbidopa, and pyridoxine in Parkinson disease: Metabolic interactions." *Archives of Neurology* 30, 6 (1974): 444-447.

[72] Wurtman, R., C. Watkins. 'Suppression of noradrenaline synthesis in sympathetic nerves by carbidopa, an inhibitor of peripheral dopa decarboxylase.' *Nature,* 265, (1977), 79–80.

[73] Shelton, R. C., F. Karoum, L. Chuang, R. J. Wyatt. "Metabolism of carbidopa to alphamethyldopamine and alphamethylnorepinephrine in rats." *Biochemical Pharmacology* 34, 20 (1985): 3623-3626.

[74] Pericas, A., A. Shafir, A. Vallribera. "Asymmetric Synthesis of L-carbidopa based on a highly enantioselective α-amination." *Organic Letters* 15, 7 (2013): 1448-1451.

[75] Asano, T., M. Moritani, M. Nakajima, S. Kotani. "Chiral lithium binaphtholate for enantioselective amination of acyclic α-alkyl-β-keto esters: Application to the total synthesis of L-carbidopa." *Tetrahedron* 73, 41 (2017): 5975-5982.

[76] Vickers, S., E. K. Stuart, H. B. Hucker. "Further studies on the metabolism of carbidopa, (minus)-L-alpha-hydrazino-3, 4-dihydroxy-alpha-methylbenzene-propanoic acid monohydrate, in the human, Rhesus monkey, dog, and rat." *Journal of Medicinal Chemistry* 18, 2 (1975): 134-138.

[77] Hely, M. A., J. G. Morris, W. G. Reid, D. J. O'Sullivan, P. M. Williamson, D. Rail, G. A. Broe, S. Margrie. "The Sydney Multicentre Study of Parkinson's disease: a randomised, prospective five year study comparing low dose bromocriptine with low dose levodopa-carbidopa." *Journal of Neurology, Neurosurgery & Psychiatry* 57, 8 (1994): 903-910.

[78] Earley, C. J., R. P. Allen. "Pergolide and carbidopa/levodopa treatment of the restless legs syndrome and periodic leg movements in sleep in a consecutive series of patients." *Sleep* 19, 10 (1996): 801-810.

[79] Leguire, L. E., G. L. Rogers, P. D. Walson, D. L. Bremer, M. L. McGregor. "Occlusion and levodopa-carbidopa treatment for childhood amblyopia." *Journal of American Association for Pediatric Ophthalmology and Strabismus* 2, 5 (1998): 257-264.

[80] Gilbert, J. A., L. M. Frederick, M. M. Ames. "The aromatic-L-amino acid decarboxylase inhibitor carbidopa is selectively cytotoxic to human pulmonary carcinoid and small cell lung carcinoma cells." *Clinical Cancer Research* 6, 11 (2000): 4365-4372.

[81] Norcliffe-Kaufmann, L., J. Martinez, F. Axelrod, H. Kaufmann. "Hyper-dopaminergic crises in familial dysautonomia: a randomized trial of carbidopa." *Neurology* 80, 17 (2013): 1611-1617.

[82] Cansino-Torres, M. A., V. Sandoval-Rodríguez, M. Sáenz-Farret, C. Zúñiga-Ramírez. "Allergy to carbidopa." *Clinical Neuropharmacology* 41, 2 (2018): 75-79.

[83] Rojo-Sebastián, A., C. González-Robles, J. G. de Yébenes. "Vitamin B6 deficiency in patients with Parkinson disease treated with levodopa/carbidopa." *Clinical Neuropharmacology* 43, 5 (2020): 151-157.

[84] Rinne, U. K., P. Mölsä. "Levodopa with benserazide or carbidopa in Parkinson disease." *Neurology* 29, 12 (1979): 1584-1589.

[85] Akpinar, S. "Treatment of restless legs syndrome with levodopa plus benserazide." *Archives of Neurology* 39, 11 (1982): 739-739.

[86] Yang, X., R. Zheng, Y. Cai, M. Liao, W. Yuan, and Z. Liu. "Controlled-release levodopa methyl ester/benserazide-loaded nanoparticles ameliorate levodopa-induced dyskinesia in rats." *International Journal of Nanomedicine* 7 (2012): 2077.

[87] Li, W., M. Zheng, S. Wu, S. Gao, M. Yang, Z. Li, Q. Min W. Sun, L. Chen, G. Xiang H. Li. "Benserazide, a dopadecarboxylase inhibitor, suppresses tumor growth by targeting hexokinase 2." *Journal of Experimental & Clinical Cancer Research* 36, 1 (2017): 1-12.

[88] Zhou, Youyou, Zunnan Huang, Juan Su, Jie Li, Shuang Zhao, Lisha Wu, JiangLing Zhang, W. Sun, L. Chen, G. Xiang, H. Li. "Benserazide is a novel inhibitor targeting PKM2 for melanoma treatment." *International Journal of Cancer* 147, 1 (2020): 139-151.

[89] Steel, M. J., J. Conner, S. Nelles, R. Kirsch. "A case of carbidopa/levodopa associated colitis mimicking inflammatory bowel disease." *Human Pathology Reports* 28 (2022): 300648.

[90] Yang, Y., W. Li Ke, G. Feng, J. Yongmei, H. Zhen, H. Fuming, G. Yunsheng, Z. Jianhong, L. Huili. Shanghai Yishengyuan Pharmaceutical Co Ltd "*Preparation method of benserazide hydrochloride*" CN104788338A 22 July 2015.

[91] Koch-Weser, J. "Hydralazine." *New England Journal of Medicine* 295, 6 (1976): 320-323.

[92] Vardanyan, R., V. Hruby. *Synthesis of Essential Drugs*. Elsevier, 200:304.

[93] Heravi, M. M., V. Zadsirjan. "Prescribed drugs containing nitrogen heterocycles: an overview." *RSC Advances* 10, 72 (2020): 44247-44311.

[94] Hofstra, A. H., "Metabolism of hydralazine: relevance to drug-induced lupus." *Drug Metabolism Reviews* 26, 3 (1994): 485-505.

[95] Israili, Z. H., P. G. Dayton. "Metabolism of hydralazine." *Drug Metabolism Reviews* 6, 2 (1977): 283-305.

[96] Zak, S. B., T. G. Gilleran, J. Karliner, G. Lukas. "Identification of two metabolites of hydralazine from human urine." *Journal of Medicinal Chemistry* 17, 3 (1974): 381-382.

[97] Reilly, C. A., S. D. Aust. "Peroxidase substrates stimulate the oxidation of hydralazine to metabolites which cause single-strand breaks in DNA." *Chemical Research in Toxicology* 10, 3 (1997): 328-334.

[98] Reece, P. A., "Hydralazine and related compounds: chemistry, metabolism, and mode of action." *Medicinal Research Reviews* 1, 1 (1981): 73-96.

[99] Segura-Pacheco, B., C. Trejo-Becerril, E. Perez-Cardenas, L. Taja-Chayeb, I. Mariscal, A. Chavez, C. Acuna, A. M. Salazar, M. Lizano, A. Duenas-Gonzalez. "Reactivation of tumor suppressor genes by the cardiovascular drugs hydralazine and procainamide and their potential use in cancer therapy." *Clinical Cancer Research* 9, 5 (2003): 1596-1603.

[100] Zambrano, P., B. Segura-Pacheco, E. Perez-Cardenas, L. Cetina, A. Revilla-Vazquez, L. Taja-Chayeb, A. Chavez-Blanco, E. Angeles, G. Cabreral, K. Sandoval, C. Trejo-Becerril, J. Chanona-Vilchis, A. Duenas-González. "A phase I study of hydralazine to demethylate and reactivate the expression of tumor suppressor genes." *BMC Cancer* 5, 1 (2005): 1-12.

[101] Taylor, A. L., S. Ziesche, C. Yancy, P. Carson, R. D'Agostino Jr, K. Ferdinand, M. Taylor, A. Kirkwood, M. Sabolinski, M. Worcel, J. N. Cohn. "Combination of isosorbide dinitrate and hydralazine in blacks with heart failure." *New England Journal of Medicine* 351, 20 (2004): 2049-2057.

[102] Bouguerne, B., N. Belkheiri, F. Bedos-Belval, C. Vindis, K. Uchida, H. Duran, M. Grazide, M. Baltas, R. Salvayre, A. Nègre-Salvayre. "Antiatherogenic effect of bisvanillyl-hydralazone, a new hydralazine derivative with antioxidant, carbonyl scavenger, and antiapoptotic properties." *Antioxidants & Redox Signaling* 14, 11 (2011): 2093-2106.

[103] El-Sherif, A. A., M. M. Shoukry, M. MA Abd-Elgawad. "Synthesis, characterization, biological activity and equilibrium studies of metal (II) ion complexes with tridentate hydrazone ligand derived from hydralazine." *Spectrochimica Acta Part A: Molecular and Biomolecular Spectroscopy* 98 (2012): 307-321.

[104] Subramanian, G., C. P B. Rajeev, C. D. Mohan, A. Sinha, T. T. T. Chu, S. Anusha, X. Huang, J. E. Fuchs, A. Bender, K. S. Rangappa, R. Chandramohanadas, Basappa. "Synthesis and in vitro evaluation of hydrazinyl phthalazines against malaria parasite, Plasmodium falciparum." *Bioorganic & Medicinal Chemistry Letters* 26, 14 (2016): 3300-3306.

[105] Visconti, M., L. Citerio, M. Borsa, and G. Pifferi. "Degradative behavior of cadralazine in aqueous solution." *Journal of Pharmaceutical Sciences* 73, no. 12 (1984): 1812-1815.

[106] Semeraro, C., L. Dorigotti, S. Banfi, C. Carpi. "Pharmacological studies on cadralazine: a new antihypertensive vasodilator drug." *Journal of Cardiovascular Pharmacology* 3, no. 3 (1981): 455-467.

[107] Voto, L., A. M. Lapidus, P. Catuzzi, F. U. Imaz, C. Zin, M. Margulies. "Cadralazine for the treatment of preeclampsia. An open, noncomparative, dose-finding pilot study." *Hypertension* 19, no. 2_supplement (1992): II132.

[108] Simonotti, L., M. Zanol, F. Parravicini, G. Pifferi. "Studies on the metabolism of cadralazine in rat." *European Journal of Drug Metabolism and Pharmacokinetics* 7, no. 1 (1982): 39-45.

[109] Schütz, H., J. W. Faigle, W. Küng, W. Theobald. "Disposition and pharmacokinetics of cadralazine and individual metabolites in man." *European Journal of Drug Metabolism and Pharmacokinetics* 10, no. 2 (1985): 147-153.

[110] McNeil, J., E. L. Conway, O. H. Drummer, L. G. Howes, N. Christophidis, William J. Louis. "Clopamide: Plasma concentrations and diuretic effect in humans." *Clinical Pharmacology and Therapeutics* 42, no. 3 (1987): 299-304.

[111] Panderi, I., M. Parissi-Poulou. "Determination of clopamide-pindolol combination in tablets by fourth-order derivative UV spectrophotometry." *International Journal of Pharmaceutics* 99, no. 2-3 (1993): 327-331.

[112] Gupta, A., M. R. Zaheer, S. Iqbal, Roohi, A. Ahmad, M. B. Alshammari. "Photodegradation and In Silico Molecular Docking Study of a Diuretic Drug: Clopamide." *ACS Omega* 7, no. 16 (2022): 13870-13877.

[113] Jucker, E., A. Lindenmann. "Über diuretisch wirksame Benzoesäure-Derivate Untersuchungen über synthetische Arzneimittel, 9. Mitteilung [*About diuretic*

benzoic acid derivatives studies on synthetic drugs, 9th part]." *Helvetica Chimica Acta* 45, no. 7 (1962): 2316-2325.

Biographical Sketches

Shubhangi Sachidanand Savale

Affiliation: Institute of Chemical Technology, Mumbai

Education: B. Pharmacy

Business Address: 2VF4+GX4, Nathalal Parekh Marg, near Khalsa College, Matunga, Mumbai, Maharashtra 400019

Professional Appointments: Pursuing a Master's in Pharmacy (Pharmaceutical Chemistry)

Honors:

1. Secured AIR 85 in GPAT Entrance Examination, 2021.
2. Secured AIR 532 in NIPER JEE Entrance Examination, 2021.
3. Secured Second Rank in Bachelor of Pharmacy degree course at Shivaji University.

Publications from the Last 3 Years:

1. Published Research Article: Ganesh D. Mote, Shubhangi S. Savale, Shubhangi S. Kharat and Aditya A. Bandgar "In Silico Molecular Modeling Study on Isatin Derivatives as Anti-Covid Agents Based on QSAR and Docking Analysis" *Der Pharma Chemica,* 2022,14(3),1-15, 10.4172/0975-413X.14.3.1-15.

Vijayendran K. K. Praneeth

Affiliation: Department of Chemistry, Kalasalingam Academy of Research and Education

Education: M. Sc., Ph. D. Chemistry

Business Address: Department of Chemistry, Kalasalingam Academy of Research and Education, Anand Nagar, Krishnankoil, Tamil Nadu

Research and Professional Experience:

- Ph. D. in Synthetic Chemistry, University of Kiel, Germany and University of Michigan, USA.
- Postdoctoral Research Fellow, University of Minnesota, USA.

Professional Appointments:

- 2011-2014; Senior Teaching and Research Associate, University of Basel, Switzerland.
- 2014-2019: Research Staff, IMS., Okazaki, Japan.
- 2019-2021: Associate Professor, School of Applied Sciences, Pratap University, Jaipur.
- 2021-Present: Associate Professor & Head, Kalasalingam Academy of Research and Education (Deemed to be University), Tamil Nadu.

Honors: Chemfab Gold Medal in M. Sc. from Pondicherry University, CSIR-JRF-NET and GATE qualified.

Publications from the Last 3 Years:

1. Bhansali, P. R., Vijayendran K. K. Praneeth, "Dehydrogenation Reactions and its pointer from nature for the synthesis of building blocks leading to valued pharmaceutical compounds" chapter in the book entitled *"Chemistry of Dehydrogenation Reactions and its Application,"* 2022 (Under Review).
2. Izu, Hitoshi, Mio Kondo, Masaya Okamura, Misa Tomoda, Sze Koon Lee, Takuya Akai, Vijayendran Praneeth, Mari Kanaike, Satoshi Kawata, and Shigeyuki Masaoka. "Precise manipulation of electron transfers in clustered five redox sites." *ChemRxiv.,* 2022 (Submitted).
3. Praneeth, Vijayendran KK, Mio Kondo, Masaya Okamura, Takuya Akai, Hitoshi Izu, and Shigeyuki Masaoka. "Pentanuclear iron catalysts for water oxidation: substituents provide two routes to

control onset potentials." *Chemical Science* 10, no. 17 (2019): 4628-4639.

4. Subramaniam, Jeevithra Dewi, Sze Koon Lee, Pondchanok Chinapang, Vijayendran KK Praneeth, Masaya Okamura, Mio Kondo, Shigeyuki Masaoka, and Pei Meng Woi. "Synthesis and structural characterization of centrosymmetric multinuclear nickel (II) complexes with neutral tetradentate N6-ligand." *Transition Metal Chemistry* 46, no. 3 (2021): 255-262.

Ganesh U. Chaturbhuj, Professor

Affiliation: Department of Pharmaceutical Sciences and Technology, Institute of Chemical Technology, N M Parekh Road, Matunga (E), Mumbai, India

Education: Ph.D. (Pharmaceutical Chemistry) University of Mumbai, Post-Doc. (Medicinal Chemistry) Northeastern University, Boston, Massachusetts, USA

Business Address: Department of Pharmaceutical Sciences and Technology Institute of Chemical Technology, N M Parekh Road, Matunga (E), Mumbai, India

Research and Professional Experience: Two years of industrial and 20 years of academic and research experience.

Professional Appointments: Professor of Pharmaceutical Chemistry

Honors: Best Teacher Award at the Institute of Chemical Technology; University Grants Commission, India Raman Post-Doctoral Fellowship.

Publications from the Last 3 Years: (2020-2022) Twenty research articles were published, and one Indian Patent applied.

1. Ganwir P, Gavali K, Chaturbhuj GU. N-(Phenylsulfonyl) Benzenesulfonamide: A New Organocatalyst for One-Pot, Solvent-Free Synthesis of Biginelli's 3, 4-Dihydropyrimidine-2 (1*H*)-

Thiones. *Polycyclic Aromatic Compounds.* 2022:1-10. https://doi.org/10.1080/10406638.2022.2067191.

2. Ganwir, P., Bandivadekar, P., Kudale, P., & Chaturbhuj, G. U. (2022). Catalyst-free, one-pot expeditious synthesis of polyhydroquinolines and 2-amino-4H-chromenes. *Research on Chemical Intermediates*, *48*(8), 3429-3447.
3. Patil NB, Chaturbhuj GU. Sulfated polyborate catalyzed rapid and efficient electrophilic thiocyanation of activated arenes. *Tetrahedron Letters.* 2022;96:153763.
4. Mali AS, Ganwir P, Chaturbhuj GU. Aluminized Polyborate: A New and Eco-friendly Catalyst for One-pot Multicomponent Synthesis of 1,3-Diaryl-3-(arylamino)propan-1-ones via Mannich Reaction. *Organic Preparations and Procedures International.* 2022:54(4), 338-345.
5. Ganwir P, Jaydeokar S, Chaturbhuj GU. Phthaloylation of amines, hydrazines, and hydrazides by N-substituted phthalimides using recyclable sulfated polyborate. *Results in Chemistry.* 2022;4:100293.
6. Ganwir P, Kale I, Chaturbhuj G. Wet copper-slag: A new and eco-friendly catalyst for Knoevenagel condensation. *Sustainable Chemistry and Pharmacy.* 2022;25:100614.
7. Mali AS, Bandivadekar P, Chaturbhuj GU. Aluminized Polyborate: A New and Eco-friendly Catalyst for the Synthesis of Symmetrical N, N'-Di (aryl/alkyl) formamidines. *Organic Preparations and Procedures International.* 2022;54(3):242-248.
8. Wijayasinghe YS, Bhansali MP, Borkar MR, Chaturbhuj GU, Muntean BS, Viola RE, et al. A Comprehensive Biological and Synthetic Perspective on 2-Deoxy-d-Glucose (2-DG), A Sweet Molecule with Therapeutic and Diagnostic Potentials. *Journal of Medicinal Chemistry.* 2022;65(5):3706-28.
9. Jadhav SS, Daware KT, Varma KS, Patil K, Patil SJ, Chaturbhuj GU, Lokhande TN, et al. A Revision of the Biginelli Reaction: A Combined Experimental and Pharmacological Investigation. *Der Pharma Chemica,* 2021, 13(10): 15-24.
10. Palav A, Misal B, Chaturbhuj G. NCBSI/KI: A Reagent System for Iodination of Aromatics through In Situ Generation of I-Cl. *The Journal of Organic Chemistry.* 2021;86(17):12467-12474.
11. Sutar P, Khedkar P, Chaturbhuj G. Sulfated Polyborate, a Novel Buffer for Low pH Mobile Phase on a Nonend Capped Stationary

Phase in Reverse Phase Liquid Chromatography. *Current Chromatography.* 2021;8(1):33-43.

12. Misal B, Palav A, Ganwir P, Chaturbhuj G. Sulfated polyborate-H_2O assisted tunable activation of N-iodosuccinimide for expeditious mono and diiodination of arenes. *Tetrahedron Letters.* 2021;74: 153154.
13. Ganwir P, Chaturbhuj G. Sulfated polyborate: A dual catalyst for the reductive amination of aldehydes and ketones by $NaBH_4$. *Tetrahedron Letters.* 2021;74:153143.
14. Palav A, Misal B, Ganwir P, Badani P, Chaturbhuj G. Rapid, chemoselective and mild oxidation protocol for alcohols and ethers with recyclable N-chloro-N-(phenylsulfonyl) benzenesulfonamide. *Tetrahedron Letters.* 2021;73:153094.
15. Rekunge DS, Mali AS, Chaturbhuj GU. One-Pot Expeditious Synthesis of 2-Amino-4, 6-(disubstituted)nicotinonitriles Using Activated Fuller's Earth as Catalyst. *Organic Preparations and Procedures International.* 2021;53(2):112-119.
16. Misal B, Palav A, Ganwir P, Chaturbhuj G. Activator free, expeditious and eco-friendly chlorination of activated arenes by N-chloro-N-(phenylsulfonyl) benzene sulfonamide (NCBSI). *Tetrahedron Letters.* 2021;63:152689.
17. Indalkar K, Malge SS, Mali AS, Chaturbhuj GU. Expeditious and Highly Efficient One-Pot Synthesis of Functionalized Imidazoles Catalyzed by Sulfated Polyborate. *Organic Preparations and Procedures International.* 2021;53(4):387-396.
18. Mali AS, Indalkar K, Chaturbhuj GU. Solvent-free, Efficient Transamidation of Carboxamides with Amines Catalyzed by Recyclable Sulfated Polyborate Catalyst. *Organic Preparations and Procedures International.* 2021;53(4):369-378.
19. Shah A, Shah AA, Nandakumar K, Rekunge D, Chaturbhuj G, Kishore A, et al. The prophylactic approach of sesamol and 3′, 4′-(methylenedioxy) acetophenone to prevent associated cardiotoxicity of doxorubicin at high dose in prostate cancer rat model. *Rasayan Journal of Chemistry.* 2021;14(3):1938-1946.
20. Mali AS, Sharma AB, Chaturbhuj GU. Sulfated Polyborate Catalyzed Selective Friedlander Annulation for Synthesis of Highly Functionalized Quinolines. *Organic Preparations and Procedures International.* 2020;52(4):297-303.

Patent:
Application No: 202121059848 A
Title of innovation: "An Improved Acid-Catalyzed Process for Synthesis of Enamies and Enaminones Based Chemical Intermediates."
Publication date: 31/12/2021

Pravin Bhansali

Affiliation: Alliance University, Department of Science, Faculty of Science and Technology, Anekal, Bengaluru-562 106, Karnataka, India

Education: Ph. D.

Business Address: Alliance University, Department of Science, Faculty of Science and Technology, Anekal, Bengaluru-562 106, Karnataka, India

Research and Professional Experience:

- PhD University of Toledo, USA.
- Postdoctoral research fellow, University of Toledo, USA.
- Postdoctoral research fellow, Johns Hopkins University, USA.

Professional Appointments:

- 08/2021-present – Professor, Alliance University, Bangalore- Taught Engineering Chemistry-I, Engineering Chemistry-II, Engineering Chemistry Laboratory.
- 11/2020-07/2021 – Associate Prof, SVKM/Dattakala Institute of Pharmacy, MS India.
- 07/2018-10/2020 – Principal, Rajesh Bhaiyya Tope/Ojas College of Pharmacy, MS, India.
- 08/2016-04/2021 – Co-director, Aspiration academy, Aurangabad, MS, India.
- 08/2017-07/2018 – Visiting professor, Government College of Pharmacy, Aurangabad, India.

- 2/2015-06/2016 – Postdoctoral Research Associate, School of Medicine, Johns Hopkins University, Baltimore, MD, USA.
- 8/2011-1/2015 – Postdoctoral Research Associate, the University of Toledo, Toledo, OH, USA.
- 8/2011-12/2011 – Organic chemistry recitation and organic chemistry laboratory instructor (Fall), Department of Chemistry, the University of Toledo, Toledo, OH, USA.
- 8/2006-8/2011 – Graduate Teaching Assistant, the Department of Medicinal and Biological Chemistry, the University of Toledo, USA. Tutored and graded various subjects including Biochemistry, Medicinal Chemistry, Medicinal Organic Chemistry Laboratory, Immunology and Microbiology Laboratories.
- 6/2002-7/2006 – Lecturer, Modern/SGRS/Allana College of Pharmacy, Pune, India. Taught Biochemistry, Organic Chemistry, and Medicinal Chemistry.
- 9/2000-4/2002 – Junior Research Fellow, Bharati Vidyapeeth's Poona College of Pharmacy, Pune, India.

Honors:

- Travelship to the 2010 Int. Pharm. Fed.'s (FIP) Pharmaceutical Sciences World Congress /Am. Ass. of Pharm. Scientists (AAPS) Annual Meeting and Exposition, New Orleans, Nov 2010.
- American Chemical Society's Division of Organic Chemistry Travel Award to attend the 239th ACS National Meeting and Expositions, San Francisco, Mar. 2010.
- The Wayne and Dorothy Hoss Fund's Award for Graduate Education to attend the Mid-Atlantic Graduate Student Symposium in Medicinal Chemistry, Detroit, July 2008.
- Travel award from the Lotus Trust, India to join the University of Toledo as a Ph. D. student, Aug. 2006.
- Successfully qualified GATE 2000-Pharm. Sci. (national level graduate aptitude test) with 98.25 percentile and 60th all over India rank.

Publications from the Last 3 Years:

Patents:

1. Ronald E. Viola, Bharani Thangavelu, Mutthamsetty; Vinay, Wang; Qinzhe, Pravin Bhansali, "Potent phthalate inhibitors of aspartate N-acetyltransferase and selective aspartate pathway inhibitors" US Patent (United States Serial No. 10449168 Issue Date: October 22, 2019) http://patft.uspto.gov/netacgi/nph-Parser?Sect1=PTO1&Sect2=HITOFF&d=PALL&p=1&u=%2Fnetahtml%2FPTO%2Fsrchnum.htm&r=1&f=G&l=50&s1=10449168.PN.&OS=PN/10449168&RS=PN/10449168.

Publications:

1. Yasanandana Supunsiri Wijayasinghe#, Pravin Bhansali#, Ronald E. Viola, Mohammad A. Kamal and Nitesh Kumar Poddar, "Natural Products: A Rich Source of Antiviral Drug Lead Candidates for the Management of COVID-19" *Current Pharmaceutical Design;* 27 (33), 2021, 3526-3550 # Equal authorship.
2. Pallavi Sharma; Arti Tyagi; Pravin Bhansali; Vineeta Singh; Shubra Pareek; Nitesh Poddar, "Saponins: Extraction, Bio-medicinal Properties and way forward to Anti-viral representatives" *Food and Chemical Toxicology;* Volume 150, April 2021, 112075 (Elsevier Publications).
3. Wijayasinghe, Yasanandana; Bhansali, Meenakshi; Borkar, Maheshkumar; Chaturbhuj, Ganesh; Muntean, Brian; Viola, Ronald; Bhansali, Pravin, "A Comprehensive Biological and Synthetic Perspective on 2-Deoxy-D-Glucose (2-DG), A Sweet Molecule with Therapeutic and Diagnostic Potential" *Journal of Medicinal Chemistry,* 2022 March 10, 65 (5), 3706–3728.

Book Chapters:

1. Jineesh AG, Jyotishkumar Parameswaranpillai, Murthy Chavali, Vijay Bahadur, Vipin Prasad, Chitrakara Hegde, Pravin R Bhansali "Automotive applications of biodegradable polymers" in book *"Biopolymers"* Submitted.

2. Pravin R. Bhansali, Vijayendran K. K. Praneeth "Dehydrogenation Reactions and its pointer from nature for the synthesis of building blocks leading to valued pharmaceutical compounds" Book: *"Chemistry of dehydrogenation reactions and its applications"* Submitted.

Chapter 2

Hydrazine Derivatives (Pyrazole, Pyridazine): A Source of Pharmaceutically Bioactive Drugs

Shubhangi S. Savale[1]
Neetu Talreja[2]
Vijayendran K. K. Praneeth[3]
Shashank M. Sonkusare[1]
Ganesh U. Chaturbhuj[1,*]
and Pravin R. Bhansali[2,†]
[1] Department of Pharmaceutical Sciences and Technology, Institute of Chemical Technology, Mumbai, Maharashtra, India
[2] Department of Science, Faculty of Science and Technology, Alliance University, Bengaluru, Karnataka, India
[3] Department of Chemistry. School of Advanced Sciences, Kalasalingam Academy of Research and Education (Deemed to be University), Krishnankoil, Tamilnadu, India

Abstract

Heterocyclic compounds derived from hydrazine (pyrazole, pyridazine, etc.) are vital in drug design and pharmaceutical research. Many structurally diverse heterocyclic derivatives (pyrazole, pyridazine, and other miscellaneous heterocyclic compounds containing *N-N* linkage) possess interesting therapeutical properties, including sedative, antidepressant, non-steroidal anti-inflammatory drugs, anticancer, uricosuric drugs, steroids, anorectic, alcohol dehydrogenase inhibitory,

* Corresponding Author's Email: gu.chaturbhuj@ictmumbai.edu.in.
† Corresponding Author's Email: pravin.bhansali@alliance.edu.in.

In: A Review of Hydrazine and Its Applications
Editor: Thomas W. Gerdes
ISBN: 979-8-88697-671-7

antipsychotic, antibiotic, anticoagulant, antibacterial, antifungal, antihistaminic, vasodilator, and many more to cite. More than 65 drugs consisting of heterocycles derived from hydrazine cater to the needs of many patients suffering from numerous diseases and disorders.

In addition to the established convenient chemistry, medicinal chemists get enticed to claim these pharmaceutically attractive moieties in their potential bioactive compounds. This marching of drugs from this heterocyclic category containing *N-N* linkage towards the century mark would be achieved sooner as many newer bioactive molecules claim to have these moieties in their structure. This chapter provides a comprehensive analysis of drugs consisting of heterocyclic nucleuses consisting of *N-N* linkage.

Introduction

Pyrazole is a five-membered heterocyclic aromatic nucleus with two nitrogen at 1 and 2 positions. It is basic with a pK_b of 11.5 (pK_a of the conjugated acid = 2.49 at 25°C), and it exhibits tautomerism based on the substituents present in the heterocyclic core. Various synthetic protocols can easily assemble it. The prominent one among these includes the reaction of α, β-unsaturated carbonyl compounds with hydrazine, which provides the formation of pyrazole. Literature exploration regarding drugs indicates that the compounds featuring pyrazole core exhibit diverse and efficient pharmacological activities (e.g., sedative, antidepressant, non-steroidal anti-inflammatory, anticancer, uricosuric, steroids, anorectic, alcohol dehydrogenase inhibitory, anti-psychotic, anticoagulant, antibacterial, antifungal, antihistaminic, vasodilator, etc.) for the treatment of various health conditions [1, 2, 3].

Similarly, pyridazine moiety is a privileged scaffold in various bioactive molecules with therapeutic properties, including anticancer, antidepressant, and antibiotic activities. Pyridazine is a six-membered heterocyclic aromatic compound containing two adjacent nitrogen atoms.

This chapter provides a comprehensive review of drugs comprising hydrazine derivatives such as pyrazole, and pyridazine moieties, including several miscellaneous drug molecules featuring *N–N* bond in the ring structure. The medicinal importance, pharmacological action, efficacy, adverse effects, structural features, metabolism studies, and synthesis reported for these drugs are compiled, analysed, and reported here. We are sure this review will update the information to propel further new directions in the drug design and development process.

Pyrazole Ring Containing Drugs

Table 1. Pyrazole ring containing drugs

Sr. No.	Drug	Class	Structure
1.	Zaleplon	Sedative	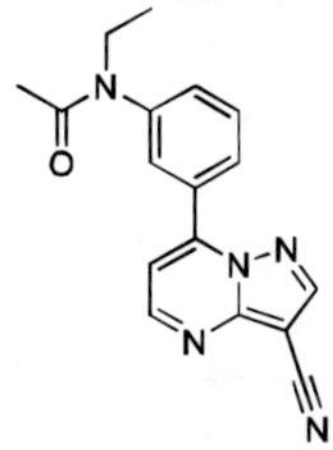
2.	Indiplon	Sedative	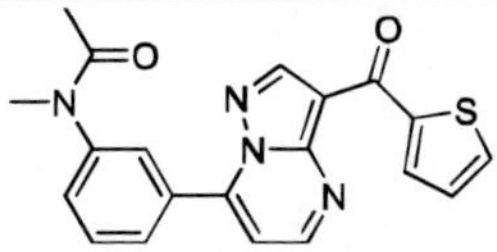
3.	Ocinaplon	Sedative	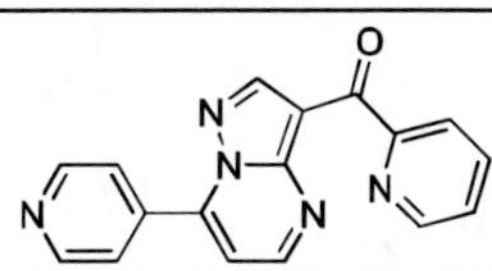
4.	Fezolamine	Antidepressant	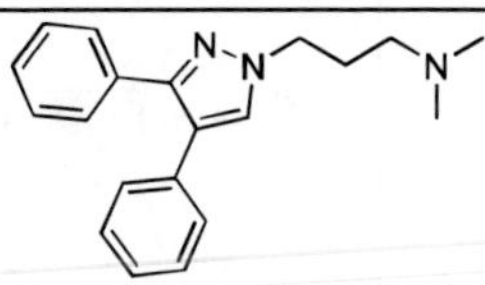
5.	Zometapine	Antidepressant	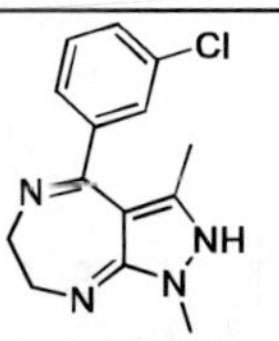
6.	CDPPB	Antipsychotic	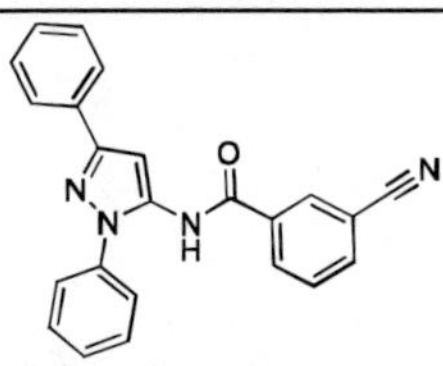
7.	Mepiprazole	Minor tranquilizer	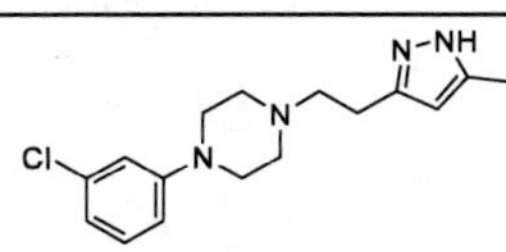

Table 1. (Continued)

Sr. No.	Drug	Class	Structure
8.	Elexacaftor	CNS	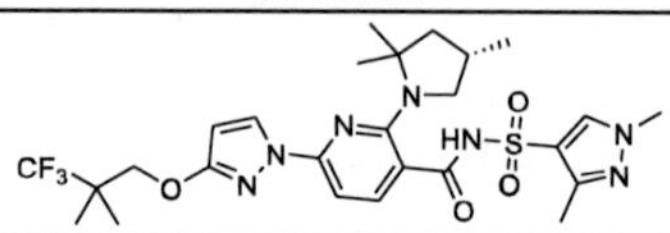
9.	Epirizole	NSAID	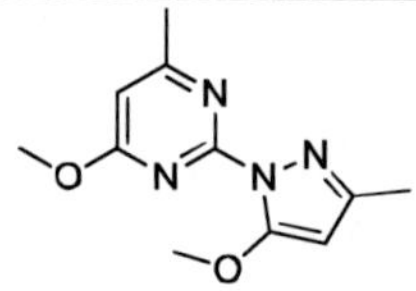
10.	Phenylbutazone	NSAID	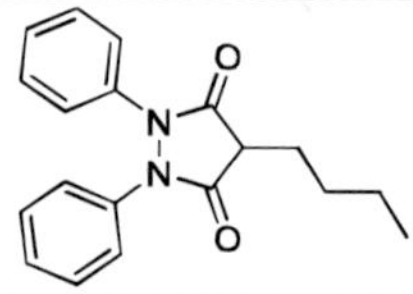
11.	Lonazolac	NSAID (Antirheumatic agent)	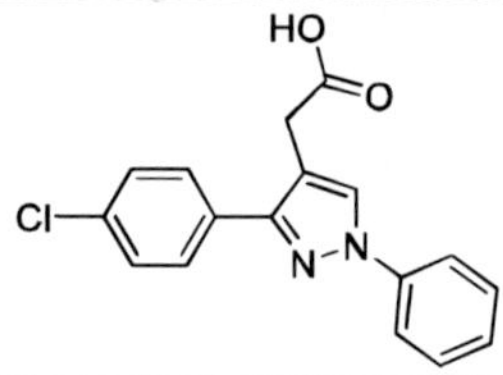
12.	Difenamizole	NSAID	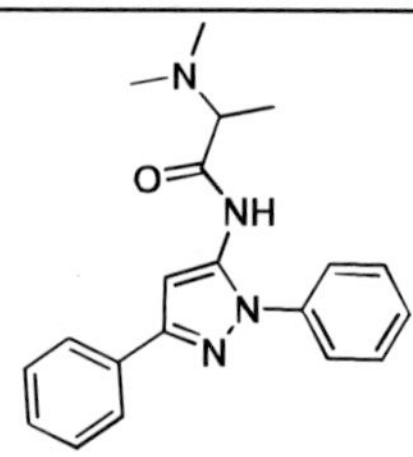
13.	Antipyrine (phenazone)	NSAID	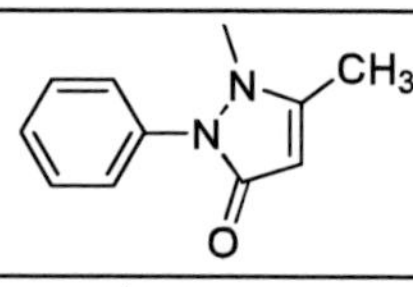
14.	Dipyrone (Novalgin)	Analgesic, antipyretic and antispasmodic	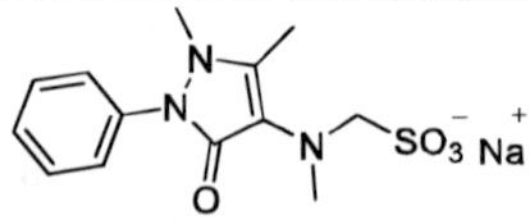
15.	Propyphenazone	NSAID	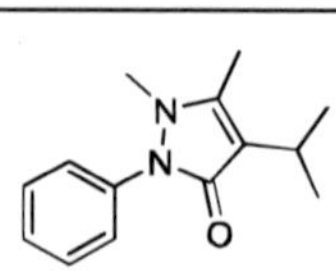

Sr. No.	Drug	Class	Structure
	Ramifenazone		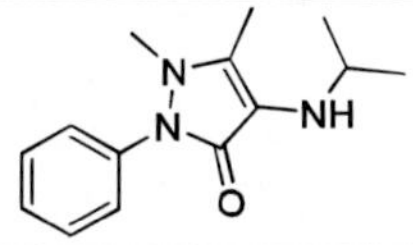
16.	Tepoxalin	NSAID	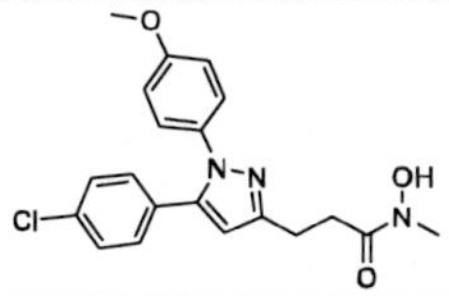
17.	SC-560	Cyclooxygenase inhibitor	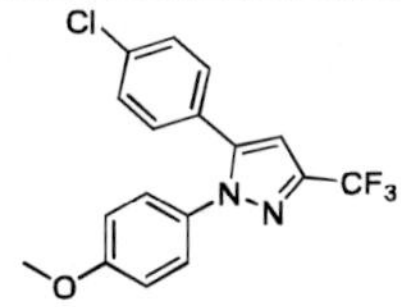
18.	Celecoxib	Cyclooxygenase 2 inhibitor	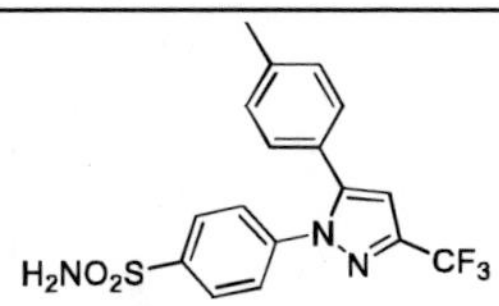
19.	Deracoxib	Cyclooxygenase 2 inhibitor	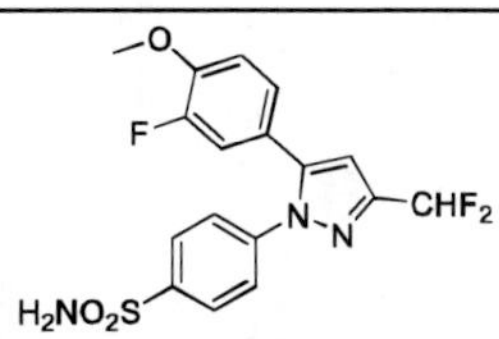
20.	Crizotinib	Anticancer	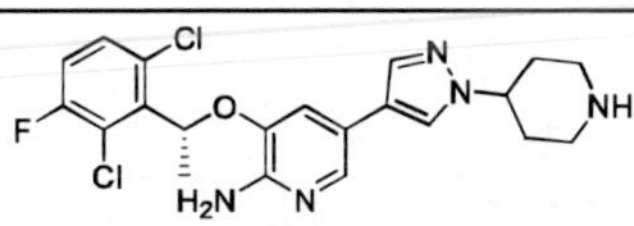
21.	Pazopanib	Anticancer	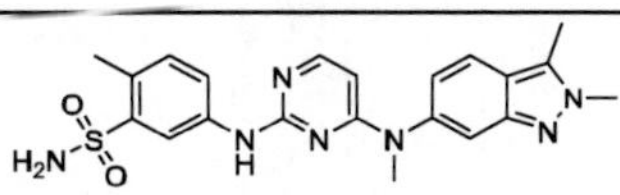
22.	Entrectinib	Anticancer	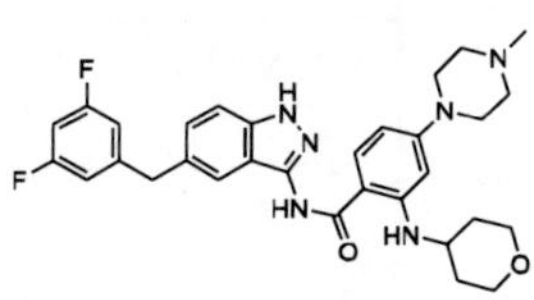

Table 1. (Continued)

Sr. No.	Drug	Class	Structure
23.	Avapritinib	Anticancer	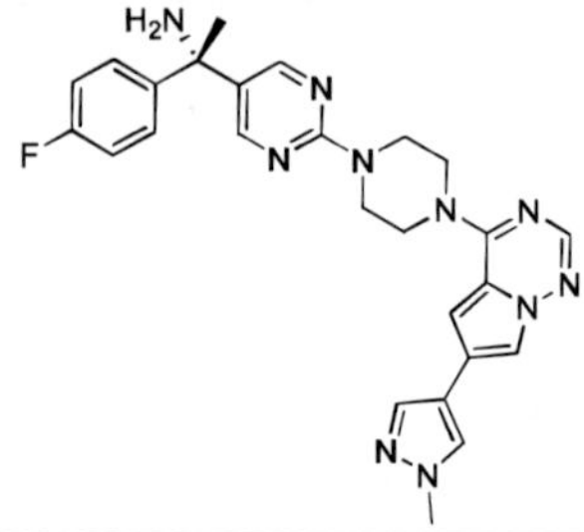
24.	Pyrazomycin (Pyrazofurin)	Anticancer	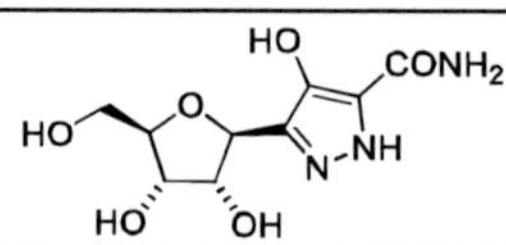
25.	Pralsetinib	Anticancer	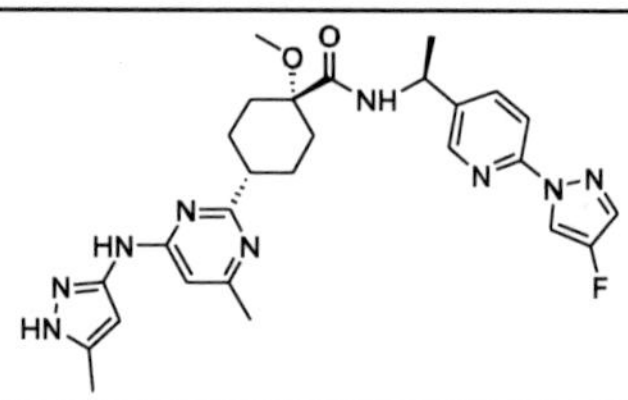
26.	Lorlatinib	Anticancer	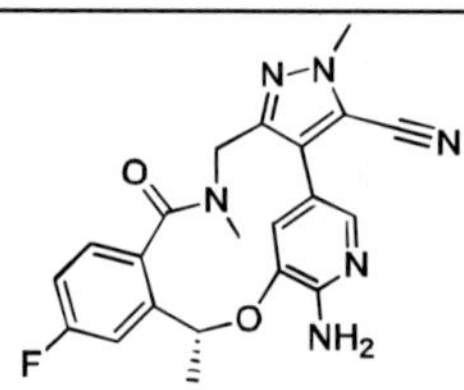
27.	Larotrectinib	Anticancer	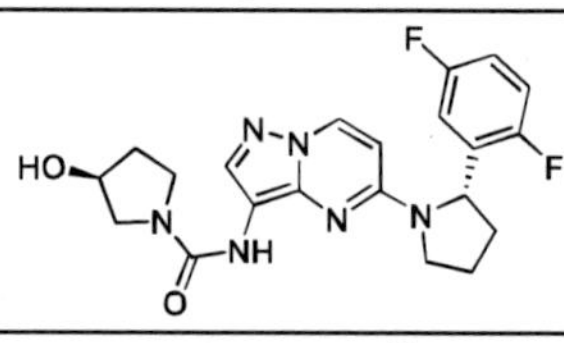
28.	Umbralisib	Philadelphia chromosome-positive chronic myeloid leukemia	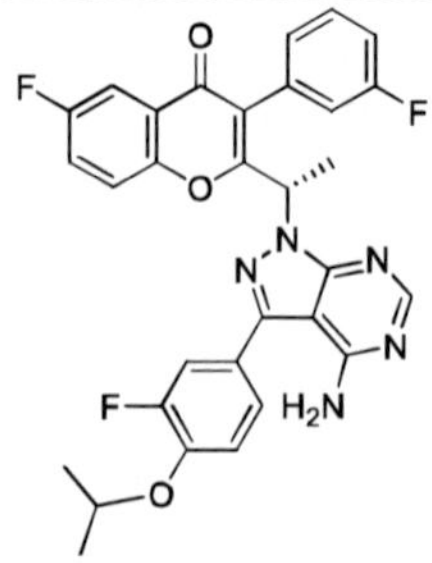

Sr. No.	Drug	Class	Structure
29.	Asciminib		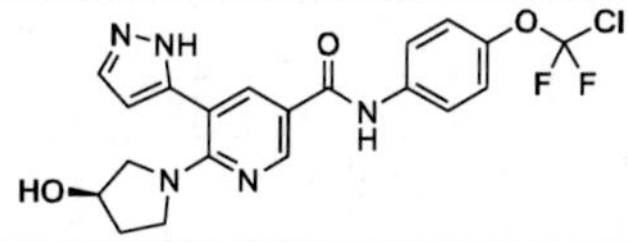
30.	Belumosudil	Chronic graft versus host disease	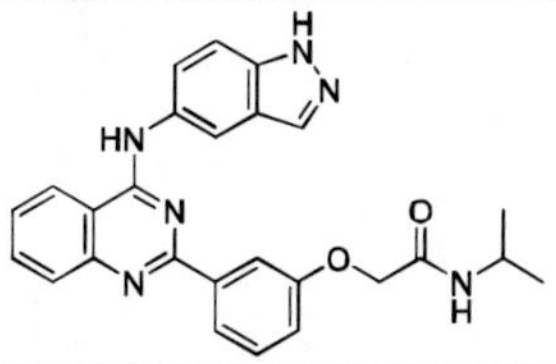
31.	Sulfinpyrazone	Uricosuric Drug (Gout treatment)	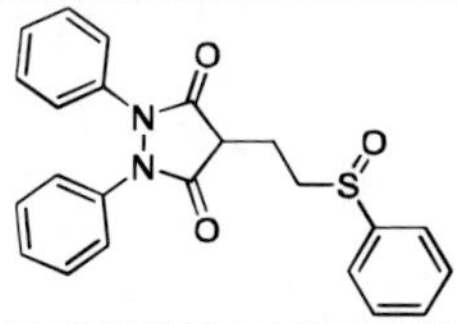
32.	Stanozolol	Steroidal	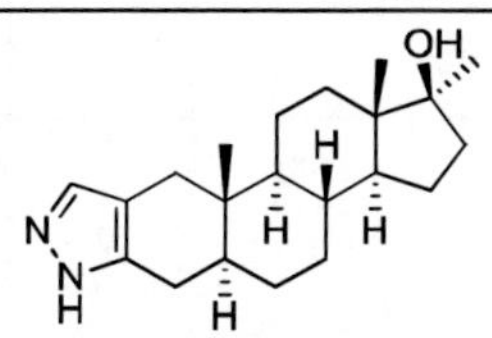
33.	Fomepizole	Alcohol dehydrogenase inhibitor	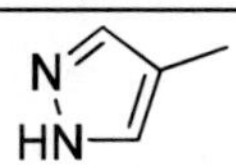
34.	Sildenafil	To treat erectile dysfunction	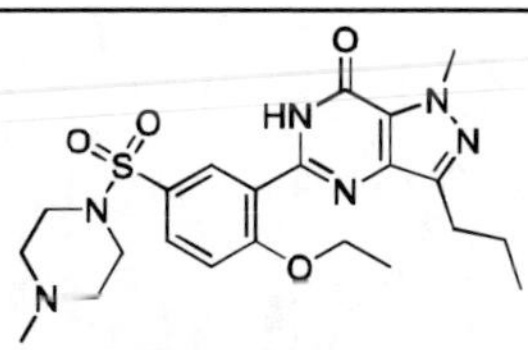
35.	Betazole	H_2-receptor agonist	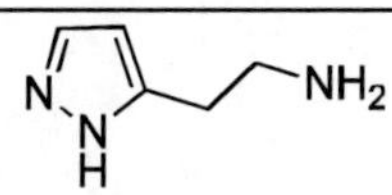
36.	AS-19	5-HT_7 receptor agonist	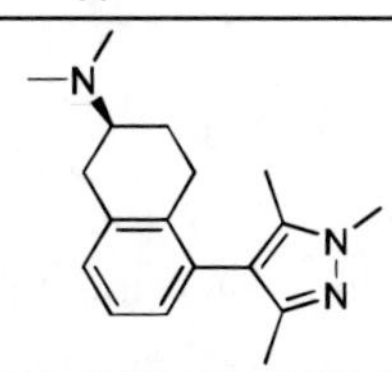

Table 1. (Continued)

Sr. No.	Drug	Class	Structure
37.	Apixaban	Anticoagulant	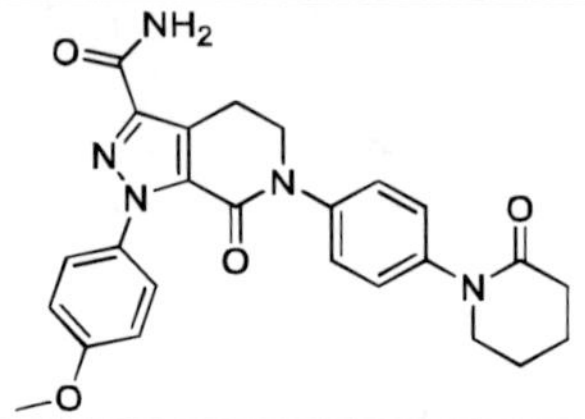
38.	Vericiguat	Chronic heart disease	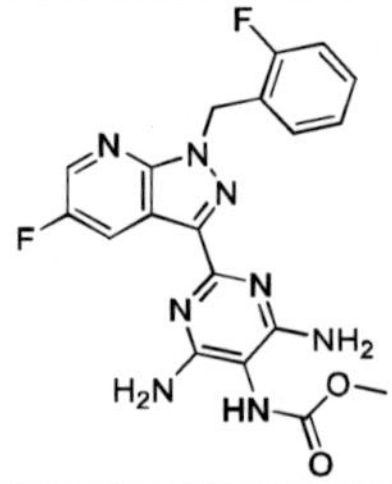

Zaleplon

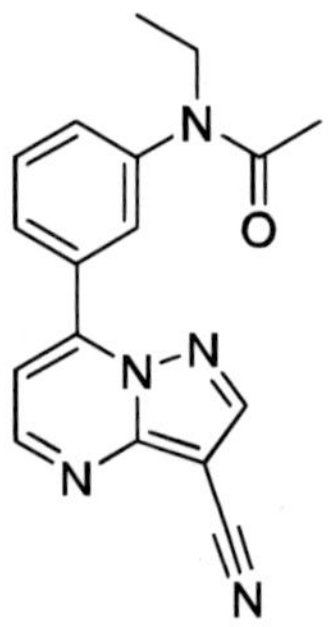

1

Zaleplon (**1**), Pyrazolopyrimidine derivative; *N*-(3-(3-cyanopyrazolo[1,5-*a*]pyrimidin-7-yl)phenyl)-*N*-ethylacetamide is a short-acting, non-benzodiazepine sedative-hypnotic drug. Also called as a non-benzodiazepine Z-class drug which include a total of five drugs (zaleplon, indiplon, eszopiclone, zolpidem, and alpidem) [4]. FDA approved zaleplon as a sedative-hypnotic drug in 1999, and the preparation of zaleplon was reported in the US patent by the American Cyanamid Company [5]. The use of barbiturates as sedative-hypnotics was diminished because of its safety issues like rapid development of tolerance, dependence, and risk of drug abuse; also, many BZD drugs have a longer half-life, cause daytime sedation, and shorter-acting BZD drugs cause rebound insomnia and day time anxiety. Z-class drugs induce sleep rapidly with lesser side effects [6]. Due to the rapid onset of action and shorter half-life, it is used in drug-facilitated crimes like sexual assault, robbery, kidnapping, and recreational abuse.

Zaleplon binds to the $GABA_A$ receptor at the receptor's α (+) and γ (−) interface to increase the frequency of Cl-channel opening. It binds at the allosteric receptor site and reduces the brain's excitability. Hence zaleplon and other Z-class drugs are commonly used to treat insomnia compared to BZD [1, 7].

Zaleplon is metabolized by the aldehyde oxidase to ring oxidation product **2** and to *N*-dealkylated product **3** by the human cytochrome CYP3A subfamily. Although cimetidine inhibits both the CYP3A and aldehyde oxidase-mediated pathways of zaleplon metabolism in the liver, interestingly, cimetidine inhibits aldehyde oxidase to more extent than CYP3A (Scheme 1) [8].

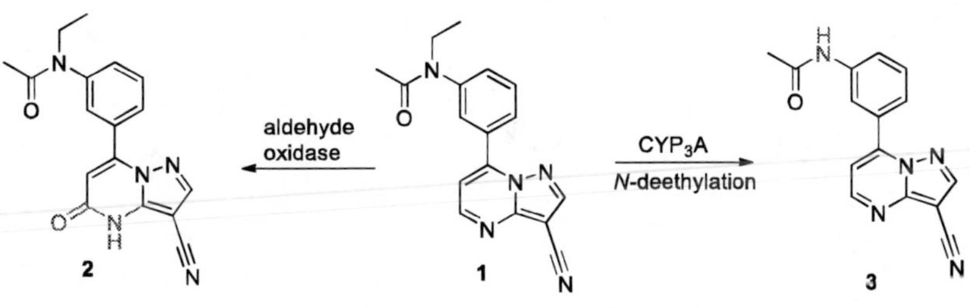

Scheme 1. Metabolic pathway of zaleplon (**1**).

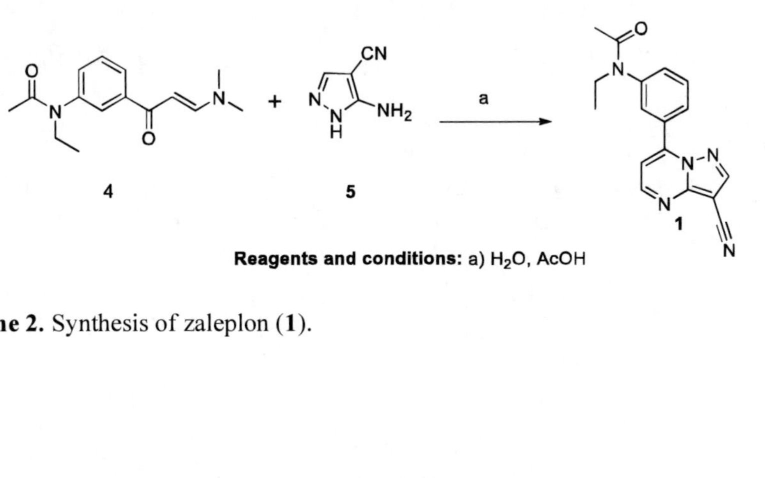

Scheme 2. Synthesis of zaleplon (**1**).

Padmanathan *et al.* developed the zaleplon synthesis method given in Scheme 2 [9]. Enaminone **4** is reacted with nitrile **5** in acetic acid and water for about 1 hr at room temperature to 70°C to generate zaleplon (**1**).

Bharathi *et al.* reported the process-related impurities in zaleplon bulk drugs, which are isolated, identified, and characterized by HPLC, LC-MS, FT-IR, and NMR (^{1}H, ^{13}C, and DEPT) techniques [10]. Rádl *et al.* reported the synthesis of *N*-[3-(3-cyanopyrazolo[1,5-*a*]pyrimidin-5-yl)phenyl]-*N*-ethyl-acetamide, a Zaleplon regioisomer which is one of the impurities formed during the synthesis of zaleplon [11].

Baraldi *et al.* synthesized a series of zaleplon analogs. Researchers have replaced the cyano group at the fifth position of pyrazolo[1, 5-a]pyrimidine moiety with different functional groups. Also, *N*-ethylacetamidophenyl moiety was replaced by (substituted) heterocycles. Determination of the bioactivity of these 43 synthesized analogs was performed to complete structure-activity relationship studies. Relative affinities of synthesized analogs for the α_1 of the $GABA_A$ receptor were done by labelling the membrane with the benzodiazepine site antagonist, flumazenil. The α_1 of the $GABA_A$ receptor was derived from the rat cerebral cortex.

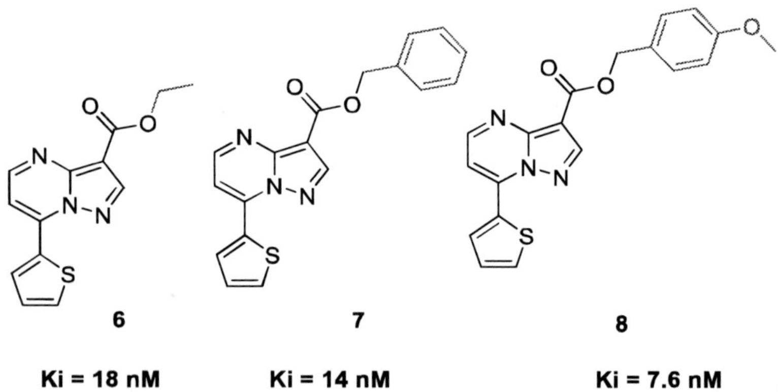

Figure 1. Prominent analog of zaleplon.

Most synthesized analogs displayed higher or equivalent activity than Zaleplon (Ki = 57.3nM). For example, a few analogs were active with Ki values of molecule **6** (18 nM), molecule **7** (14 nM), and molecule **8** (7.6 nM), as shown in Figure 1 [12]. Although Behbehani *et al.* described the synthesis of various analogs of Zaleplon, researchers did not publish their biological activity [13].

Indiplon

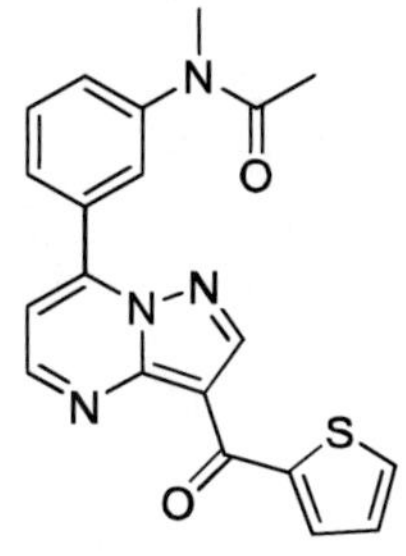

9

Indiplon (**9**), a pyrazolopyrimidine derivative, *N*-methyl-*N*-(3-(3-(thiophene-2-carbonyl)pyrazolo-[1,5-*a*]pyrimidin-7-yl)phenyl)acetamide is a short-acting Z-class non-benzodiazepine class drug approved for the treatment of insomnia. It acts by modulating the α_1 subunit of the GABA receptor complex to induce sleep. Compared with zaleplon, indiplon is 10 times more selective to the α_1 subunit. Due to the lesser side effects, indiplon formulations like immediate release (IR) and modified release (MR) were developed. The immediate release form (IR) was designed for sleep onset problems, and the modified release form (MR) was developed to help patients stay asleep throughout the night. The FDA approved indiplon-IR 5 and 10 mg in May 2006 [14].

Burgos *et al.* reported that indiplon could modulate platelet function that affects the cardiovascular system. They studied the interaction of the A2A receptor with indiplon using Schrodinger (Schrodinger LLC, New York 2016) software, platelet aggregation assay, and measurement of P-selectin surface expression. This study concluded that indiplon had a remarkable activity inhibiting platelet aggregation by activating the A2A receptor [15].

Metabolism of indiplon is shown in Scheme 3. Shimizu *et al.* reported the study on the human arylacetamide deacetylase (AADAC) enzyme involved in the hydrolysis of various drugs like indiplon. Indiplon has an *N*-acetyl group which is readily hydrolyzed by the *N*-esterase enzyme to *N*-deacetylindiplon **10** in human liver microsomes (HLMs). They investigated the human AADAC can catalyze the hydrolysis of indiplon. It also gets metabolized by CYP3A/5 to *N*-demethylindiplon **11** [16].

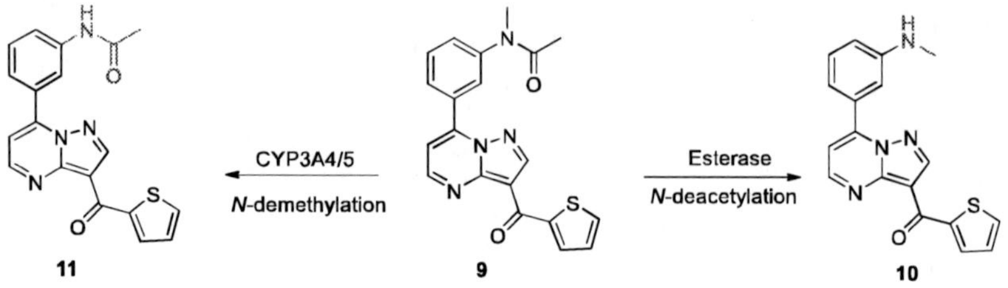

Scheme 3. Metabolic pathway of indiplon (**9**).

Czarnik *et al.* submitted a US patent application for deuterium-enriched indiplon, pharmaceutically acceptable salt forms. Glenmark generics filed a patent for the synthesis of indiplon (**9**) (Scheme 4) by *N*-alkylation of enaminone **13,** followed by condensation with amino pyrazole **14** [17].

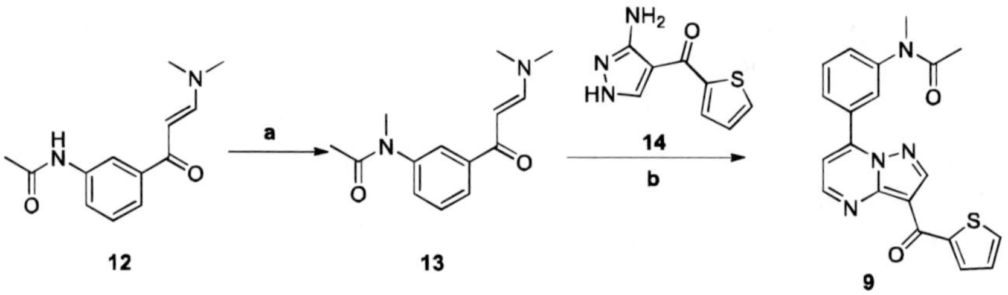

Reagents and conditions: a) CH_3I/NaH, DMF; b) (3-amino-1*H*-pyrazol-4-yl)(thiophen-2-yl)methanone, AcOH/reflux

Scheme 4. Synthesis of indiplon (**9**).

Ocinaplon

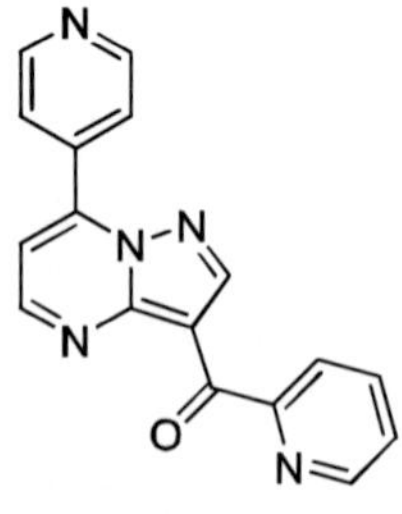

15

Ocinaplon (**15**), pyridin-2-yl(7-(pyridin-4-yl)pyrazolo[1,5-*a*]pyrimidin-3-yl)methanone, pyrazolopyrimidine derivative belongs to the Z-class benzodiazepine drugs but used as an anxiolytic drug instead of sedative-hypnotic. It has a low affinity toward the benzodiazepine binding site of the $GABA_A$ receptor. The mechanism of action of ocinaplon is that it selectively modulates the $GABA_A$ receptor isoforms involved in the mediation of anxiety. During the clinical trials, ocinaplon demonstrated anxiolytic efficacy in phase II clinical trials for generalized anxiety disorder (GAD) [18]. Ocinaplon may activate multiple $GABA_A$ isoform. It may help to treat patients with GAD as demonstrated by Lippa *et al.* [19]. The new use of ocinaplon in the treatment of organophosphate exposure was reported in a US patent by Helton *et al.* [20].

Berezhnoy *et al.* reported the pharmacological properties of a metabolite of ocinaplon (Scheme 5). Ocinaplon is converted to DOV 315090, a 4'-*N*'-oxide metabolite formed after oxidative biotransformation. DOV 315090 also acts as a $GABA_A$ receptor modulator, like a parent compound that may participate in pharmacological action. Its *in-vitro* activity was evaluated on recombinant $\alpha_1\beta_2\gamma_2S$, $\alpha_2\beta_2\gamma_2S$, and $\alpha_3\beta_2\gamma_2S$ receptors have a slight difference from ocinaplon. However, its affinity for $\alpha_5\beta_2\gamma_2S$ is two-fold lower than ocinaplon [21].

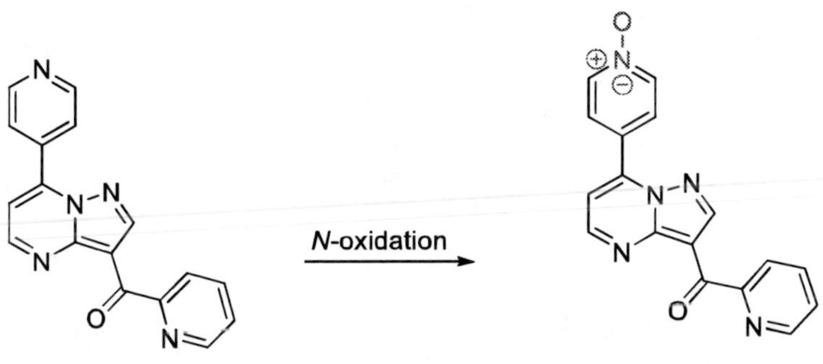

15

DOV 315090

Scheme 5. Metabolic pathway of ocinaplon (**15**).

Synthesis of ocinaplon (Scheme 6) performed by reacting ethyl picolinate **16** with acetonitrile in toluene under the basic condition to obtain compound **17**, which was then followed by aldol condition using DMF-DMA to get compound **18**, which undergoes cyclization with aminoguanidine (**19**) to yield compound **20**. Product (**20**) of earlier step is condensed with enaminone **21** under acidic condition to produce ocinaplon (**15**) [22].

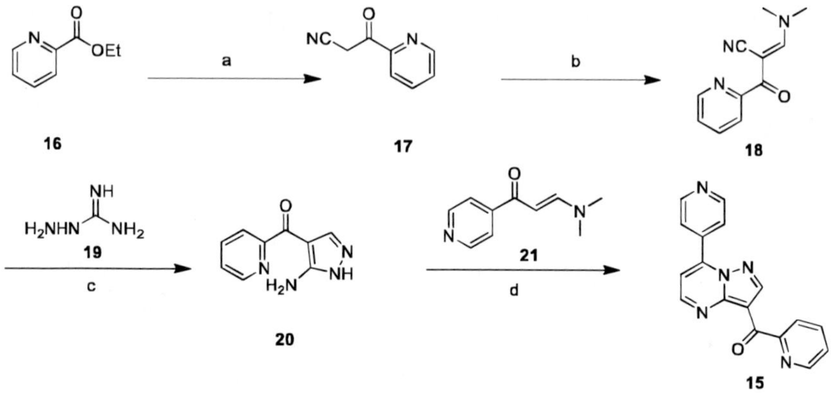

Reagents and Conditions: a) MeCN, NaOMe/toluene, heat; b) toluene, DMF/DMF-DMA; c) water, heat; d) HOAc, 12 h, reflux

Scheme 6. Synthesis of ocinaplon (**15**).

Fezolamine

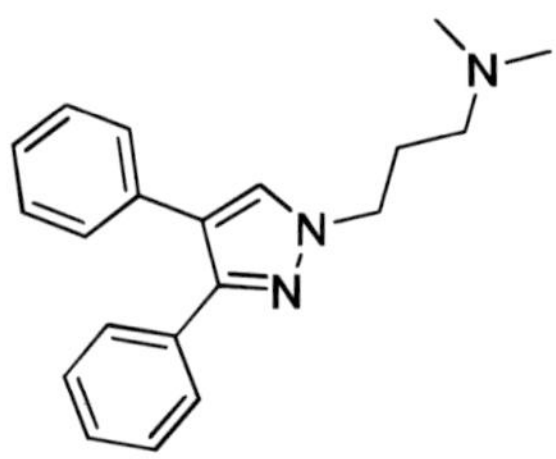

22

Fezolamine (**22**), 3-(3,4-diphenyl-1*H*-pyrazol-1-yl)-*N,N*-dimethylpropan-1-amine, is an antidepressant drug with selectivity in blocking synaptosomal uptake of [^{3}H]norepinephrine [23]. Reports showed that fezolamine is an effective antidepressant with minimal anticholinergic and sedative effects. Nausea, headache, constipation, and dry mouth were the common side effects of the drug [24].

Human CYP450 enzymes metabolize fezolamine into reactive metabolites. Dealkylated metabolites are primary metabolites formed in humans; fezolamine is converted to the mono-dealkylated metabolite and further to the di-dealkylated metabolite (Scheme 7) [25].

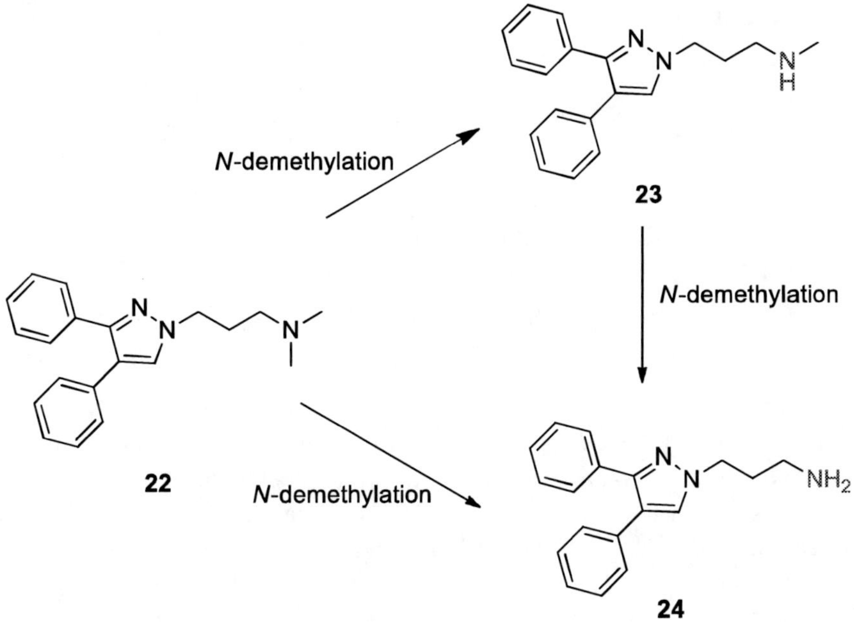

Scheme 7. Metabolic pathway of fezolamine (**22**).

Synthesis of fezolamine is shown in Scheme 8. The base-catalyzed Michael addition of diphenylpyrazole **25** to acrylonitrile **26** produces l-(2-cyanoethyl)-3,4-diphenylpyrazole (**27**). Compound **27** gives fezolamine upon reaction with dimethylamine (**28**) under hydrogenation condition [26].

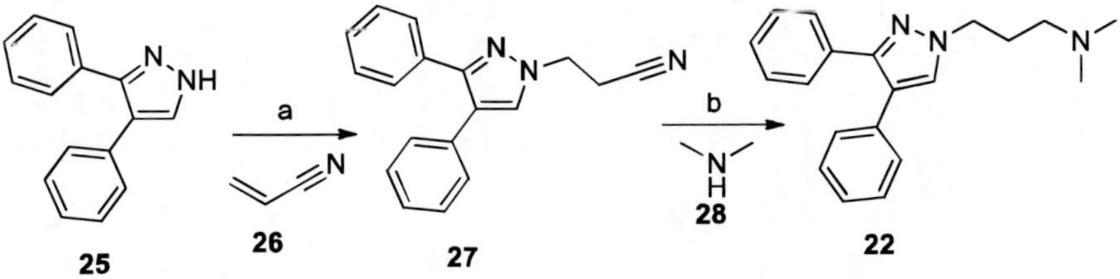

Reagents and conditions: a) Ethanol, KOH; b) Ethanol, H_2/Pd

Scheme 8. Synthesis of fezolamine (**22**).

Zometapine

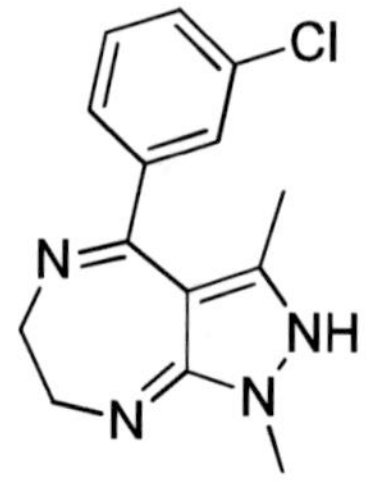

29

Zometapine (**29**), 4-(3-chlorophenyl)-1,3-dimethyl-1,6,7,8-tetrahydropyrazolo-[3,4-*e*][1,4]-diazepine, a pyrazolodiazepine derivatives resembles to benzodiazepine class of drug. It acts as an antidepressant. According to the clinical study by Katz *et al.*, the use of zometapine drug could reverse chronic stress-induced behavioral depression in humans after three weeks of use of zometapine [27].

CDPPB

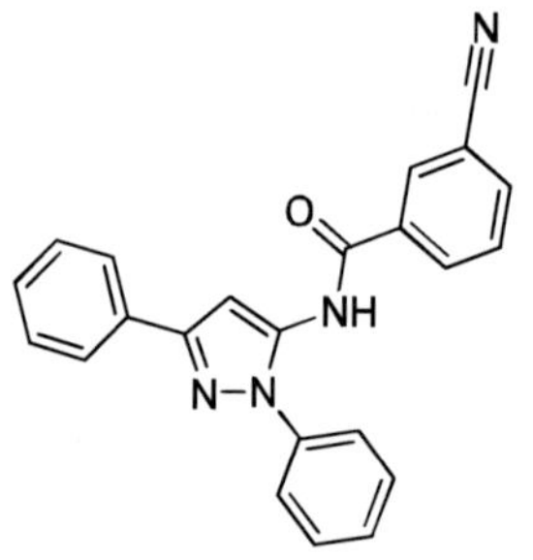

30

CDPPB (**30**), 3-cyano-*N*-(1,3-diphenyl-*1H*-pyrazol-5-yl)benzamide is an antipsychotic agent. It is a potent, selective positive modulator of metabotropic glutamate receptor subtype 5. It can easily cross the blood-brain barrier and act on GluR5, which may produce behavioral effects [28]. Uslaner *et al.* reported that CDPPB activating the mGluR5 could produce a change in markers of synaptic plasticity and a cognition-enhancing effect. This study concluded that CDPPB could increase the phosphorylation of GluR1 [29].

CDPPB is also effective in conditioned and unconditioned fear in a population with post-traumatic stress disorder (PTSD) [30].

Pharmacokinetic studies in rats showed that CDPPB had a plasma half-life of 4.4 h. Metabolite profiling studies in rats, dogs, and humans indicated that the primary metabolic pathway of CDPPB involves oxidation and glucuronidation of the diphenyl pyrazole moiety of the drug [28].

The CDPPB (**30**) synthesis was reported from benzoyl acetonitrile **31**, which is refluxed with phenylhydrazine to obtain pyrazole derivative **32** followed by acylation with cyanobenzoyl chloride in DCM or THF at low temperature to yield CDPPB (**30**) (Scheme 9) [31].

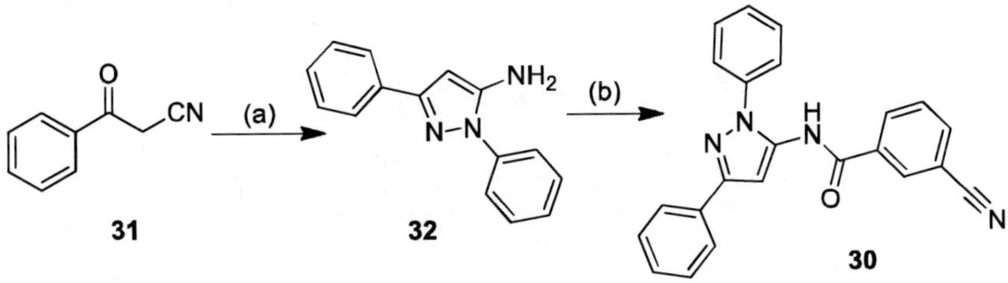

Reagents and conditions: (a) phenylhydrazine, reflux, 2 h; (b) cyanobenzylchloride, DCM or THF, 0-10 °C

Scheme 9. Synthesis of CDPPB (**30**).

Mepiprazole

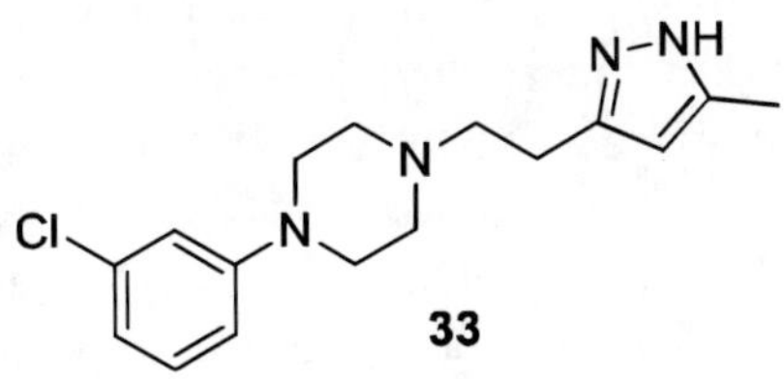

Mepiprazole (**33**) is 1-(3-chlorophenyl)-4-[2-(5-methyl-*1H*-pyrazol-3-yl)ethyl]piperazine. Presently, Mepiprazole is only used as a tranquilizer in Spain [32]. Studies showed that the drug was beneficial in treating irritable bowel syndrome [33].

The major metabolite of mepiprazole was *m*-chlorophenyl piperazine (*m*-CPP) [34]. Lawlor *et al.* reported the effect of the metabolite *m*-CPP on blood pressure [35]. The studies showed that the administration of 0.08 mg/Kg of *m*-CPP produced a marginal increase in blood pressure. Studies by Gijsman *et al.* showed that m-CPP does not alter heart rate [36]. *m*-CPP also does not induce sedation. However, studies on human subjects showed that after 8-12 h of administration of m-CPP, side effects such as headache or migraine were reported.

Elexacaftor

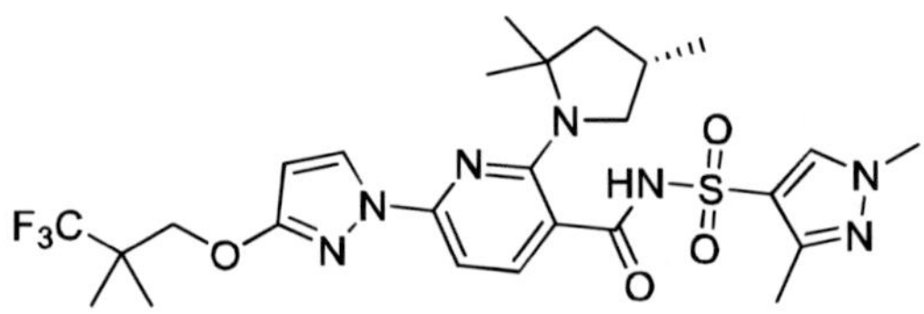

34

Elexacaftor (**34**), (*S*)-*N*-((1,3-dimethyl-1*H*-pyrazol-4-yl)sulfonyl)-6-(3-(3,3,3-trifluoro-2,2-dimethylpropoxy)-1*H*-pyrazol-1-yl)-2-(2,2,4-trimethylpyrrolidin-1-yl)nicotinamide, is a CFTR corrector that efficiently increases the amount of the CFTR protein and controls the cystic fibrosis (CF) disease.

The primary active metabolite of elexacaftor (ELX) was found to be M23-ELX. Around 87% of the administered elexacaftor excreted in faces as its metabolites, with only 0.23% found in the urine [37]. CYP3A4/5 was the chief enzyme that catalyzed the metabolic pathway of elexacaftor. The metabolite M23-ELX exhibits similar potency as elexacaftor.

Synthetic route of elexacaftor (**34**) shown in Scheme 10. 2,6-dichloronicotinic acid **35** is protected as its *t*-Bu ester **36**. Nucleophilic aromatic substitution with **37** occurred using potassium carbonate and DABCO (15 mol %) as bases in DMF at room temperature, followed by Boc group was cleaved using HCl in *i*-PrOH, carboxylic acid **38** was activated with CDI in THF and then coupled with sulfonamide **39** using DBU to furnish imide **40**. In the final step, the nucleophilic aromatic substitution of pyrrolidine **41** with **40** was carried out in a DMSO/DEE mixture at 120°C using potassium carbonate as the base forms elexacaftor (**34**) [38].

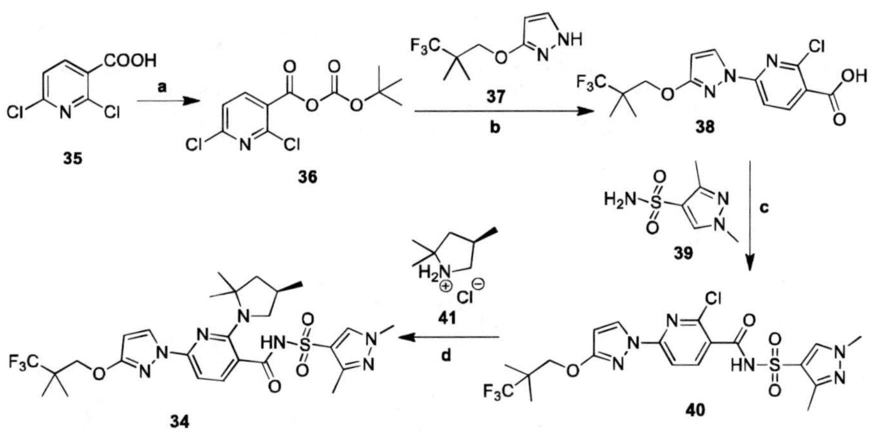

Reagents and conditions: a) $(Boc)_2O$, DMAP, THF; b) i) K_2CO_3, DABCO, DMF, ii)HCl, *i*Pr-OH, reflux; c) CDI, DBU, THF; d) K_2SO_4/DMSO/DEE, 120 °C.

Scheme 10. Synthesis of elexacaftor (**34**).

The preclinical study suggested that the elexacaftor combined with other molecules like ivacaftor and tezacaftor shows promising effects on modulation/correction of the CFTR gene [39, 40, 41, 42].

Epirizole

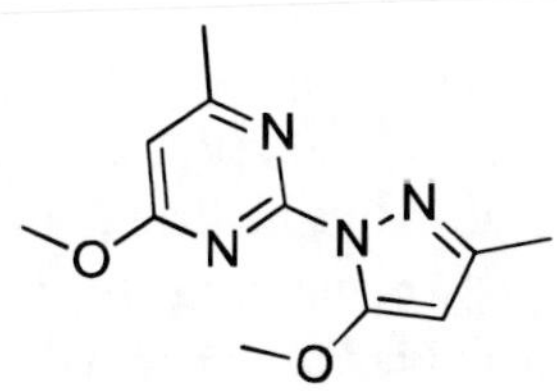

42

Epirizole (**42**), 4-methoxy-2-(5-methoxy-3-methyl-1*H*-pyrazol-1-yl)-6-methylpyrimidine is a NSAID drug used to treat musculoskeletal disorders and inflammation in Japan. It is also written as meperizole in some articles. It is a duodenal ulcerogenic drug that shows pathogenic lesions in the duodenum and stimulates gastric acid secretion on oral administration. Epirizole may cause stimulation of gastric acid secretion by changing the concentration of HCO_3^- in the duodenum. Therefore, it is used to study drugs inhibiting acid secretion activity, like pantoprazole [43].

When combined with indomethacin, epirizole may increase its bioavailability more than individual drugs [44]. Onoa *et al.* studied the effect of Pt(II) and Pd(II) complexes with the bi-heterocyclic ligand epirizole on tumor cells; however, it shows less activity than cisplatin [45].

Phenylbutazone

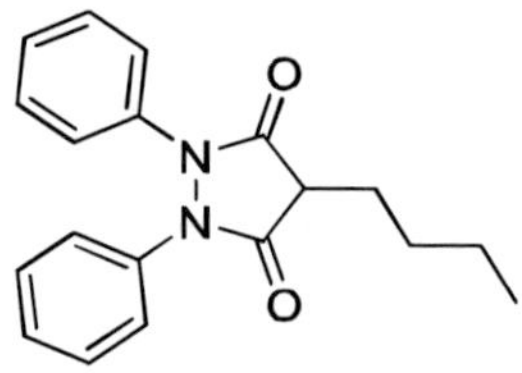

43

Phenylbutazone (**43**), 4-butyl-1,2-diphenylpyrazolidine-3,5-dione, a pyrazolone derivative is a NSAID drug with analgesic and antipyretic activity. It is used for acute pain and muscular–skeleton disorder, especially for the treatment of rheumatoid arthritis [46].

In 1949, phenylbutazone, for the first time, was used with aminopyrine (an anti-inflammatory drug) in equal concentration with the intention that it may serve as a solvent for another drug. However, these studies have proved that phenylbutazone has analgesic and anti-inflammatory activity. At that time, cortisones were used for treating chronic rheumatoid arthritis, but this would only treat symptoms and not cure the disease. Introducing phenylbutazone was a miracle in the medical field as it would treat rheumatoid disease. It is also called as a "wonder drug" [47].

There are different trajectories behind this drug. It is used in both animals and humans to treat pain and inflammation. In 2013, many patients with rheumatoid arthritis suffered from severe side effects and even death. As a result, this drug was banned, but some rheumatoid arthritis physicians opposed it as it positively impacted the disease [24].

In 2008, a European country faced a meat adulteration scandal; FDA banned the use of phenylbutazone in animals being slaughtered for human consumption. We cannot fix the safe level of phenylbutazone in the animal because of its bone marrow toxicity [48].

Phenylbutazone gets metabolized through an oxidative pathway (Scheme 11). Ring oxidation of phenylbutazone forms oxyphenbutazone (**44**), an active metabolite, while γ-hydroxyphenylbutazone **45** formed by γ–oxidation of the aliphatic chain is an inactive metabolite because of the lack of interactions at the active site with protein. Loss of lipophilic character might be the contributory factor in the inactivity of the γ-hydroxyphenylbutazone. As oxyphenbutazone is an active metabolite, it also influences the drug's toxicity because it has a high electron density [49]. Toxicity is related to its mechanism of action as it is COX-I and COX-II non-selective inhibitor. The significant side effects are gastric ulcers, thrombosis, and chronic renal failure [32, 50].

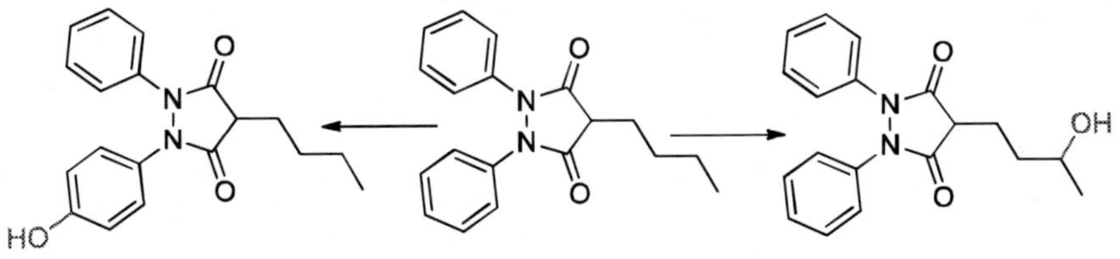

44 **43** **45**

Scheme 11. Metabolic pathway of phenylbutazone (**43**).

The synthesis of phenylbutazone (Scheme 12) in one-step was reported. It involves condensation of diester **46**, diphenylhydrazine (**47**), and bromobutane (**48)** at 70-90 °C under basic conditions, followed by ring closure of pyrazolidinedione at 120-150 °C to yield phenylbutazone (**43**) [51].

46 **47** **48** **43**

Reagents and conditions: a) i. NaOEt, EtOH, 70 °C, 3 h; ii. 120 - 150 °C, 2 h

Scheme 12. Synthesis of phenylbutazone (**43**).

Due to restrictions on phenylbutazone and its toxic side effects, Borges *et al.* designed and evaluated a newer series of phenylbutazone derivatives. The toxicity of this drug is due to ring hydroxylation; hence the aromatic ring is substituted with chlorine showing promising activity. Compounds **49, 50, 51,** and **52** (Figure 2) show good activity amongst all designed compounds [32].

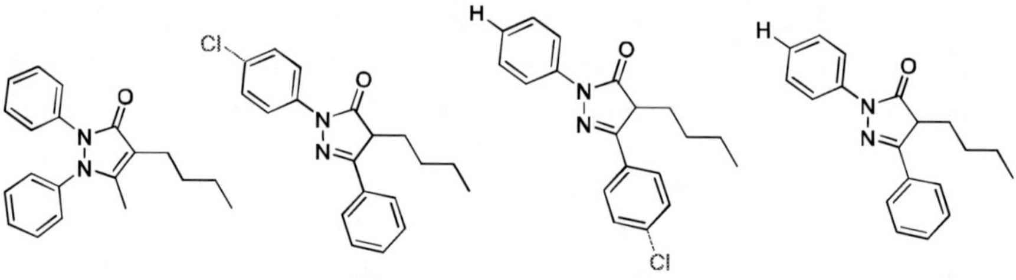

49 50 51 52

Figure 2. Prominent analog of phenylbutazone.

Lonazolac

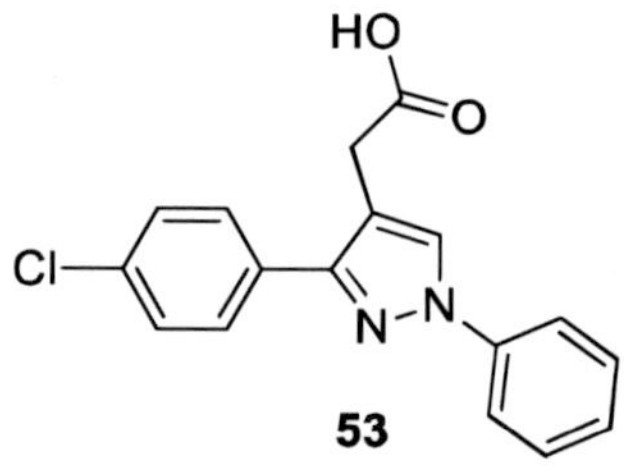

53

Lonazolac (**53**), 2-(3-(4-chlorophenyl)-1-phenyl-1*H*-pyrazol-4-yl)acetic acid, is an aryl acetic acid derivative of pyrazole categorized under the class of NSAID drugs. It has anti-inflammatory and analgesic activity. Lonazolac is marketed in its calcium-coordinated form as Lonazolac Ca.

The mechanism of action of lonazolac calcium inhibits COX-1 and COX-2 enzymes resulting in an antipyretic, analgesic, and anti-inflammatory action. It also has anti-platelet aggregation action; however, it is less than COX inhibition activity [52]. Vinge *et al.* reported the activity of lonazolac by inhibition of the generation of thromboxane and platelet aggregation. Indoprofen and lonazolac were compared for this study revealing both drugs with the same potency toward platelet aggregation after oral administration [53].

Friedrich *et al.* reported the procedure for the analysis of metabolites of lonazolac by the HPLC method. Formation of hydroxyllonazolac **54** was catalyzed by CYP 2C9, which further undergoes sulfonation by arylsulfotransferase to lonazolac sulphate **55** as shown in Scheme 13. This metabolite is separated by HPLC and detected by fluorescence [53].

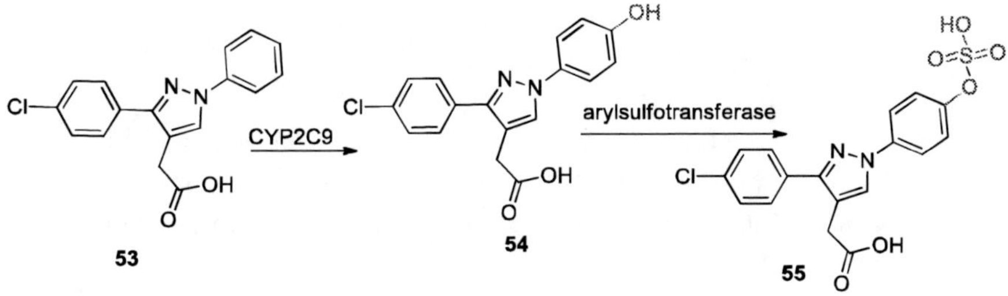

Scheme 13. Metabolic pathway of lonazolac (**53**).

The synthesis of lonazolac is shown in Scheme 14. Lonazolac was synthesized by cross-condensation of *N*-phenylhydrazone **56** with conjugated hydrazone **57**, resulting in the intermediate **58**, which is further hydrolyzed to yield Lonazolac (**53**) [54].

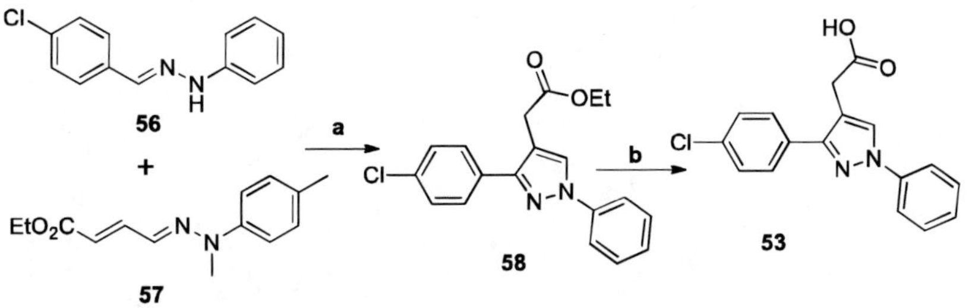

Reagents and conditions: a) MsOH, CH_2Cl_2, rt, 6 hr, 37%; b) LiOH, THF/H_2O, rt, 3 hr, 89%

Scheme 14. Synthesis of lonazolac (**53**).

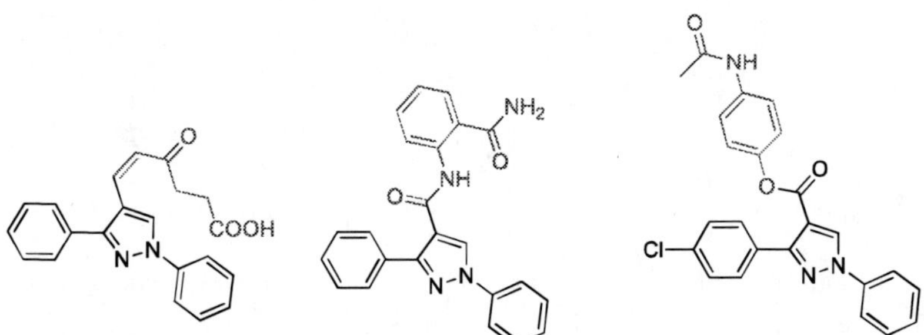

Figure 3. Prominent analog of lonazolac.

Harras *et al.* designed, synthesized, and evaluated the non-acidic analogs of lonazolac. 1,3,4-trisubstituted pyrazole derivatives displayed good selectivity towards COX-2 and good anti-inflammatory activity *in-vitro* and *in-vivo*. *In-silico* docking study also confirms the same results. Compounds **62**, **63**, **64**, **65**, and **66**, as shown in Figure 4, possess good selectivity to COX-2 enzyme with potent anti-inflammatory activity. These compounds are less ulcerogenic than the standard drug indomethacin [56].

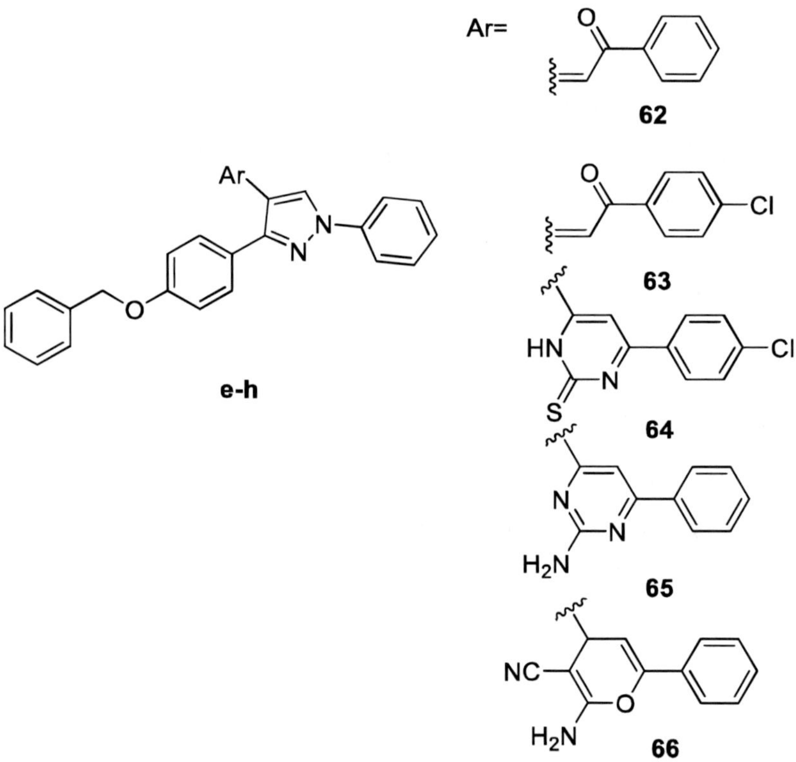

Figure 4. Prominent non-acidic 1,3,4–trisubstituted pyrazole analog of lonazolac.

Abdellatif *et al.* also designed, synthesized, and evaluated the non-acidic 1,3,4–trisubstituted pyrazole for their selectivity towards COX-2, anti-inflammatory activity, and ulcerogenic profile. bisaminosulphonyl substituted derivatives show good potency satisfying all the safety data [57].

Difenamizole

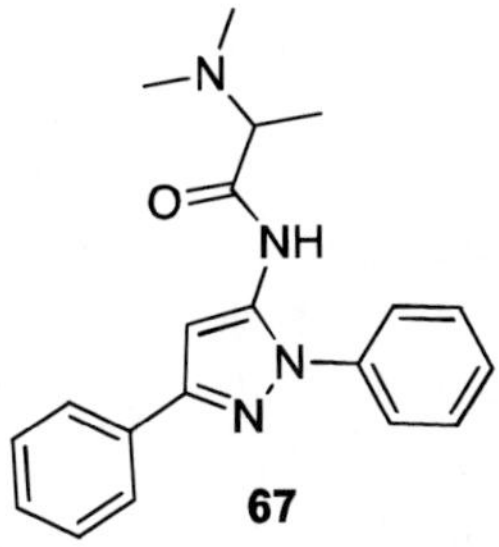

67

Difenamizole (**67**), 2-(dimethylamino)-*N*-(1,3-diphenyl-1*H*-pyrazol-5-yl)propanamide is an NSAID class of drug, but it is demonstrated that it may act by the central nervous system.

Difenamizole also acted as an analgesic and relaxed abnormal muscles while showing no action on the normal muscle [58]. Nabeshima *et al.* reported the study of the action of difenamizole on catecholamines in the brain. This study's result shows that difenamizole acts by depressing the dopaminergic receptor by decreasing the release of catecholamines from monoaminergic neurons. It inhibits dopaminergic receptors to a greater extent compared to norepinephrinergic receptors [59]. It also inhibits monoaminoxidase activity in the mouse brain [60]. Monoamine oxidase reduces the secretion of neurotransmitters such as norepinephrine and dopamine in the brain. Decreased levels of these neurotransmitters in the brain can lead to depression and anxiety. Difenamizole was shown to effectively enhance dopamine secretion in the mouse brain and showed an inhibitory effect on potassium-induced dopamine release [60].

Pyrazolone-Based Drugs

Antipyrine (Phenazone), Dipyrone, Aminophenazone, Melubrin

Pyrazolone unit is one of the key structural motifs present in various marketed pharmaceutical drugs. The first pyrazolone-based antipyretic, analgesic, and anti-inflammatory drug successfully marketed was phenazone (**68**) [61]. Antipyrine (Phenazone), the first approved pyrazolone-containing drug, is still widely used as an anti-inflammatory, analgesic, and antipyretic agent [62]. The introduction of the dimethylamino group at the 4th position led to the new compound called aminophenazone (**69**), 4-(dimethylamino)-1,5-dimethyl-2-phenyl-1,2-dihydro-3*H*-pyrazol-3-one, is also called as pyramidone. SAR studies revealed that the presence of the dimethylamino group in pyramidone increases its activity by three times compared to the parent phenazone. The three primary metabolites of aminophenazone found were: a) 4-methyl aminoantipyrine, b) 4-aminoantipyrine, and c) 4-acetyl aminoantipyrine. Other metabolites of aminophenazone found were 4-hydroxyantipyrine, 4-formyl antipyrine, rubazonic acid, and methyl rubazonic acid [63]. Among the metabolites, 4-methyl aminoantipyrine was the primary metabolite found in a study conducted in rats [64].

To produce the drug in a soluble and injectable form with enhanced activity, Melubrin-sodium (**70**), sodium ((1,5-dimethyl-3-oxo-2-phenyl-2,3-dihydro-1*H*-pyrazol-4-yl)amino)methanesulfonate was prepared and marketed. Melubrin-Sodium (**70**) contains a sodium methane sulphonate group on side chain nitrogen to increase water solubility. Melubrin (**70**) was less successful in the market since the later-arrived dipyrone (**71**) (novalgin) showed improved activity and solubility compared to melubrin (**70**).

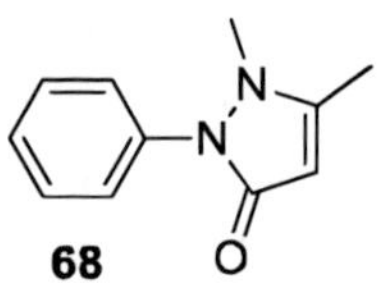

68

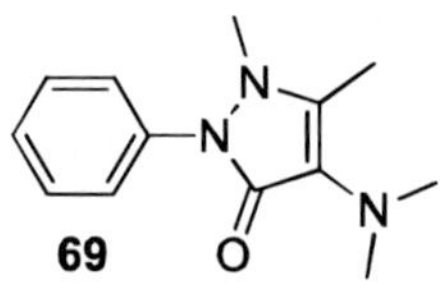

69

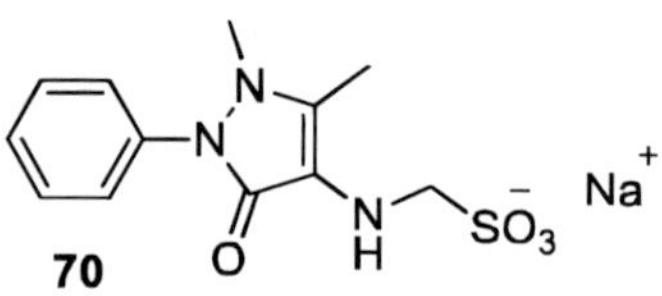

70

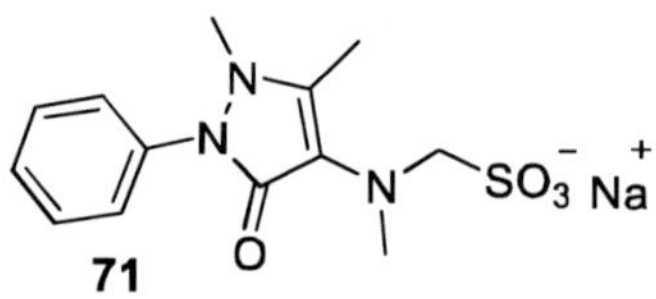

71

Antipyrine (Phenazone)

Phenazone (**68**), 1,5-dimethyl-2-phenyl-1,2-dihydro-3*H*-pyrazol-3-one also known as antipyrine.

After discovering antipyrine and its multiple roles in therapeutics, various pyrazolone analogs have been synthesized to improve the activity of the parent antipyrine. Many FDA-approved pyrazolone-based drugs are currently available in the market. Multiple studies showed that pyrazolone derivatives were also effective as anti-tubercular, antimicrobial, anti-tumor, lipid-lowering, and antihyperglycemic agents [65].

Ludwig Knorr synthesized Antipyrine in 1883 from the reaction of 3-methyl-1-phenyl-1*H*-pyrazol-5(4*H*)-one **72** with MeI in methanol [66]. An improved synthetic method for the synthesis of antipyrine was later demonstrated by K. Krohn *et al.* in 1989, as shown in Scheme 15. In this work, 3-methyl-1-phenyl-1*H*-pyrazol-5(4*H*)-one **72** was allowed to react with Me_2SO_4 at 160 °C for 5 h to yield antipyrine (**68**) in 71% yield [67].

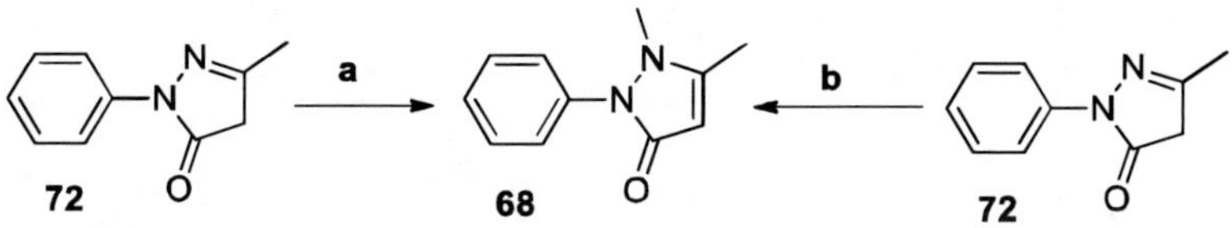

Reagents and condtions: a) MeI, MeOH; b) Me_2SO_4, 160 °C, 5h, 71%

Scheme 15. Synthesis of antipyrine (**68**).

Dipyrone

Dipyrone (**71**), ((1,5-dimethyl-3-oxo-2-phenyl-2,3-dihydro-1*H*-pyrazol-4-yl) (*N*-methyl)amino) methane sulfonate is an effective analgesic and antipyretic drug. Dipyrone can treat pain caused by smooth muscle spasms. Dipyrone also acts as a non-steroidal anti-inflammatory drug (NSAID) [68]. The sulphonic acid moiety in dipyrone helps the drug immensely enhance water solubility compared to the aminopyrine molecule.

Dipyrone can be used safely for patients suffering from chronic obstructive pulmonary disease (COPD) since dipyrone does not induce a bronchodilation effect [69]. R. Zapf *et al.* studied the possible interference of dipyrone with anticoagulants such as phenprocoumon and ethyl biscoumacetate [70]. The possible interaction of the drug with the anticoagulant can lead to a severe haemorrhage condition. The study was carried out by measuring the prothrombin ratio in healthy male subjects after administering dipyrone. The study revealed that the anticoagulant activity of phenprocoumon and ethyl biscoumacetate is not affected by the oral dose of dipyrone.

Dipyrone is non-enzymatically hydrolyzed to active metabolite **73** in the gastrointestinal tract. Further metabolized in the liver by CYP enzymes to the second active metabolite **75** by demethylation and the first inactive metabolite **74** by oxidation (Scheme 16). Also converted to another inactive metabolite **76** via *N*-acetylation [71, 72].

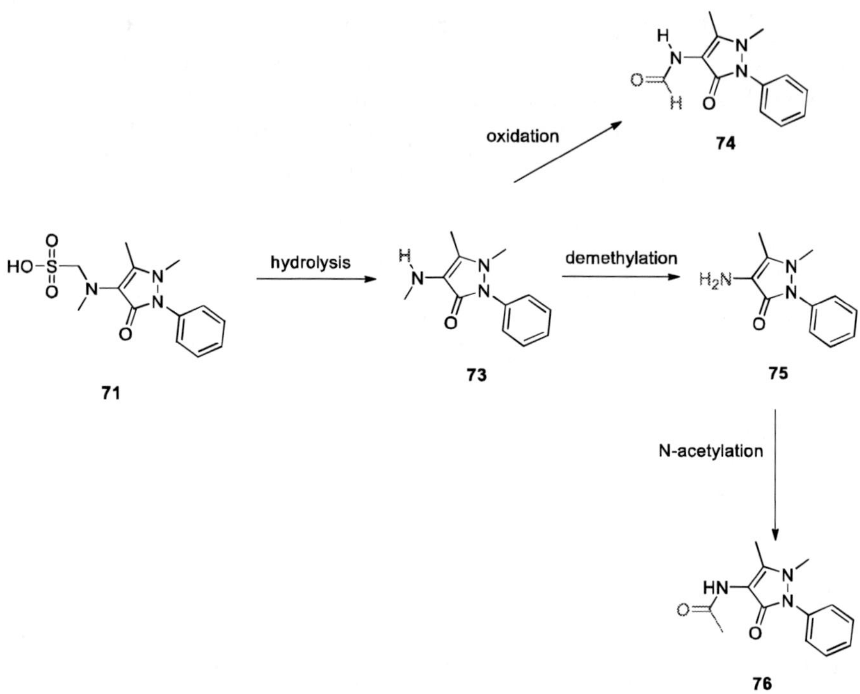

Scheme 16. Metabolic pathway of dipyrone.

Synthesis of dipyrone (Scheme 17) was done by reacting phenylhydrazine (77) and ethyl acetoacetate **78** to obtain intermediate **79**. Further, intermediate **79** was alkylated by Friedel-Crafts reaction using methyl chloride and Lewis acid to get phenazone **80**. Nitrosation, followed by reduction and alkylation of **80** were performed to obtain **81**, **82,** and **83** respectively. Formation of dipyrone (**71**) by Mannich-reaction was done using formaldehyde and sodium hydrogen sulphite [73].

The significant adverse effect of using dipyrone is agranulocytosis [74]. In the case of pregnant women and new-born infants is limited. In one such study by Bar-Oz *et al.*, dipyrone metabolites were found in breast milk in the concentration up to 4.3 mg/L and in the urine of the breastfed infant up to 3.2 mg/L [75]. G. Rizzoni *et al.* reported cyanotic episodes in one infant following the dipyrone dose for the first time. The FDA banned dipyrone in the USA due to the potential cause of drug-induced agranulocytosis. Nevertheless, many countries in Africa, Asia, and Europe use the drug due to its excellent antipyretic and analgesic properties.

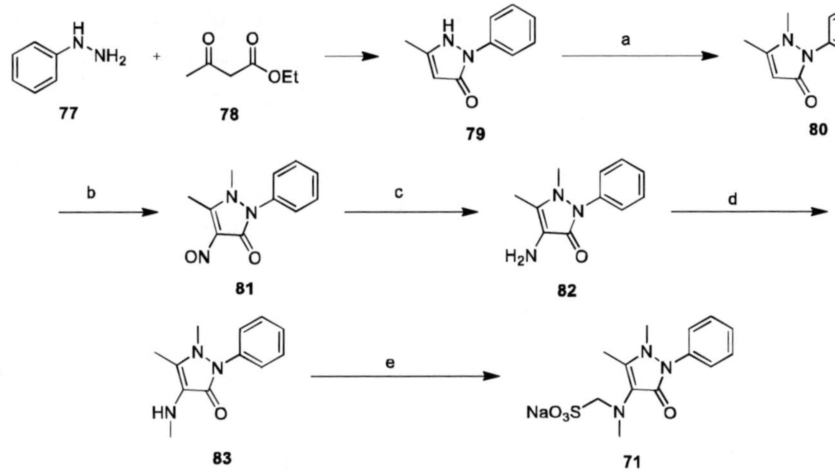

Reagents and conditions: a) $AlCl_3$, CH_3Cl; b) HNO_2; c) [H] ; d) $AlCl_3$, CH_3Cl; e) HCHO, $NaHSO_3$

Scheme 17. Synthesis of dipyrone (**71**).

Propyphenazone and Ramiphenazone

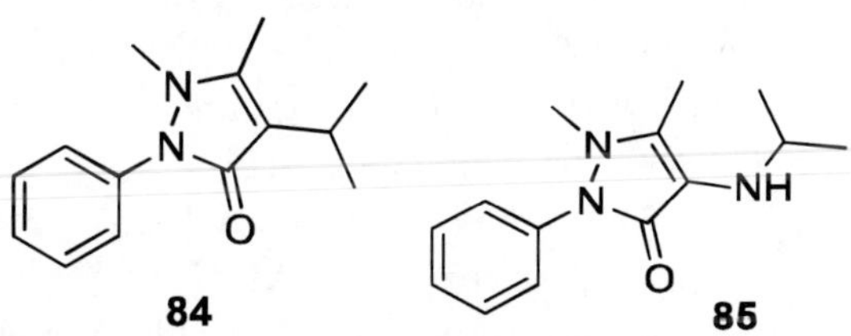

Propyphenazone (**84**), 4-isopropyl-1,5-dimethyl-2-phenyl-1*H*-pyrazol-3(2H)-one, and Ramiphenazone (**85**), 4-(isopropylamino)-1,5-dimethyl-2-phenyl-1*H*-pyrazol-3(2H)-one, are pyrazolone-based compounds widely used as analgesic and anti-inflammatory drugs. Sarridon, a combination of propyphenazone (150 mg), paracetamol (250 mg), and caffeine (50 mg), showed excellent pain relief for patients with moderate and severe acute dentoalveolar pain. Propyphenazone showed much better analgesic activity than placebo therapy. Prolonged intake of sarridon caused drug-induced hepatitis in a patient with no history of liver disease [76]. Also, prolonged use of these anti-inflammatory drugs can cause gastrointestinal, renal, and cardiovascular issues [77].

The foremost metabolic route of propyphenazone (**84**) is demethylation, and the main metabolite is enolglucuronide of *N*-(2)-demethylpropy-phenazone, whereas 3-aminopyrine is completely absorbed after administration [78].

Tepoxalin

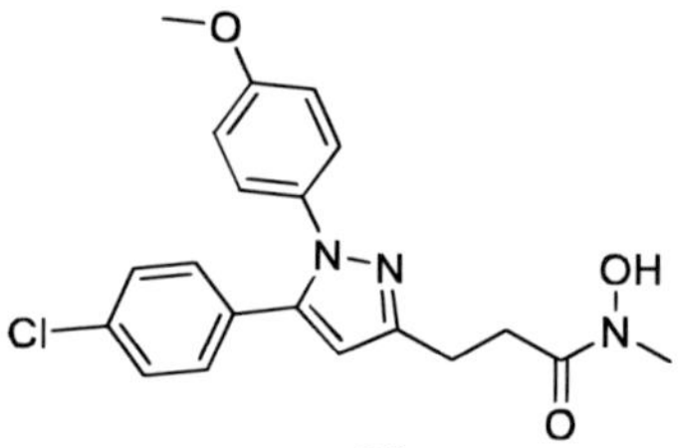

86

Tepoxalin (**86**), 3-(5-(4-chlorophenyl)-1-(4-methoxyphenyl)-4,5-dihydro-1*H*-pyrazol-3-yl)-*N*-hydroxy-*N*-methylpropanamide, pyrazole derivative has dual action of cyclooxygenase/lipoxygenase inhibition. It is primarily used to treat pain and inflammation. Tepoxalin is a prostaglandin–H-synthase-1 inhibitor. Many NSAIDs inhibit only cyclooxygenase enzyme, but tepoxalin inhibits both the rate-limiting enzymes that are cyclooxygenase and peroxidase non-competitively. It is known that NSAIDs require an acidic group in their structure to interact with the enzyme's active site. Tepoxalin contains hydroxamic acid instead of carboxylic acid. Hydroxamic acid of tepoxalin additionally provides peroxidase enzyme activity [79].

FDA has approved tepoxalin for use along with chemotherapy. Clinical trials show that tepoxalin given with chemotherapy may restore sensitivity and improve drug response in patients with ATP binding cassette subfamily B member 1 (ABCB1)-overexpressing breast cancer. ABCB1 is a gene that encodes for P-glycoprotein, which prevents intracellular accumulation of cytotoxic drugs [80]. Zhou *et al.* studied the *in-vivo* effect of tepoxalin on neutrophile migration. It was shown that tepoxalin blocks the upregulation of Mac-1 (it is a key adhesion molecule that facilitates cancer progression and mediates the adhesion of tumor cells to the endothelium of blood vessels) and inhibits the neutrophile migration into inflammatory sites in murine skin stimulated by TNF-α or LPS. Researchers also demonstrated that tepoxalin is a potent inhibitor of NFxB-induced transcription *in-vitro*. NFxB is a signalling pathway involved extensively in cancer development and progression [81].

The metabolism pathway of tepoxalin is shown in Scheme 18. After oral administration tepoxalin is converted to RWJ-20142, a carboxylic acid metabolite. It has cyclooxygenase-inhibiting activity but does not have lipoxygenase activity [82].

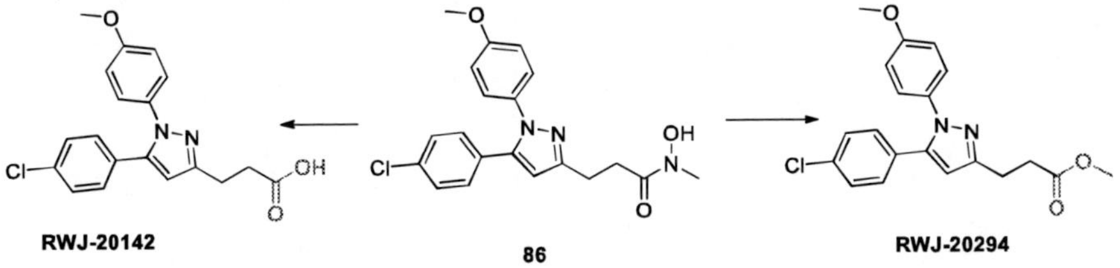

Scheme 18. Metabolic pathway of tepoxalin (**86**).

The synthesis of tepoxalin is shown in Scheme 19. Tepoxalin was synthesized from substituted dioxo-hexanoic acid **87** and treated with acetic anhydride to obtain lactone **88**. Compound **88** was added to *N*-methyl hydroxylamine hydrochloride in methylene chloride containing triethylamine to generate acetylated amine **89** followed by stirring at room temperature to yield hydroxamic acid derivative **90**. Compound **90** was reacted with 4-methoxyphenylhydrazine hydrochloride under the basic condition to obtain tepoxalin (**86**) [83].

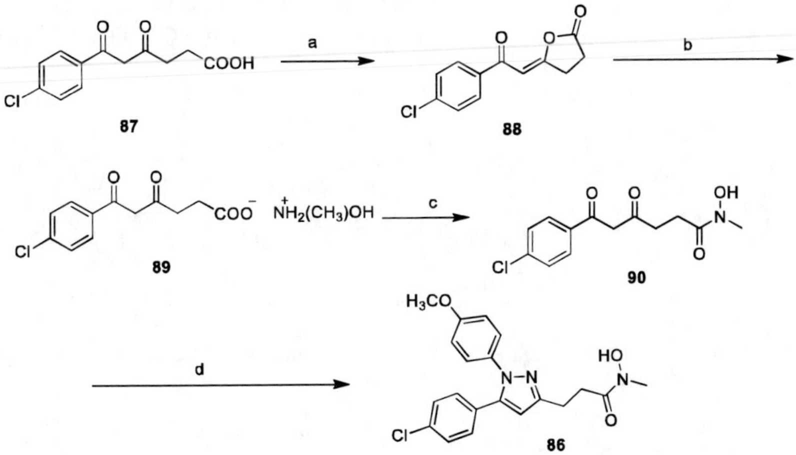

Reagents and conditions: a) Ac_2O; b) Et_3N/CH_2Cl_2, $NH(CH_3)OH$ .HCl; c) heat; d) 4-Methoxy phenylhydrazine hydrochloride, EtOH / Na_2CO_3

Scheme 19. Synthesis of tepoxalin (**86**).

SC-560

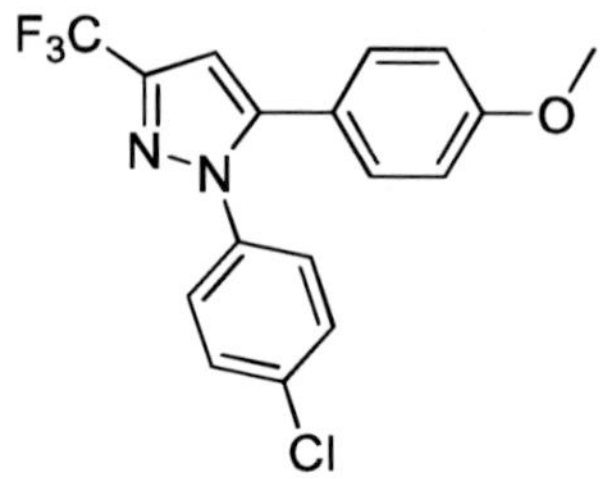

91

SC-560 (**91**), 1-(4-chlorophenyl)-5-(4-methoxyphenyl)- 3-trifluoromethyl pyrazole is a highly selective and potent cyclooxygenase-1 (COX-1) inhibitor used as an anti-inflammatory and analgesic drug [84]. The drug consists of a pyrazole with two aryl rings and a trifluoromethyl group. SC-560 also shows anti-tumor activity [85].

SC-560 was synthesized starting from 1,1,1-trifluoro-4-(4-methoxyphenyl)but-3-en-2-one (**92**) as the substrate (Scheme 20). Treatment of **92** with *p*-toluene sulphonyl hydrazide in the presence of NaOAc resulted in the formation and isolation of pyrazole derivative **93** in 90% yield. Finally, the Cu-catalyzed reaction was successfully applied in the reaction of **93** with 1-chloro-4-iodobenzene **94** as the substrate to produce SC-560 in 77% yield [86].

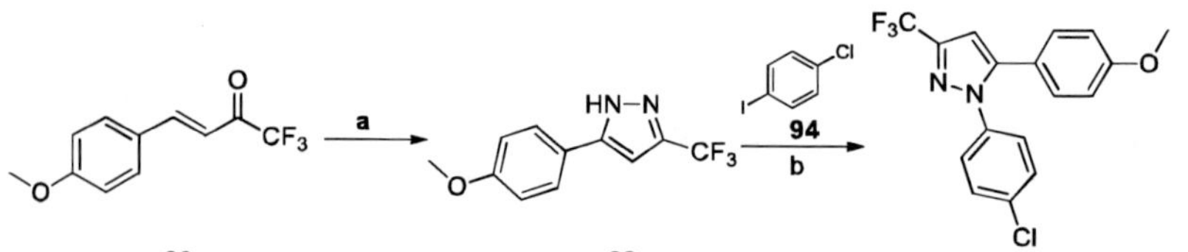

Reagents and conditions: a) $TsNNH_2$,NaOAc, EtOH, 80 °C; b) 1-chloro-4-iodobenzene, CuI, K_2CO_3, 1,4-dixoane, 150 °C, 77%

Scheme 20. Synthesis of SC-560 (**91**).

Celecoxib

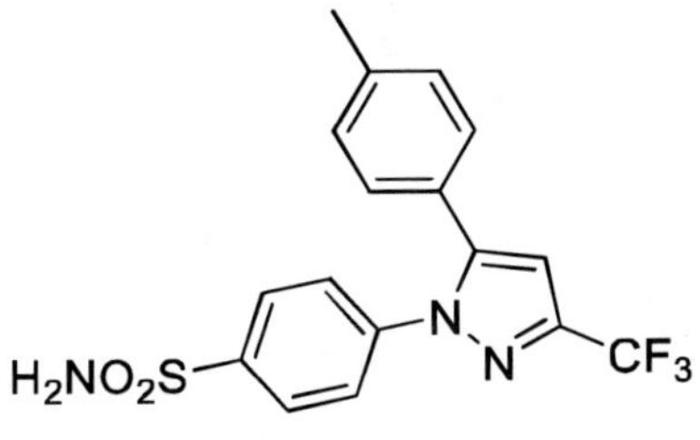

95

Celecoxib (**95**), 4-(5-(*p*-tolyl)-3-(trifluoromethyl)-1*H*-pyrazol-1-yl) benzenesulfonamide, is a non-steroidal anti-inflammatory drug (NSAID), which selectively inhibits cyclooxygenase-2 enzyme (COX-2) [87]. Celecoxib is used in the treatment of osteoarthritis and rheumatoid arthritis.

Patients treated with celecoxib (25 mg- 100 mg daily) showed much less gastrointestinal issues than those treated with other non-steroidal anti-inflammatory drugs. Celecoxib also showed drug independence with no reports on the formation of an ulcer in patients treated with the drug compared to patients administrated with ibuprofen 3 X 800 mg daily [88]. Additionally, the drug exhibited zero effect on collagen-induced platelet aggregation; however, it caused a little effect on potassium balance in the patients treated with diuretics.

The celecoxib is metabolized through methyl hydroxylation to produce hydroxycelecoxib, catalyzed by cytochrome P450 family 2 subfamily C member 9 (CYP2C9) and cytochrome P450 3A4 (CYP3A4). However, CYP3A4 has a minor role (less than 25%) in catalyzing the reaction. Next, hydroxycelecoxib is oxidized to produce carboxycelecoxib and then conjugate with glucuronic acid using uridine 5'-diphospho-glucuronosyltransferase to produce 1-O-glucuronide [89].

Celecoxib was synthesized in a two-step process (Scheme 21). First, trifluoro-*p*-tolyl-1,3-butanedione **98** was synthesized by the reaction of *p*-methyl acetophenone **96** and trifluroacetyl imidazole **97** in the presence of sodium bis(trimethylsilyl)amide. Treatment of the formed product with 4-hydrazinylbenzenesulfonamide **99** produced celecoxib (**95**) [90].

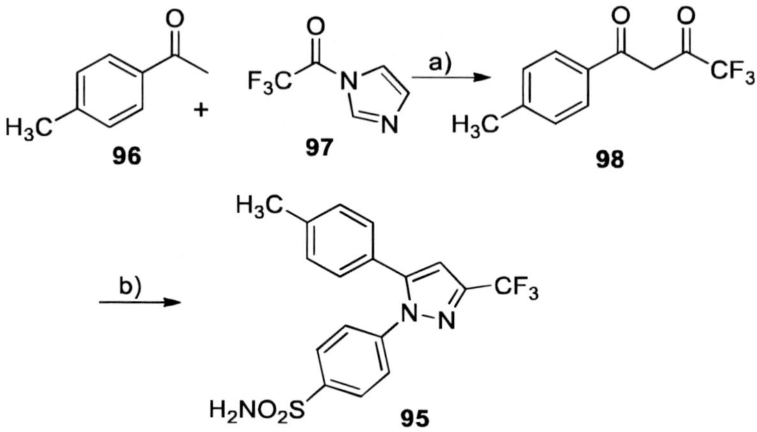

Reagents and conditions: a) $NaN(TMS)_2$; b) 4-hydrazinylbenzenesulfonamide (**99**)

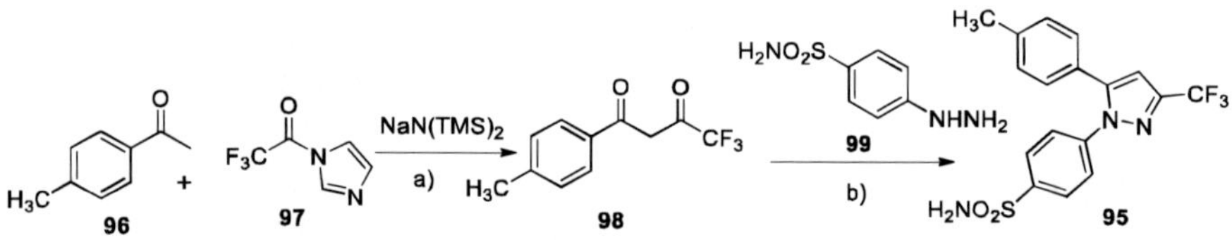

Reagents and conditions: a) $NaN(TMS)_2$; b) 4-hydrazinylbenzenesulfonamide

Scheme 21. Synthesis of celecoxib (**95**).

Deracoxib

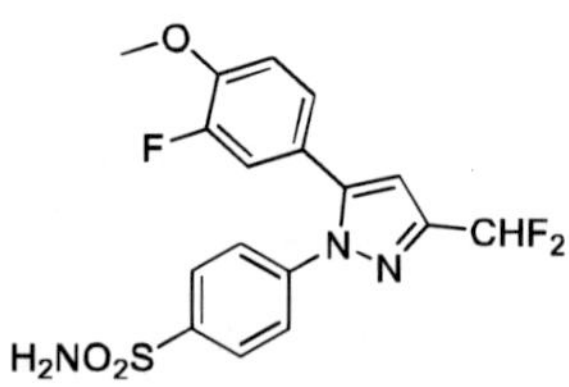

100

Deracoxib (**100**), 4-(5-(3-fluoro-4-methoxyphenyl)-3-(difluoromethyl)-1*H*-pyrazol-1-yl) benzenesulfonamide, structurally resembles to celecoxib, also has NSAIDs activity. The mechanism of action of deracoxib is it selectively inhibits the COX-2 enzyme and acts as an anti-inflammatory. Deracoxib significantly suppresses the concentration of PGE2 in blood.

Spyra *et al.* studied that deracoxib acts as a TRPV3 modulator, which potentiate the 2-APB calcium signal. The fluoromethyl group present in the structure may potentiate this activity. Deracoxib shows 4-fold more activity as a transient receptor potential vanilloid-3 (TRPV3 is a Ca^{2+} permeable nonselective cation channel) modulator than other coxib derivatives [91]. Deracoxib causes a decrease in platelet aggregation to a small extent, while other coxibs may not cause this side effect [92]. Deracoxib is approved for treating perioperative orthopedic pain and chronic pain associated with arthritis in dogs [93]. 90% is metabolized in the liver and excreted in faeces as 3-4 metabolite [94].

Deracoxib is synthesized from fluoroalkylated enaminones **101** (Scheme 22). Fluoroalkylated enaminones react with sulfonamide **102** and get converted to deracoxib (**100**) reported by Wan *et al.* [95].

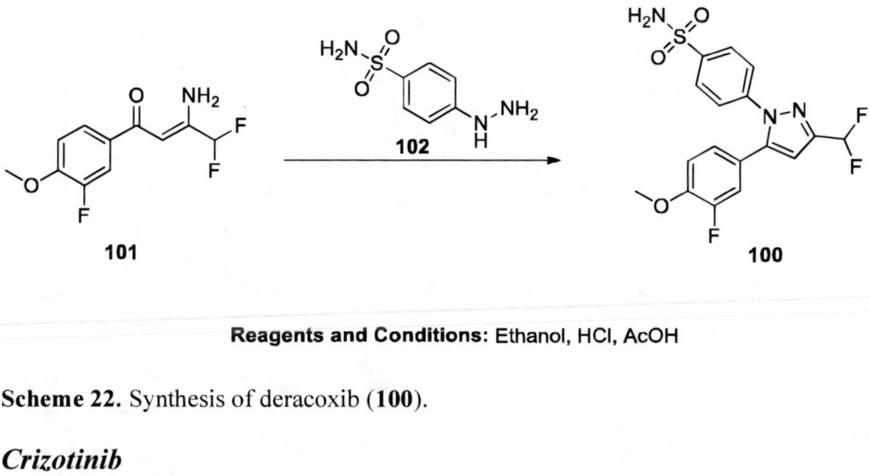

Reagents and Conditions: Ethanol, HCl, AcOH

Scheme 22. Synthesis of deracoxib (**100**).

Crizotinib

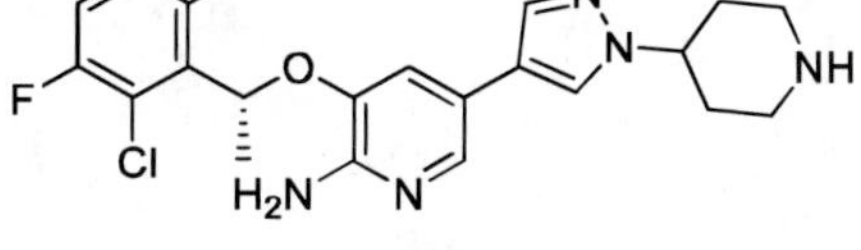

103

Crizotinib (**103**), (*R*)-3-(1-(2,6-dichloro-3-fluorophenyl)ethoxy)-5-(1-(piperidin-4-yl)-1*H*-pyrazol-4-yl)pyridin-2-amine is an inhibitor of anaplastic lymphoma kinase (ALK), a type of tyrosine kinases responsible for cancer growth in anaplastic-large cell lymphomas, lung cancer, and neuroblastoma.

Crizotinib was the first approved ALK inhibitor drug to treat the conditions mentioned above. Crizotinib showed excellent activity in the treatment of lung cancer. Choi *et al.* demonstrated that the cancer cells resisted crizotinib through secondary mutation in the kinase domain of anaplastic lymphoma kinase [96]. Two mutations (C1156Y and L1196M) have been identified, which cause resistance to the drug. Sun *et al.* used molecular docking studies to disclose the mechanism of crizotinib resistance by cancer cells. In this work, molecular docking studies suggested that C1156Y mutation induced conformation changes in the protein's binding cavity, resulting in the decreased binding affinity for crizotinib [97].

Crizotinib is mainly metabolized by CYP450 enzyme isoforms CYP3A4 and CYPP3A5 in the liver. Phase I metabolism of crizotinib includes oxidation forming amide **104**, and O-dealkylation products **105**, **107**, and phase II metabolism occurs via glutathione conjugation leading to compound **106**, **108** or sulphate conjugation (Scheme 23) [98].

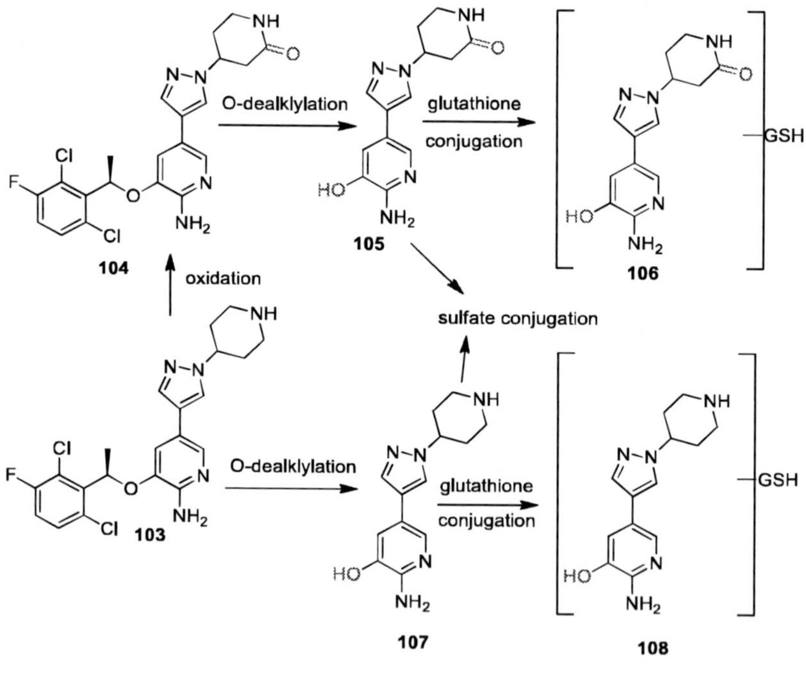

Scheme 23. Metabolic pathway of crizotinib (**103**).

In 2014, Li *et al.* developed a novel approach for synthesizing crizotinib (Scheme 24) [99]. One of the critical steps in the reaction is the asymmetric reduction of 2,6- dichloro-5-fluoro acetophenone **109** to the corresponding chiral alcohol **110**. Mitsunobu coupling of the chiral alcohol **110** with the Boc-protected substrate **111** was carried out to yield product **112** in 50% yield. The next step involves Suzuki-Miyaura coupling to give the doubly Boc-protected crizotinib **114** in 85% yield. Finally, the removal of Boc groups afforded crizotinib (**103**) [99].

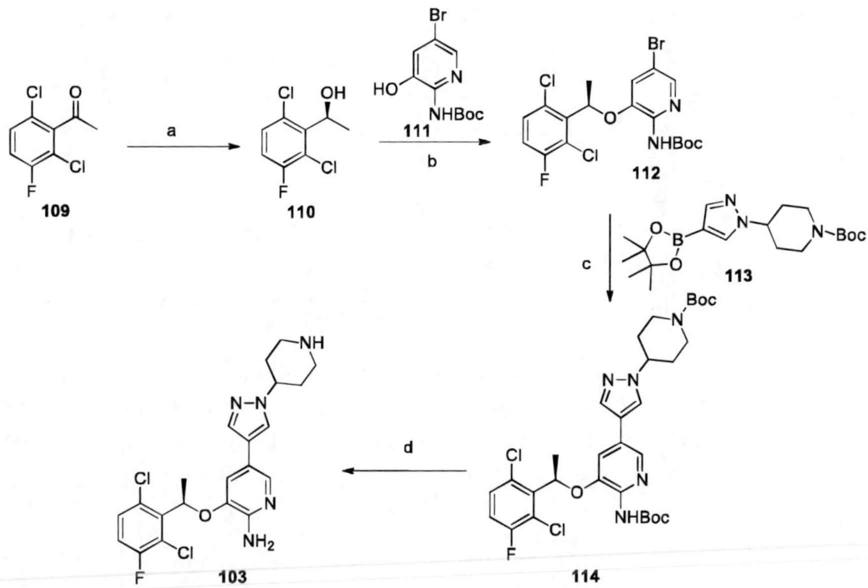

Reagents and conditions: a) Cat. Ir-spiro complex / H_2, ^tBuOK, EtOH; **b)** Ph_3P, DIAD, THF, 20-30 °c, 50%; **c)** Suzuki-Miyaura coupling, 60 °C, 3 h, 85%; **d)** HCl/EtOH, DCM, 20–30 °C, 2 h, 87%.

Scheme 24. Synthesis of crizotinib (**103**).

Pazopanib

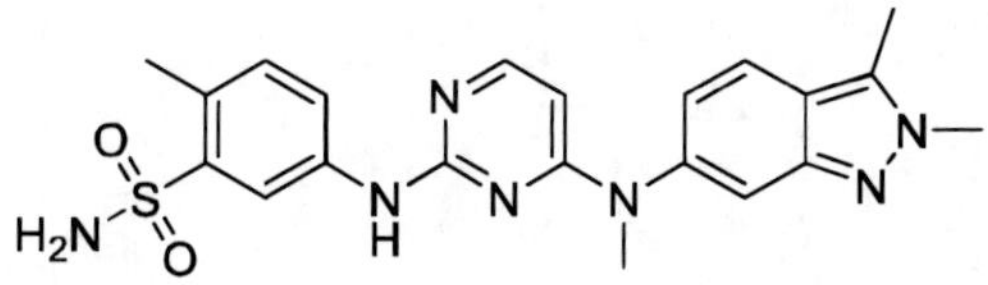

115

Pazopanib (**115**), 2-methyl-5-((4-(-*N*-(2,3-dimethyl-2H-indazol-6-yl)(*N*-methyl)amino)pyrimidin-2-yl)amino)benzenesulfonamide, an indazole derivative, is a tyrosine kinase inhibitor of VEGFR-1,2,3 and platelet-derived growth factor receptor-α,β VEGFR is a vascular endothelial growth factor, and its receptors regulate vasculogenesis and angiogenesis. It is used as an anti-cancer agent. Frankel *et al.* discussed the phase 2 study of pazopanib drug's use to treat osteosarcoma metastatic to the lung [100].

Nelson *et al.* reported the new use of pazopanib to treat neuroepithelial tumors. Pazopanib is co-administered with topotecan and showed significant potency in treating neuroepithelial tumors. Because of co-administration, efficacy and exposure of topotecan are enhanced. Pazopanib showed moderate selectivity against fibroblast growth factor receptor 1-3 (FGFR1-3) protein involved in the formation, survival, and movement of nerve cells in several brain areas [101].

Hereditary haemorrhagic telangiectasia (HHT) is an angiogenic disorder with severe gastrointestinal bleeding leading to anemia. As pazopanib, an angiogenesis inhibitor shows improvement in a patient with HHT disorder, the case study concluded that it is safe to use pazopanib in severe bleeding in HHT [102].

Metabolic pathway of pazopanib shown in (Scheme 25). Pazopanib gets metabolized in the liver microsome by the enzyme CYP3A4 to oxidative metabolites ((GSK1268992, GSK1268997, GSK1071306, and GW700201). GSK1268997 shows the same *in-vitro* potency comparable to pazopanib [103].

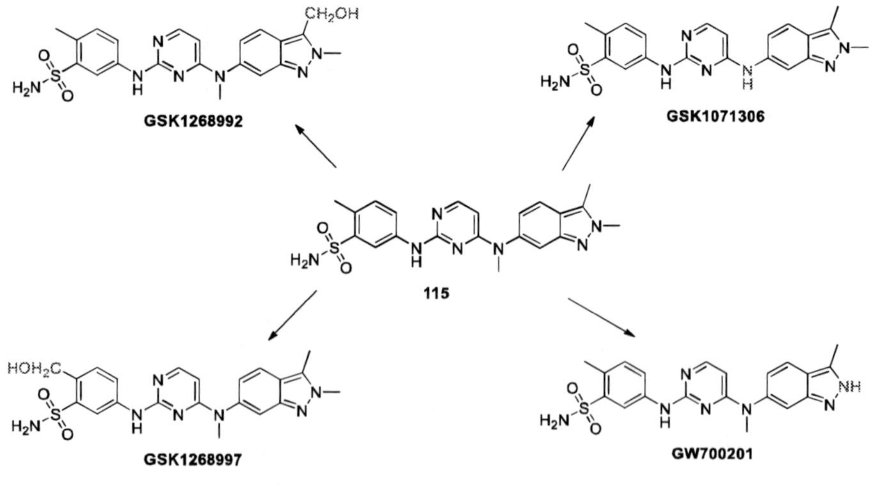

Scheme 25. Metabolic pathway of pazopanib (**115**).

Masui *et al.* reported a case study on the adverse effects of pazopanib after long-term use. Pazopanib is used to treat metastatic renal cell carcinoma (RCC). Pazopanib shows adverse effects like dysgeusia, hand-food syndrome, fatigue, hyperalbuminemia, liver dysfunction, and bone marrow dysfunction. However, after long-term use of pazopanib, the patient was diagnosed with lymphangiectasia [104].

Xue *et al.* designed, synthesized, and evaluated the novel pazopanib derivatives. Studied for the dual HDAC and VEGFR inhibitors activity for treating multi-targeted anti-cancer activity. The study result showed that **116**, **117**, and **118** (Figure 5) possess the potent activity and exhibited antiangiogenic activity compared to pazopanib [105].

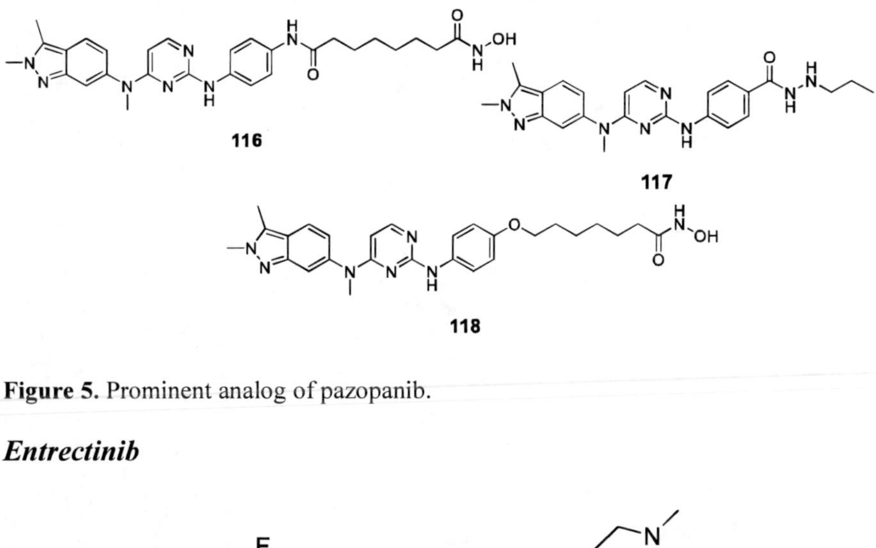

116

117

118

Figure 5. Prominent analog of pazopanib.

Entrectinib

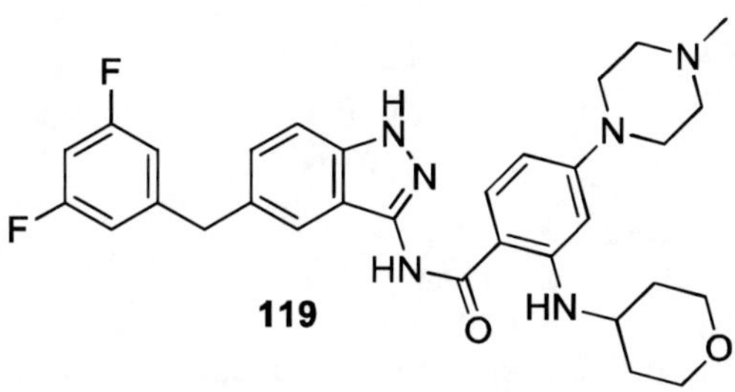

119

Entrectinib (**119**), *N*-(5-(3,5-difluorobenzyl)-1*H*-indazol-3-yl)-2-((tetrahydro-2*H*-pyran-4-yl)amino)-4-(4-methylpiperazin-1-yl)benzamide, an indazole derivative have anti-cancer activity. It is a selective adenosine triphosphate competitive inhibitor of tyrosine kinase A (TRKA), tyrosine kinase B

(TRKB), tyrosine kinase C (TRKC), ROS proto-oncogene 1, and anaplastic lymphoma kinase (ALK). USFDA approved it in 2019 to treat NTRK fusion-positive solid tumors in adults and paediatrics. NTRK is a neurotrophic tropomyosin receptor kinase gene fusion that leads to abnormal proteins called TRK fusion protein, which may cause cancer cells to grow [106].

It is used to treat primary and metastatic brain diseases as it can cross the blood-brain barrier (BBB) and act as a TRK inhibitor [107].

Entrectinib is an orally administered anti-cancer drug. It is mainly metabolized by CYP3A4 and in small amounts by CYP2C3 and CYP1C to its number of metabolites. **124** is an active metabolite. *N*-demethylation formed **124**. It further gets oxidized to **120** and **122**, while **121** is the oxidized product of entrectinib. **126** is a glucuronide metabolite (Scheme 26) [108].

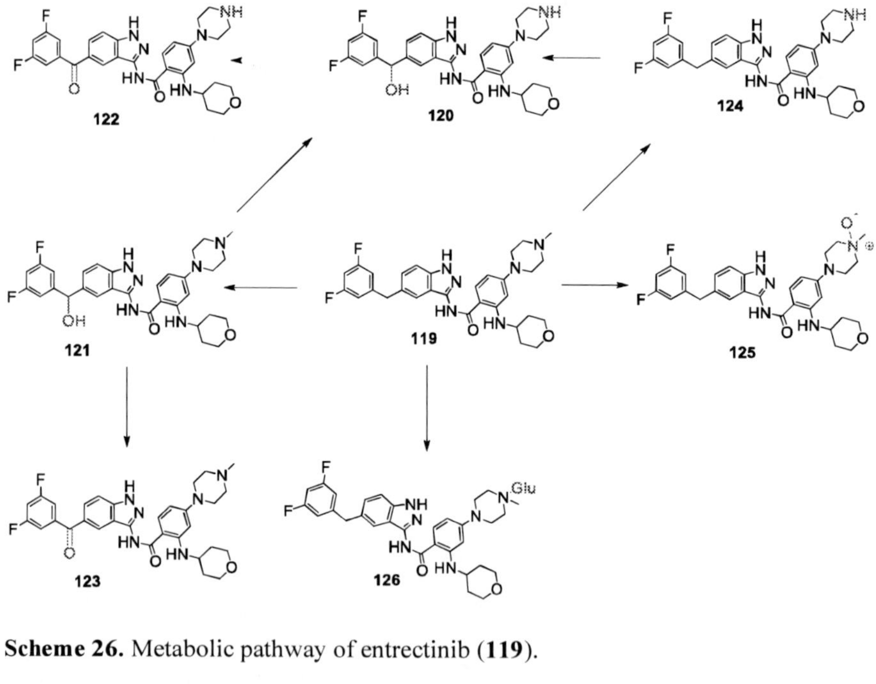

Scheme 26. Metabolic pathway of entrectinib (**119**).

The synthetic route for entrectinib (Scheme 27) was published by Liang *et al.* The reaction involves the conversion of acid **127** to acyl chloride **128**, which gets coupled with amine **129** which finally leads to the formation of amide, entrectinib (**119**) [109].

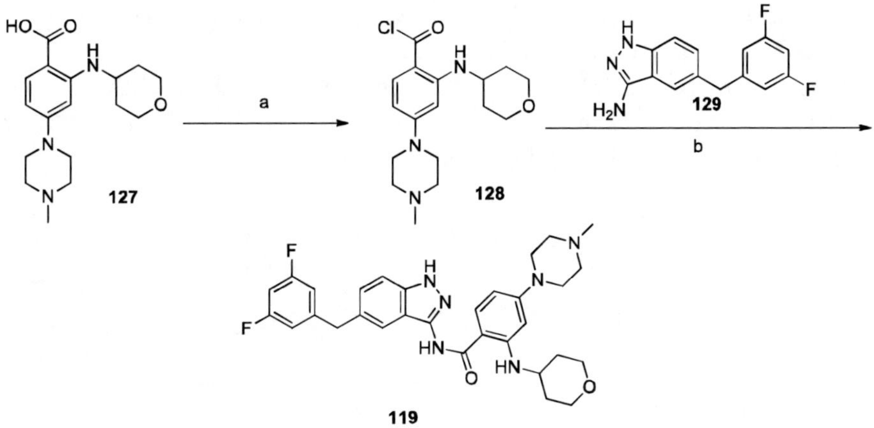

Reagents and conditions: a) $(COCl)_2$, DCM, DMF; b) (1) DIPEA, THF, 20 °C; (2) MeOH, TEA

Scheme 27. Synthesis of entrectinib (**119**).

Avapritinib

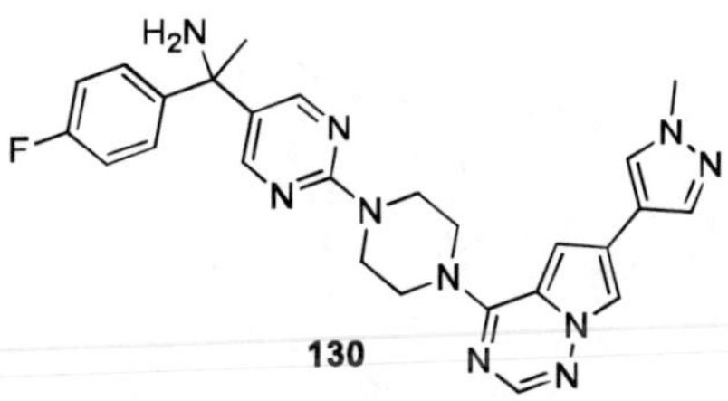

Avapritinib (**130**), 1-(4-fluorophenyl)-1-(2-(4-(6-(1-methyl-1*H*-pyrazol-4-yl)pyrrolo[2,1-*f*][1,2,4]triazin-4-yl)piperazin-1-yl)pyrimidin-5-yl)ethan-1-amine is an orally active potent selective tyrosine kinase inhibitor against a mutant form of KIT (mutation in KIT pro-oncogene are the most common genetic changes associated with gastrointestinal stromal tumors) and platelet-derived growth factor α (PDGFRA). Avapritinib is used to treat patients with advanced PDGFRA D842V-mutant gastrointestinal stromal tumors. Mutations in PDGFRA that cause GI stromal tumors are generally characterized by substituting aspartic acid (D) with Valine (V), commonly known as D842V [110]. Wu *et al.* reported the effect of avapritinib on the multidrug-resistant cancer cell. Study shows that the drug can re-sensitize multi-drug resistance cancer cells at nontoxic concentrations. Hence it may be administrated with a chemotherapeutic agent to increase the efficacy [111]. In 2020, FDA approved avapritinib for gastrointestinal stromal tumor [112].

The synthesis of avapritinib started from compound **131**, which was reacted with L-DBTA using ethanol and water in the presence of acetic acid to obtain compound **132** (Scheme 28). Further, compound **132** was treated with **133** using DIPEA in THF to afford condensed product **134** and 20% KOH in DCM and water to afford avapritinib (**130**) [113].

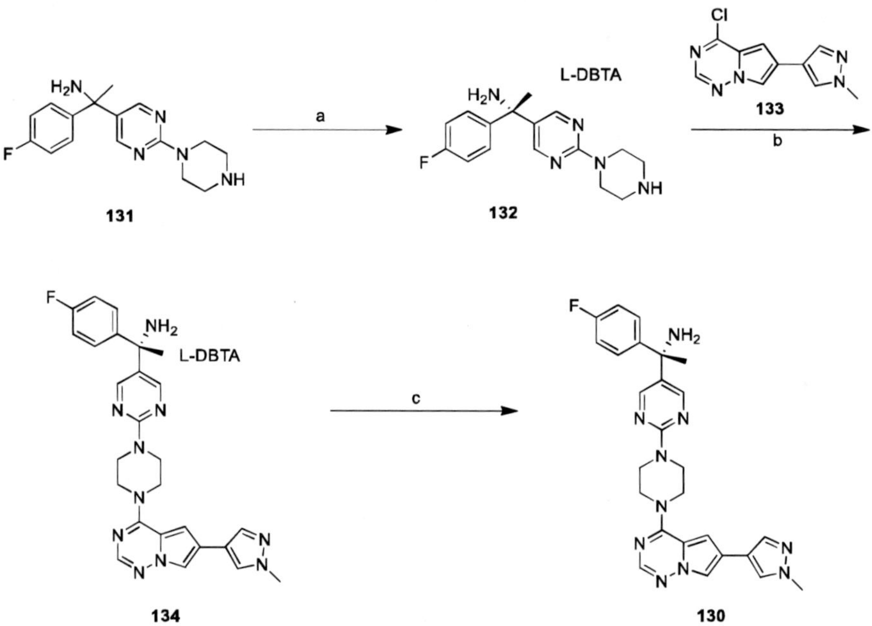

Reagents and conditions: a) L-DBTA, EtOH, H_2O,CH_3COOH; b) THF, DIPEA; c) 20% KOH, DCM, H_2O

Scheme 28. Synthesis of avapritinib (**130**).

Pyrazomycin

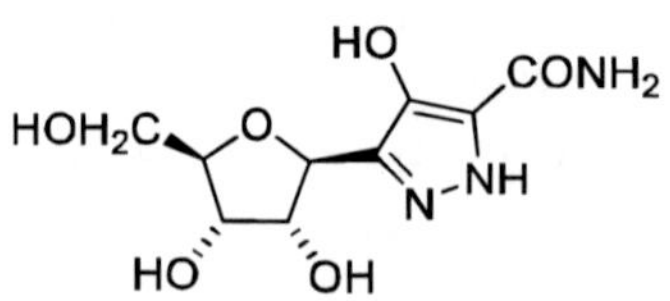

135

Pyrazole-containing scaffolds in many natural products are of immense importance due to their application in drug discovery [114]. Pyrazole-based drugs are used in many therapeutic agents such as antiviral, antibiotic, anti-tumor, and anti-inflammatory agents [2]. Pyrazomycin is an important class of natural products among the various pyrazole-based drugs. Pyrazomycin (**135**), 3-((2*S*,3*R*,4*S*,5*R*)-3,4-dihydroxy-5-(hydroxymethyl)tetrahydrofuran-2-yl)-4-hydroxy-1*H*-pyrazole-5-carboxamide exhibits broad-spectrum antiviral and antitumor activity against viruses [115]. Pyrazomycin was first isolated from the strain of Streptomyces *candidus* fermentation. The α-anomer of pyrazomin, called pyrazomycin B, was also isolated from the same fermentation.

In 1972, Sorm *et al.* developed a synthetic procedure for preparing Pyrazomycin (Scheme 29) [116]. The synthesis started with the preparation of α-ketoacid ester (**137**) from c-glycoside (**136**) using ozonolysis. Treatment of **137** with benzylhydrazinoacetic acid afforded the hydrazone derivative (**138**). The formed hydrazone derivative was converted to **139** using acetic anhydride, followed by the addition of diazomethane in methanol. Finally, the debenzylation reaction of **139** yields pyrazomycin (**135**) in 58% yield. Pyrazomycin exhibited excellent activity against myxovirus, rhabdovirus, and retrovirus. SAR studies showed that the -OH group attached to the pyrazole ring is indispensable for the marked activity of pyrazomycin [2].

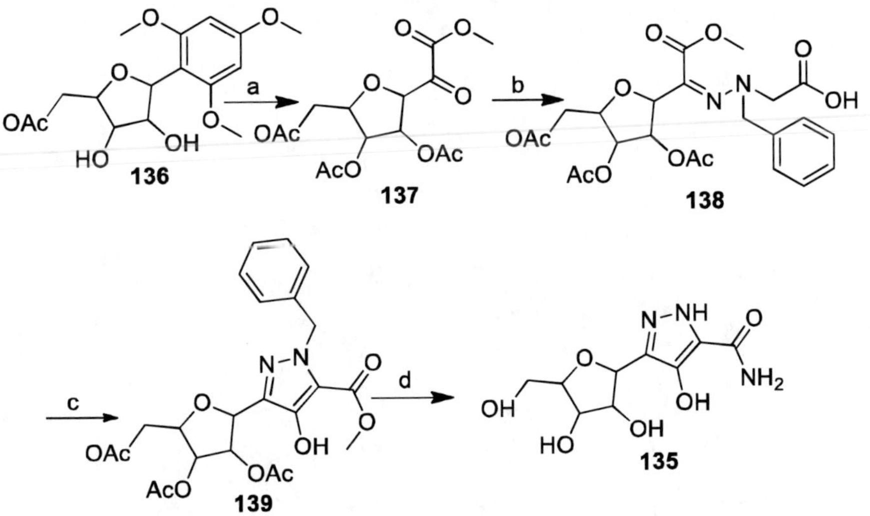

Reactions and conditions: a) ozonolysis, ethylacetate, -25°C; b) benzylhydrazinoacetic acid, methanol; c) i) acetic anhydride ,100°C, 8h, ii) CH_2N_2, NaOH, methanol; d) i) NH_3/methanol, 8h, ii) debenzylation, Pd/H_2, CH_3OH

Scheme 29. Synthesis of pyrazomycin (**135**).

Pralsetinib

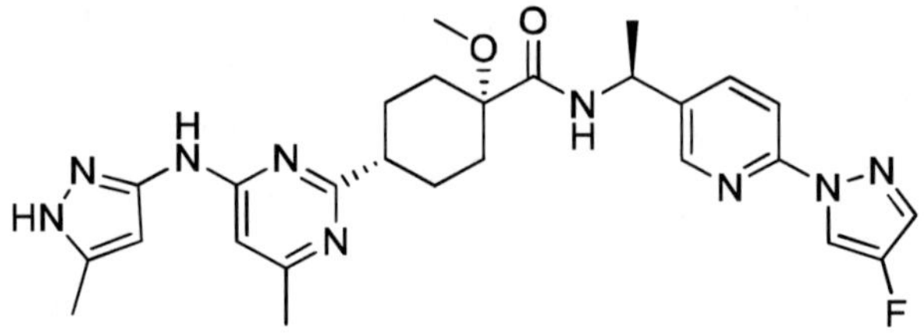

140

Pralsetinib (**140**), (1*S*,4*R*)-*N*-((*S*)-1-(2-(4-fluoro-1*H*-pyrazol-1-yl)pyridin-5-yl)ethyl)-1-methoxy-4-(6-methyl-4-((5-methyl-1*H*-pyrazol-3-yl)amino)pyrimidin-2-yl)cyclohexanecarboxamide is also a tyrosine kinase Inhibitor that targets mutation of receptor tyrosine kinase (RET) fusion and mutation with multikinase inhibitors. Pralsetinib showed anti-tumor activity in numerous RET tumor models, mainly intracranial tumors. The RET can have tumorigenic potential via downstream pathways that lead to uncontrolled cell growth. The mode of action of the pralsetinib is particularly inhibiting the RET and RET fusion with mutation, thereby controlling/inhibiting cancer cell proliferation.

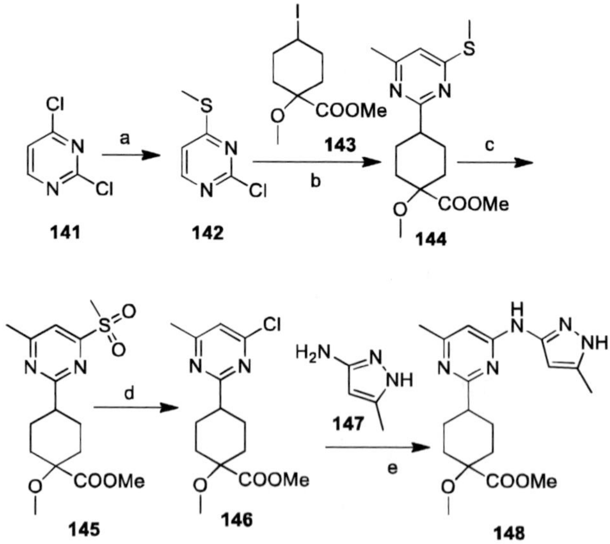

Scheme 30. (Continued).

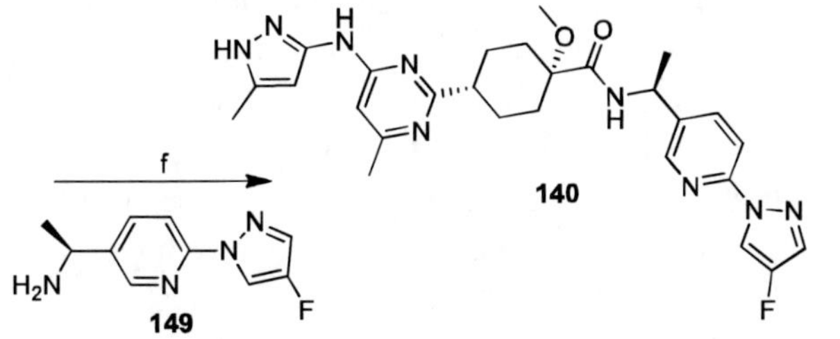

Reagents and conditions: a) MeSNa, THF, rt; b) Zn, THF, $PdCl_2$,DMA, 80 °C; c) *m*-CPBA, DCM; d) i. AcOH, 80 °C, ii. $POCl_3$, 100 °C; e) $Pd_2(dba)_3$, *t*-BuXPhos, KOAc, dioxane,100 °C; f) i. NaOH, MeOH,ii. PyBop, DIPEA, DMF,iii. Chiral resolution

Scheme 30. Synthesis of pralsetinib (**140**).

As shown in Scheme 30, the synthesis of pralsetinib started from 2,4-dichloro-6-methylpyrimidine **141**, which was reacted with sodium methanethiol to get compound **142**, which undergoes cross-coupling reaction with **143** to yield compound **144**. Further, **144** was oxidized with *m*-CPBA in DCM to get sulfone **145**. Sulfone **145** was hydroxylated with acetic acid and subsequent treatment with $POCl_3$ to obtain compound **146** followed by a reaction with 5-methyl-1*H*-pyrazol-3-amine **147** to yield compound **148**. Treatment of compound **148** with **149** in the presence of sodium hydroxide with further condensation and chiral resolution afforded pralsetinib (**140**) [117].

Lorlatinib

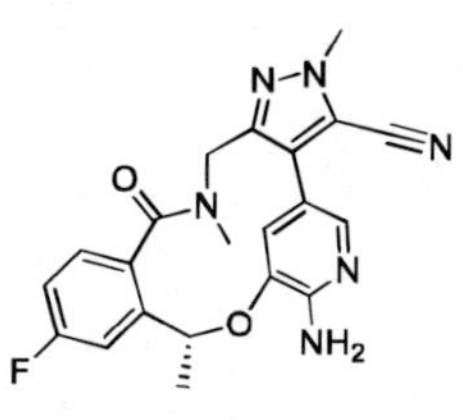

150

Lorlatinib (**150**), (16*R*)-19-amino-13-fluoro-4,8,16-trimethyl-9-oxo-17-oxa-4,5,8,20-tetrazatetracyclo[16.3.1.$0^{2,6}$.$0^{10,15}$]docosa-1(22),2,5,10(15),11,13,18,20-octaene-3-carbonitrile is a third-generation drug that inhibits anaplastic lymphoma kinase (ALK) that has activity against non-small cell lung cancer

(NSCLC). Lorlatinib shows greater organized and intracranial effectiveness than crizotinib in treating NSCLC in ALK-positive patients. Lorlatinib has anti-tumor activity against single/compound ALK resistance after the progression of the disease (first and second-generation ALK-TKI) [118]. Ceritinib, crizotinib, alectinib, brigatinib, and lorlatinib are approved drugs for treating NSCLC with ALK-positive. Usually, lorlatinib was developed mainly to target resistant mutants (first and second-generation ALK-TKI). The preclinical data suggested that lorlatinib can potentially inhibit cell growth of ALK mutants resistant (like L1196M, G1269A, and G1202R) and ALK phosphorylation [119].

Lorlatinib is metabolized by phase I metabolism via dealkylation, *N*-oxidation and hydroxylation, generally involving CYP enzymes (Scheme 31). Further, the active metabolites are converted to inactive metabolites by glucuronide conjugation [120].

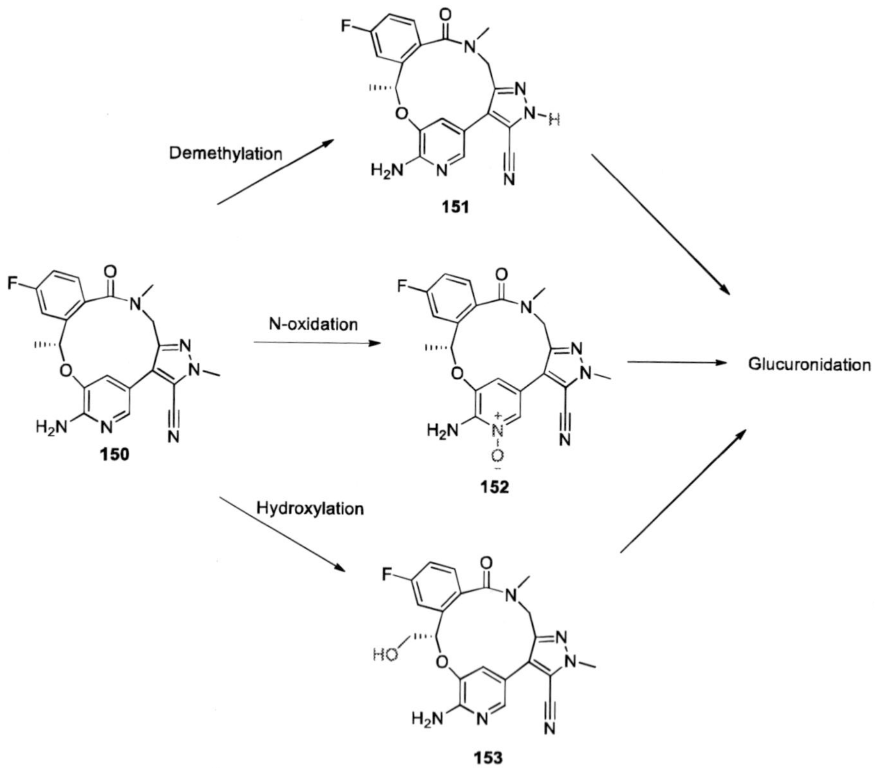

Scheme 31. Metabolic pathway of lorlatinib (**150**).

The Lorlatinib was synthesized through direct arylation/intramolecular Suzuki coupling reaction (Scheme 32). Compound **154** coupled with Boc-protected pyrazole **155** to form intermediate **156**, which undergoes cyclization to lorlatinib (**150**) [126].

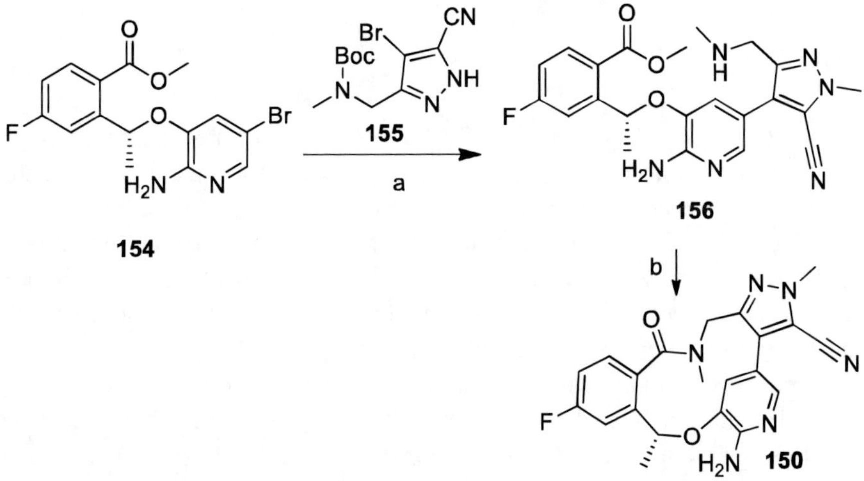

Reagents and conditions: a) i. NaOH, MeOH, H_2O, ii. 4 ml NaOH in dioxane, MeOH, 40 °C, 2.5 h; b) 1.4 equiv of HATU, 10 equiv of DIPEA, DMF, 0 °C, 30 min, slow reverse addition

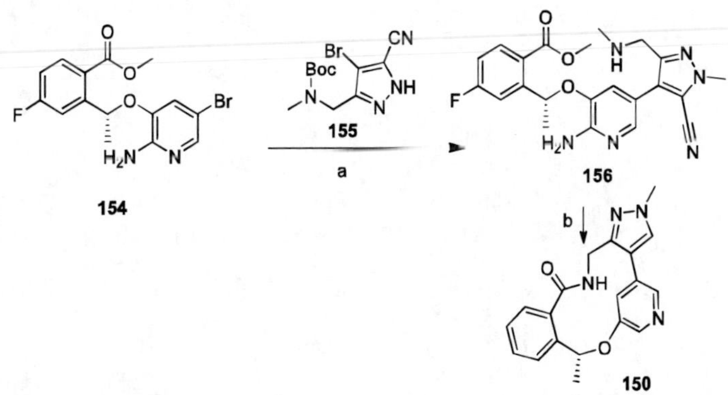

Reagents and conditions: a) i. NaOH, MeOH, H_2O, ii. 4 ml NaOH in dioxane, MeOH 40 °C, 2.5 h; b) 1.4 equiv of HATU, 10 equiv of DIPEA, DMF, 0 °C, 30 min, slow reverse addition

Scheme 32. Synthesis of the lorlatinib (**150**).

Larotrectinib

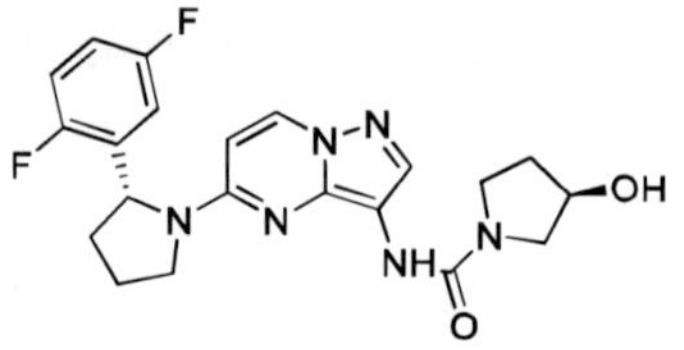

157

Larotrectinib (**157**), (*S*)-*N*-(5-((*S*)-2-(2,5-difluorophenyl)pyrrolidin-1-yl)pyrazolo[1,5-*a*]pyrimidin-3-yl)-3-hydroxypyrrolidine-1-carboxamide is a central nervous system active tropomyosin receptor kinase (TRK) inhibitor that has been approved for the treatment of TRK fusion-positive cancer based on its anti-tumor efficacy clinical trials phase I and II. Moreover, larotrectinib shows 100-fold selectivity against kinase and non-kinase targets. The mode of action blocks the ATP-binding site of TRK receptors [121, 122]. The Larotrectinib (100 mg twice daily for adults) is administered continuously for twenty-eight days. The data suggested that approximately 79% of patients show a complete response. The adverse effects are vomiting, high alanine and aspartate aminotransferase, and decreased neutrophil count were observed [123]. The above-mentioned clinical trials suggest that larotrectinib is highly sensitive to TRK gene fusion of solid tumors.

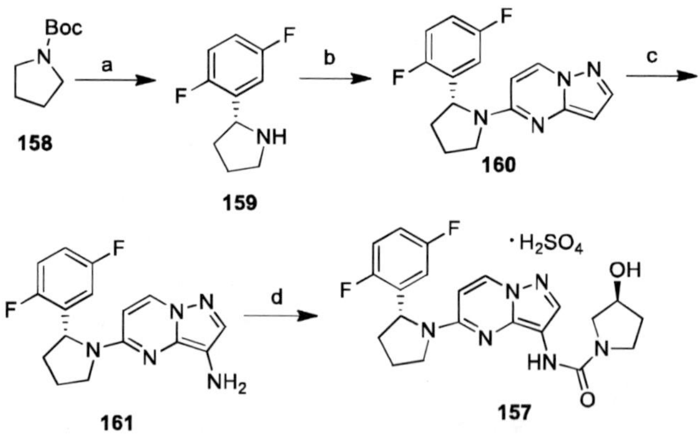

Reagents and conditions: a) i) 2-bromo-1,4-difluorobenzane, s-BuLi, $ZnCl_2$, $Pd(OAc)_2$, -78 °C to rt, ii) 4N HCl, 1,4-dioxane; b) 5-chloropyrazolo[1,5-a]pyrimidine, n-butanol, DIPEA, 160 °C; c) TFA, HNO_3, Zn powder, MeOH/DCM, sat. aq. NH_4Cl; d) (*S*)-pyrrolidinin-3-ol, DCM, CDI, H_2SO_4, MeOH

Scheme 33. Synthesis of larotrectinib (**157**).

The synthesis of larotrectinib (Scheme 33) was reported from *N*-Boc-pyrrolidine **158** with 2-bromo-1, 4-difluororbenzene using alkyl lithium to obtain **159**. Boc deprotection under acidic medium followed by a reaction with 5-chloropyrazolo[1,5-a]pyrimidine to obtain **160**. Further nitration and reduction of **160** to afford **161** followed by a reaction with (*S*)-pyrrolidinin-3-ol and treatment with sulfuric acid to yield larotrectinib (**157**) [126].

Umbralisib

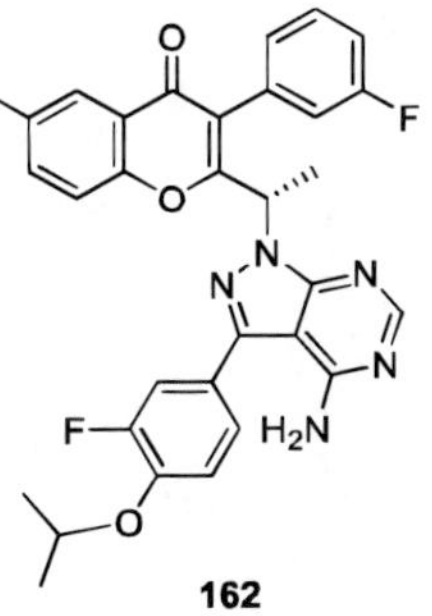

162

Umbralisib (**162**), (*S*)-2-(1-(4-amino-3-(3-fluoro-4-isopropoxyphenyl)-1*H*-pyrazolo[3,4-*d*]pyrimidin-1-yl)ethyl)-6-fluoro-3-(3-fluorophenyl)-4*H*-chromen-4-one, is a phosphatidylinositol 3-kinase delta (PI3KΔ) and casein kinase 1 epsilon (CK1ε) inhibitor. Umbralisib is approved for treating marginal zone lymphoma (MZL) and follicular lymphoma (FL). The mode of action of umbralisib is mainly via interrupting PI3KΔ pathways. The PI3KΔ pathways are essential for signalling β-cell receptors that lead to lymphoma progression. Moreover, umbralisib obstructs other pathways, particularly casein kinase pathways of lymphoma. The preclinical and clinical trials showed that umbralisib has effectively treated MZL and FL lymphoma by interrupting PI3KΔ pathways. Lunning *et al.* used Ublituximab and umbralisib molecules for patients with chronic lymphocytic lymphoma/non-Hodgkin lymphoma. The overall response rate was 46%, with no disease progression up to 20 months [124]. Fowler *et al.* studied umbralisib in patients with MZL and FL. The study included 208 patients with MZL and FL with 800 mg/day until the progression of the disease. The data suggested a 47.1% response rate and tumor reduction of 86.4% with no progression of the disease up to 2.7-4.6 months. Umbralisib can potentially treat MZL and FL with no significant toxic effect [125].

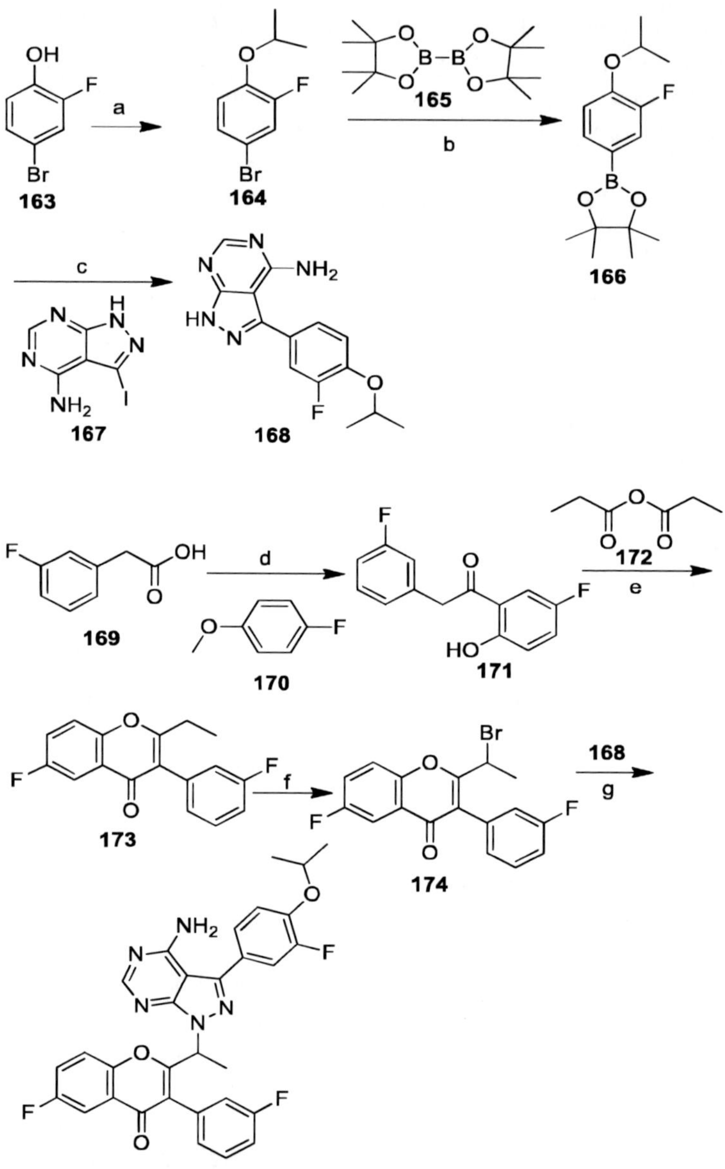

Reagents and conditions: a) *i*PrOH, PPh_3, DIAD,THF, 45 °C to reflux; b) $Pd(dppf)Cl_2$, CH_2Cl_2, KOAc, Dioxane, 80 °C; c) $Pd(PPh_3)_4$, Na_2CO_3,DMF/H_2O/EtOH, 80 °C; d) $C_2O_2Cl_2$, DCM,DMF, 0 °C, $AlCl_3$, 0 °C to rt; e) TEA, reflux; f)NBS, AIBN, CCl_4, 80 °C ; g) K_2CO_3, DMF, rt

Scheme 34. Synthesis of umbralisib (**162**).

The synthesis of umbralisib (Scheme 34). started by introduction of isopropyl group by Mitsunobu-process on 4-bromo-3-fluorophenol **163** to obtain ether **164**. Boronic ester **165** was formed from ether **164** via Suzuki-Miyaura coupling reaction. Further subsequent Suzuki-Miyaura reaction between **165** and **166** catalyzed by Pd-tetrakis(triphenylphosphine) was carried out to obtain compound **168**. The second synthon compound **174** was synthesized from fluorophenylacetic acid **169** treated with oxalyl chloride, $AlCl_3$, and 4-fluoroanisole **170** to obtain Friedel-Craft acylation product **171**. The cyclization of **171** with propionic anhydride **172** was carried out to obtain **173** under basic conditions, followed by benzylic bromination to yield bromide **174**. Compound **174** was reacted with compound **168** under the alkaline condition to afford umbralisib **162** [126].

Asciminib

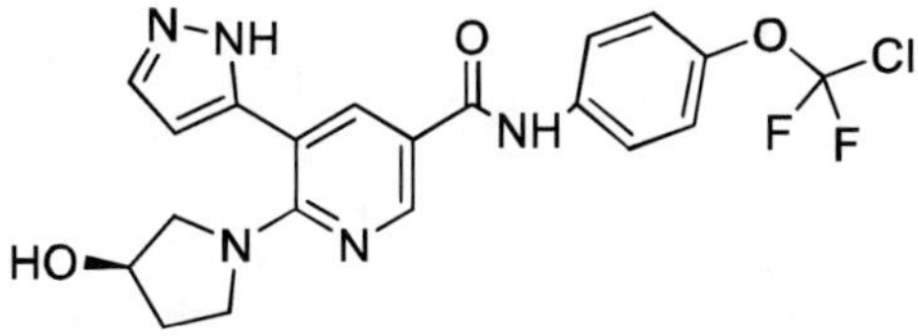

175

Asciminib **(175),** (*R*)-*N*-(4-(chlorodifluoromethoxy)phenyl)-6-(3-hydroxy-pyrrolidin-1-yl)-5-(1*H*-pyrazol-5-yl)nicotinamide, targets ABL myristoyl pocket (STAMP), used to treat Philadelphia chromosome-positive chronic myeloid leukaemia (Ph+CML) patients. Asciminib is an inhibitor of ABL1 kinase of the oncogenic fusion gene (BCR-ABL1) protein. The BCR-ABL1 protein help in the proliferation of CML. Usually, the progression of CML is mainly due to the translocation of the Ph chromosome, which develops between BCR and ABL1 genes. The fusion protein BCR-ABL1 elevates tyrosine kinase that helps the CML's proliferation. Asciminib binds the ABL1 protein myristoyl pocket (fusion protein portion) and prevents oncogenic activity [127, 128]. Asciminib might be used for the treatment of Ph+CML with T315I mutation.

CYP450 enzymes are the main metabolizing enzymes for asciminib (Scheme 35). Direct phase II metabolism via glucuronidation produces glucuronide-conjugated inactive metabolite **176**. Phase I metabolism involves O-dealkylation to phenolic metabolite **177**, hydrolysis of amide group to acid metabolite **178,** and oxidation of hydroxyl group to ketone **179**. Oxidative ring opening also is one of the metabolizing pathways leading to the formation of carboxylic acid metabolite **180** [129].

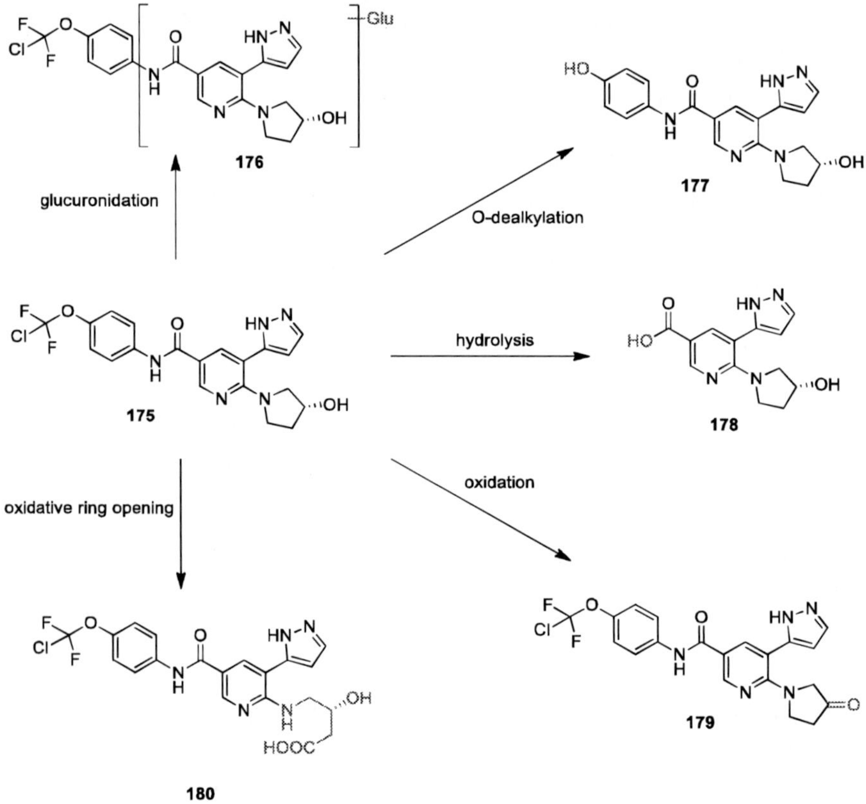

Scheme 35. Metabolic pathway of asciminib (**175**).

The synthesis of asciminib **(175)** started by a reaction of halogen-substituted nicotinic acid **181** and substituted aniline **182** upon activation of carboxylic acid to functionalize to amide **183**. Further, amide **183** was treated with chiral pyrrolidine-3-ol **184** to obtain **185**. Compound **185** with boronic ester **186** undergoes a Suzuki-Miyaura coupling reaction to obtain compound **187** followed by deprotection of tetrahydropyran protecting group to furnish asciminib **175** (Scheme 36) [130].

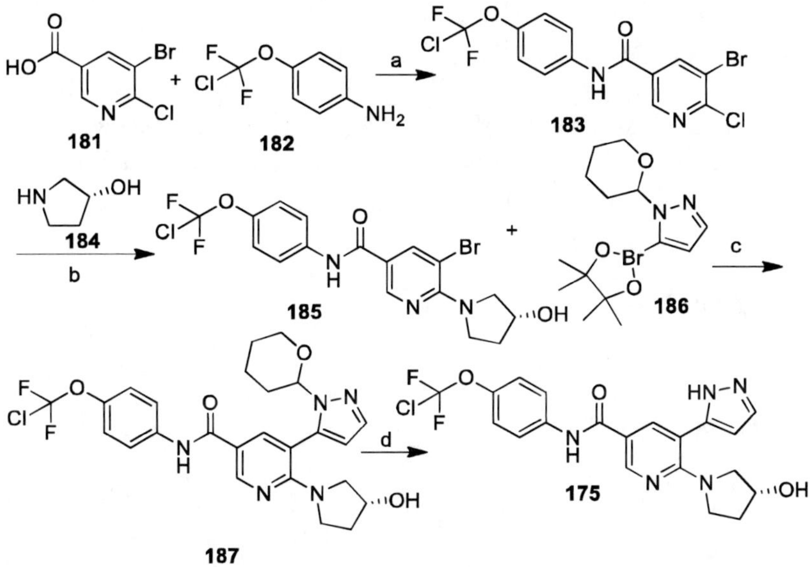

Reagents and conditions: a) 1. $SOCl_2$, DMF, toluene, 80 °C, 2. DIPEA, THF, 16 °C to rt; b) *i*PrOH, 140 °C; c) $Pd(PPh_3)_4$, K_3PO_4, toluene, 110 °C; d) TFA, DCM, 10 °C to rt

Scheme 36. Synthesis of asciminib (175).

Belumosudil

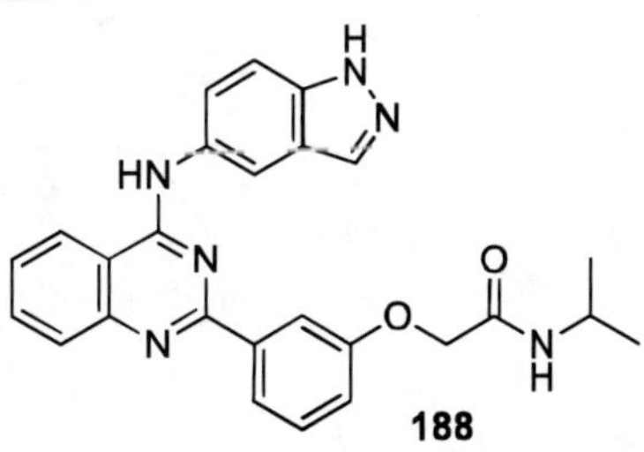

Belumosudil (**188**), 2-(3-(4-((1*H*-indazol-5-yl)amino)quinazolin-2-yl)phenoxy)-*N*-isopropylacetamide, is a Rho-kinase inhibitor drug used in the treatment of chronic graft-versus-host disease (cGVHD). US FDA first approved this drug in 2021 to treat cGVHD [131]. Belumosudil may cause harm to the foetus if given to pregnant women. Breastfeeding is not advised in patients administrated with belumosudil due to the adverse effect, which can cause harm to babies [132].

Belumosudil metabolism (Scheme 37) mainly involves CYP enzymes, i.e., CYP3A4, CYP2C8, CYP2C19, and CYP3A5. CYP catalyzes phase I metabolism involving aromatic ring hydroxylation, O-dealkylation, *N*-dealkylation, and hydrolysis of amide bond to form acid metabolite, i.e., the active metabolite. Phase II metabolism via glucuronidation involves the enzyme UGT1A9 to form inactive conjugated metabolite of belumosudil in humans [133].

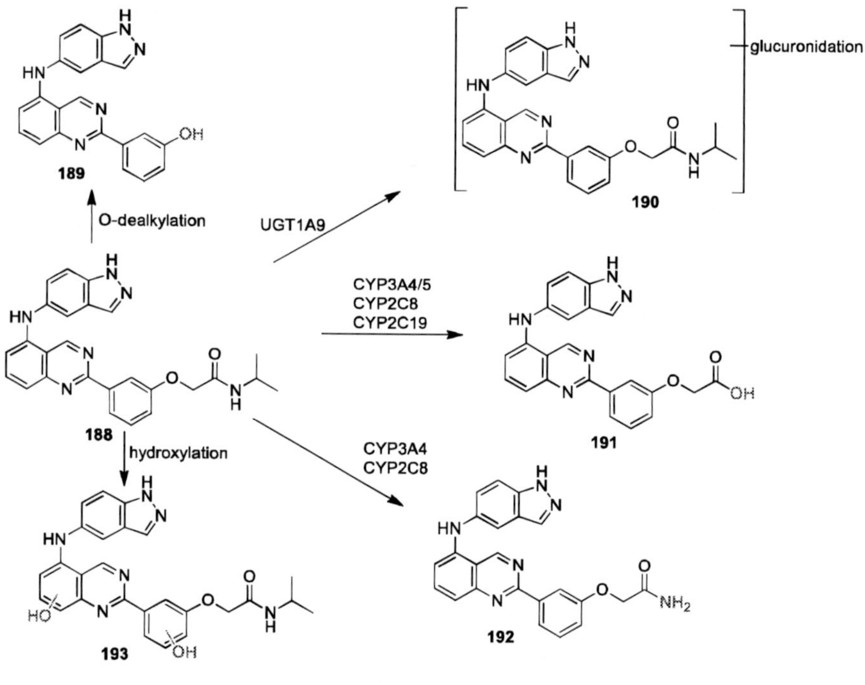

Scheme 37. Metabolic pathway of belumosudil (**188**).

Sulfinpyrazone

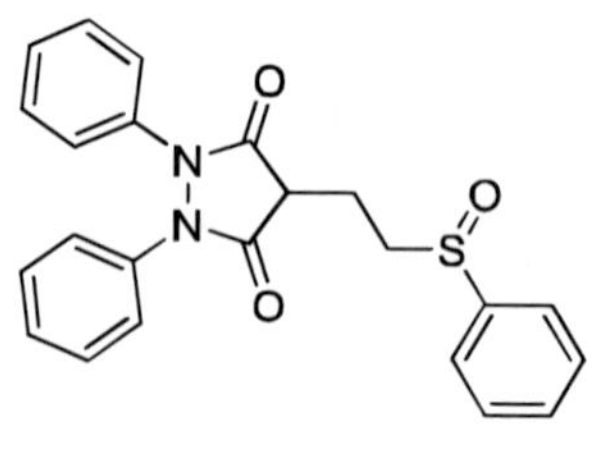

194

Sulfinpyrazone (**194**), 1,2-diphenyl-4-(2-(phenylsulfinyl)ethyl)-3,5-pyrazolidinedione, a phenylbutazone analog acts as a uricosuric drug. Sulfinpyrazone is the only drug used as a uricosuric drug for the treatment of gout. During the trials, it was observed that it affects chronic hemodialysis. Also, the antiarrhythmic activity of sulfinpyrazone in the dog was reported. It has antiplatelet aggregation activity. The anturane reinfarction trial research group has conducted clinical trials for sulfinpyrazone to treat myocardial infarction. All the activities other than uricosuric are because of its metabolites. Its sulphide and hydroxysulfide metabolites are responsible for platelet-inhibiting activity. These *p*-hydroxy, sulfone, sulphide, and *p*-hydroxysulfide metabolites in urine and plasma were detected by the HPLC method by Lentjes *et al.* [134].

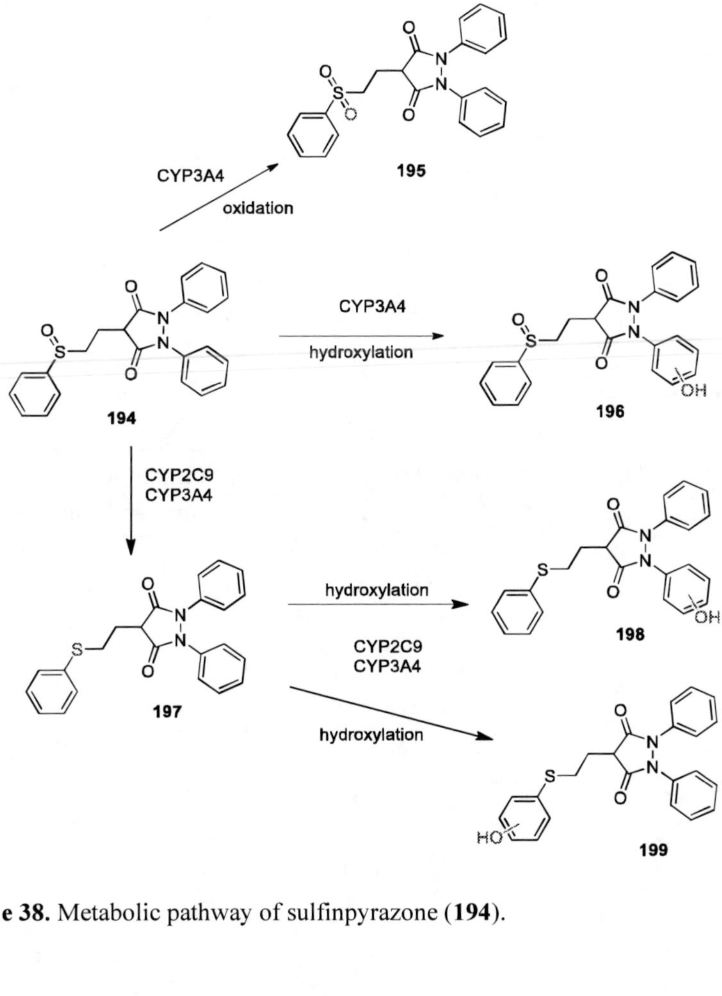

Scheme 38. Metabolic pathway of sulfinpyrazone (**194**).

CYP450 enzymes are involved in the metabolism of sulfinpyrazone. CYP3A4 and CYP2C9 are major metabolizing enzymes in humans. CYP3A4 enzyme is involved in the formation of oxidized and hydroxylated metabolites. CYP2C9 enzyme involved in the formation of reduced thioether metabolite further converted to inactive hydroxylated metabolite (Scheme 38) [135].

Synthesis of sulfinpyrazone (Scheme 39) was done by condensing *N,N'*-diphenylhydrazine **200** with substituted malonic ester **201** to produce pyrazolidinedione **202** followed by oxidation of thiol ether by hydrogen peroxide in acetic acid to obtain compound **194** [136].

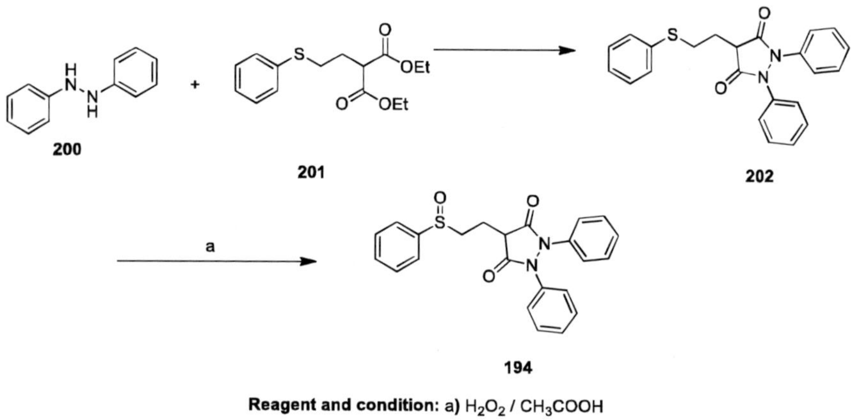

Scheme 39. Synthesis of sulfinpyrazone (**194**).

It has several other effects related to its anti-platelet activity like, inhibitory effect on mural thrombus formation, inhibition of humoral immune responses, inhibition of human platelet prostaglandin synthesis, fibrinolytic activity, and endothelial cell protection [137].

Stanozolol

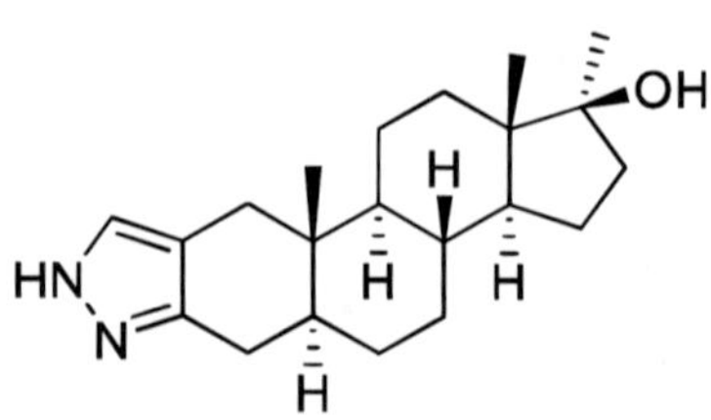

203

Stanozolol (**203**), (1*S*,3a*S*,3b*R*,5a*S*,10a*S*,10b*S*,12a*S*)-1,10a,12a-trimethyl 1,2, 3,3a,3b,4,5,5a,6,8,10,10a,10b,11,12,12a-hexadecahydrocyclopenta[5,6] naphtho[1,2-*f*]indazol-1-ol is a non-aromatic androgen, a testosterone derivative has more anabolic activity than androgenic activity. It treats diseases like turner's syndrome, aplastic anemia, and hereditary angioedema. Adjuvant for treating the vascular disorder and growth failure [138]. The misuse of stanozolol for body-building leads to serious side effects like severe flank pain, acute kidney failure, and thrombosis affecting both kidneys [139]. Stanozolol may disrupt the estrous cycle in mice, and its recovery was dose-dependent [140].

Alves *et al.* performed a randomized trial to know the use of stanozolol in the osteoarthritis model. The study shows that stanozolol is safer and has good potency to treat osteoarthritis [141].

Liu *et al.* studied that stanozolol in combination with decitabine, improves the health of the progression-free survival of patients with high-risk myelodysplastic syndrome. Due to the low price and good safety profile, it will be very economical to use with decitabine [142].

Ozcagli *et al.* reported that the study of administration of stanozolol with exercise leads to a decrease in the telomerase enzyme activity causing liver aging. In this study, the metabolites of stanozolol were identified by LC-MS. As most anabolic steroids are used as doping agents in sports, stanozolol is also used as a 17α-alkylated derivative of testosterone [143]. Stanozolol is also found in the body hair of an anabolic steroid-abused person, as per the case study performed by Kintz *et al.* [144]. In the absence of functional LDL receptors, stanozolol leads to the development and progression of atherosclerosis.

Stanozolol is metabolized in humans to mainly hydroxylated derivatives. Hydroxylation occurs at different positions to produce active metabolites (Scheme 40) [145].

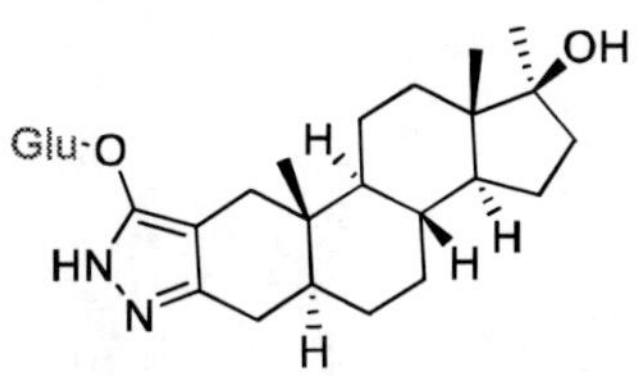

207

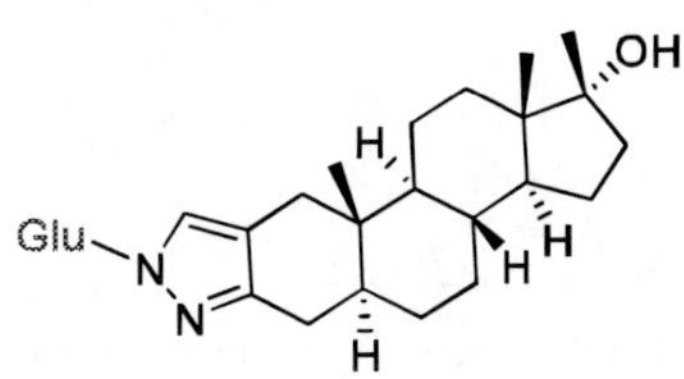

208

Figure 6. Metabolic products of stanozolol.

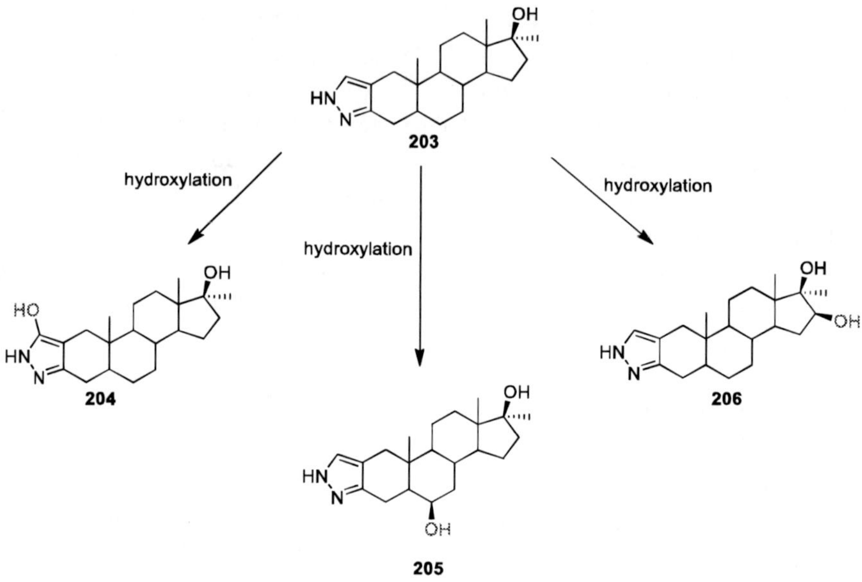

Scheme 40. Metabolic pathway of stanozolol (**203**).

Stanozolol undergoes Phase II biotransformation (Figure 6) to 3'-hydroxy-stanozolol-glucuronide **207** and 17-epi-Stanozolol-*N*-glucuronide **208**. These are detected by the zebrafish water tank (ZWT) model [146]. Stanozolol-*N*-glucuronide metabolites in human urine were beneficial for anti-doping analysis.

Fomepizole

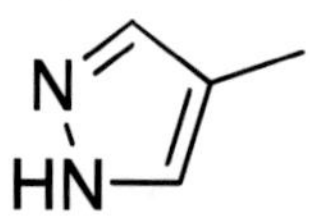

209

Fomepizole (**209**), 4-methyl pyrazole, is a competitive inhibitor of aldehyde dehydrogenase (ADH), which has >8000 times higher affinity than ethanol. Hence it is used to treat alcohol poisoning.

The mechanism of action of fomepizole is that it binds to the zinc element in the ADH enzyme, nicotinamide-adenine dinucleotide coenzyme, and inhibits the binding of alcohol to the enzyme [147]. The case report presented at the North American Congress of Clinical Toxicology Annual Meeting provided information on fomepizole's use to treat disulfiram-ethanol reactions. As disulfiram-ethanol reactions cause histamine-like reactions because of the accumulation of acetaldehyde. Fomepizole is an effective blocker of ADH that leads to inhibition of the formation of acetaldehyde [148].

Lepik *et al.* reported a case study of the administration of fomepizole during dialysis to treat ethylene glycol poisoning. It was observed that fomepizole might cause bradycardia and hypotension if administered intravenously during hemodialysis. But post, hemodialysis administration is well tolerated [149].

Fomepizole is excreted through urine via zero-order metabolism as <5% of the drug is excreted unchanged. Metabolism of fomepizole through CYP2E1 enzyme yields 4-carboxypyrazole **210** and 4-hydroxymethyl-pyrazole **211**, both metabolites undergo phase II transformations to form glucuronide conjugation (Scheme 41) [150, 151].

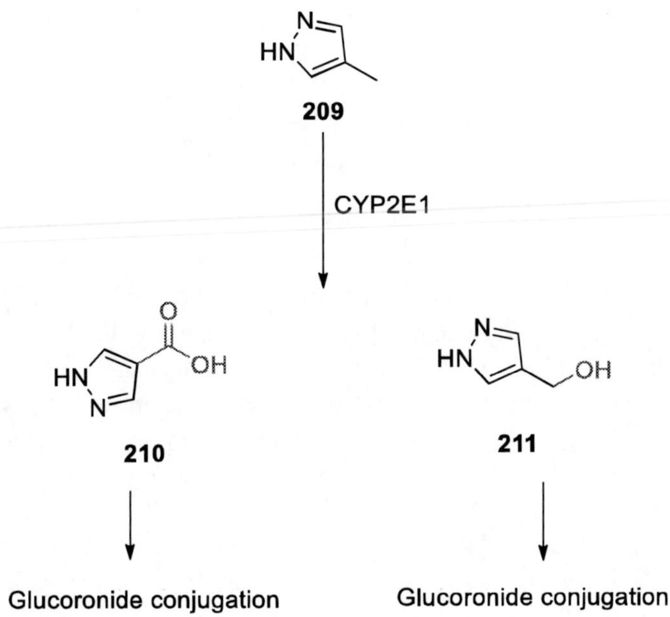

Scheme 41. Metabolic pathway of fomepizole (**209**).

The synthesis of fomepizole is depicted in Scheme 42. Ether **212** was reacted with triethylorthoformate **213** to generate diacetal **214,** which upon reaction with hydrazine sulphate produced fomepizole (**209**) [152].

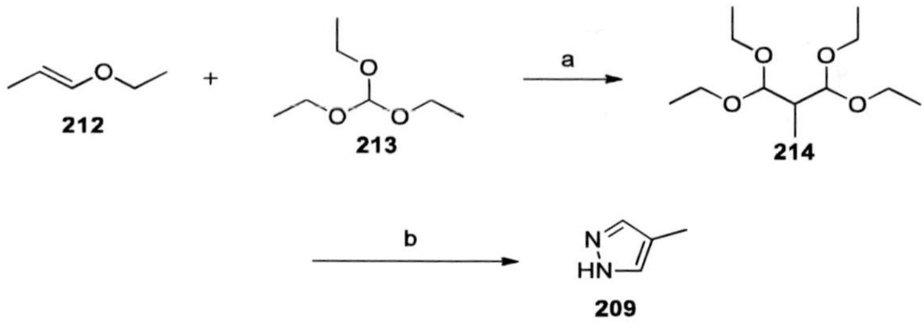

Reagents and conditions: a) EtO.BF_3 ; b) $H_2NNH_2.H_2SO_4$

Scheme 42. Synthesis of fomepizole (**209**).

Sildenafil

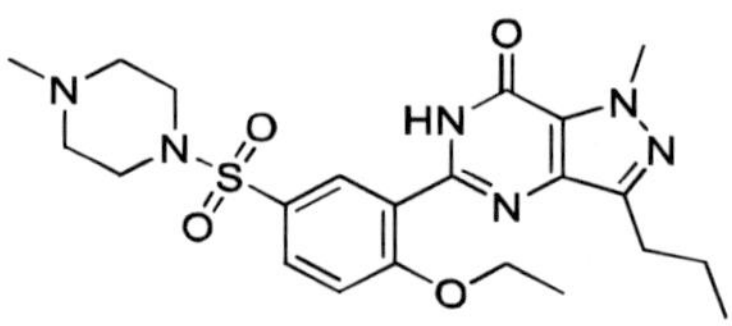

215

Sildenafil (**215**), 5-(2-ethoxy-5-((4-methylpiperazin-1-yl)sulfonyl)phenyl)-1-methyl-3-propyl-1,6-dihydro-7*H*-pyrazolo[4,3-*d*]pyrimidin-7-one is a drug used to treat erectile dysfunction in males. The mechanism of action of this drug involves selective inhibition of phosphodiesterase-5. It is useful in endothelial dysfunction in hypertensive states, which is achieved by increasing NO, reducing oxidative stress and COX-1 prostanoids, and improving the cGMP/PKG signaling [153].

Assemat *et al.* isolated sildenafil and its 14 derivatives from sexual enhancer dietary supplements. Sexual enhancer dietary supplements are adulterated with sildenafil and its derivatives for their effectiveness. These adulterants are isolated and detected by NMR [154].

Sildenafil is metabolized in the liver microsomes by oxidative metabolic pathways and forms hydroxylated metabolites (Scheme 43) [155].

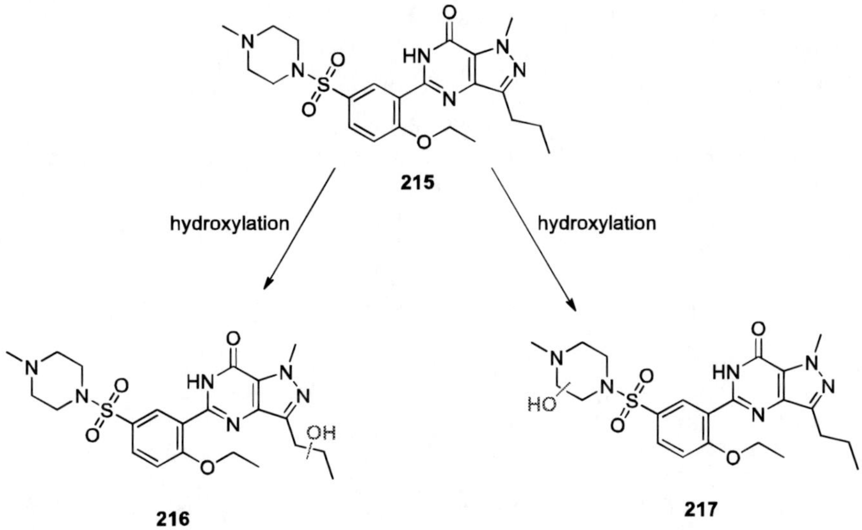

Scheme 43. Metabolic pathway of sildenafil (**215**).

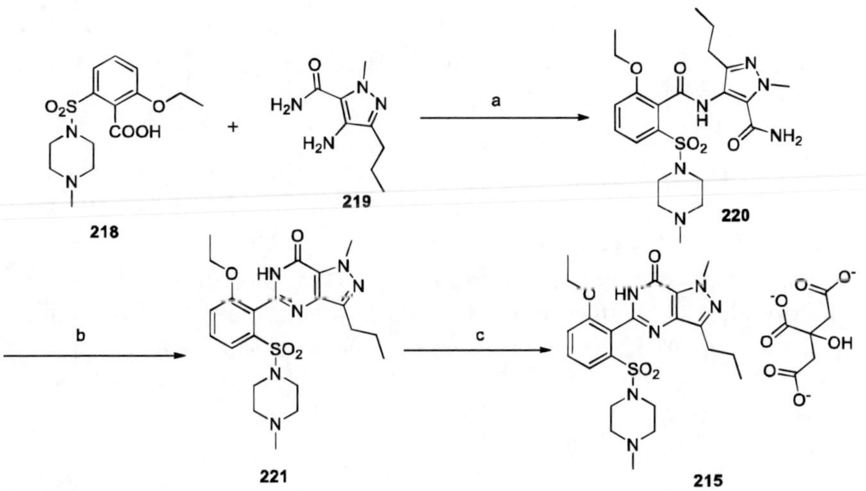

Reagents and conditions: a) EtOAc, N,N'-carbonyldiimidazole ,90% ; b) KOBu,BuOH ; c) citric acid 2-butanone, 99%

Scheme 44. Synthesis of sildenafil citrate (**215**).

The first-time developed method of synthesis of sildenafil is not scalable. Dunn *et al.* developed a new method of synthesis of Sildenafil citrate (Scheme 44). They used a greener approach to this multi-step route which does not produce toxic by-products and yields a high amount of final product. Pfizer received a green chemical technology award for the process development of sildenafil citrate by UK award 2003 [156].

Betazole

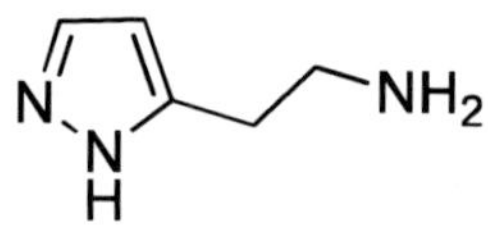

222

Betazole (**222**), 2-(1*H*-pyrazol-3-yl)ethan-1-amine is an H_2-receptor antagonist used to treat duodenal and gastric ulcers and some special conditions like Zollinger-Ellison disease. Betazole functions by blocking the H_2 receptor site, thereby controlling/decreasing the acid secretion by the stomach. A study by I. M. Spitz *et al.* demonstrated that betazole directly stimulates or induces gastrointestinal polypeptide, an inhibitory hormone responsible for the decreased secretion of gastric acid (HCl), without the mediation of gastric HCl [157].

G. Jones *et al.* reported the two-step synthesis of betazole in 1953. In this reaction, treatment of γ-pyrone **223** and hydrazine **224** in the ratio of 1:2 afforded 3-pyrozoleacetaldehyde hydrazone **225** quantitatively in the first step. Catalytic hydrogenation using Raney nickel and H_2 at 90°C and under 1500 lb. pressure resulted in the formation of betazole with 80% yield. The synthetic route for betazole formation is depicted in Scheme 45 [158].

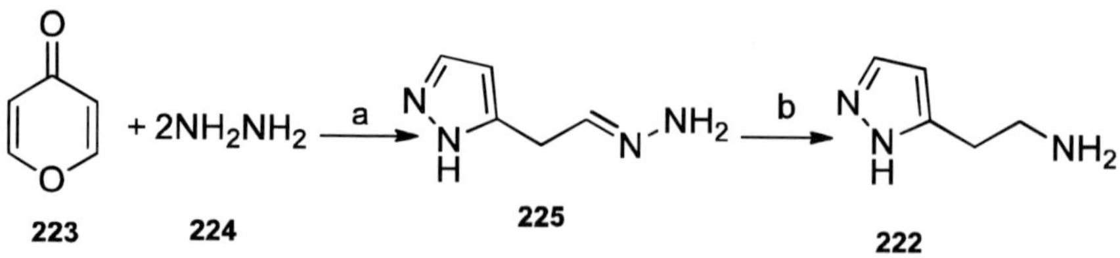

Reagents and conditions: a) methanol; b) Ni / H_2, methanol, 90 °C, 1500 lb.

Scheme 45. Classical synthesis of betazole (**222**).

Recently, Paradisi *et al.* have developed an alternative approach for preparing betazole using a multi-enzymatic catalytic system (Scheme 46). This method is a greener and more sustainable way to produce betazole. In this work, the conversion of 2-(1*H*-pyrazol-3-yl)ethanol **226** to the corresponding amine **222** product was achieved using three enzymes: a) horse liver alcohol dehydrogenase enzyme (HLADH), b) ω-transaminase from Halomonas *elongate* (HEWT), and c) NADH oxidase from Lactobacillus *pentosus* (LpNOX). A high yield (75%) for betazole was obtained using the multienzymes approach [159].

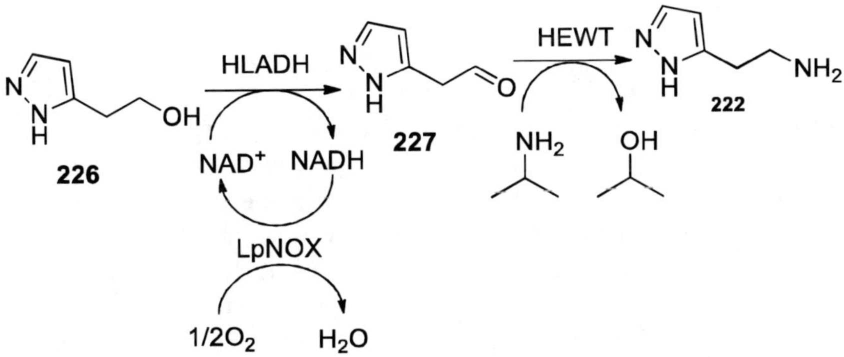

Scheme 46. Greener synthesis of betazole (**222**) using multienzymes approach.

AS-19

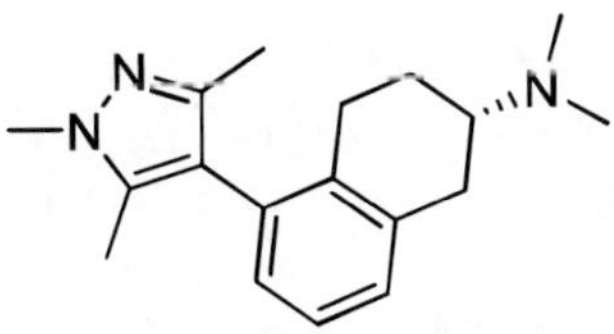

228

AS-19 (**228**), pyrazole derivative, (2*S*)-(+)-8-(1,3,5-trimethylpyrazolin-4-yl)-2-(dimethylamino)tetralin. AS-19 is a potent agonist for the 5-hydrohytryptamine (5-HT) receptor, a key receptor in memory formation in the brain [160]. *Ex-vivo* studies by G. Perez-García *et al.* observed that AS-19 reverses memory loss induced by scopolamine and dizocilpine to improve long-term memory acquisition [160].

Apixaban

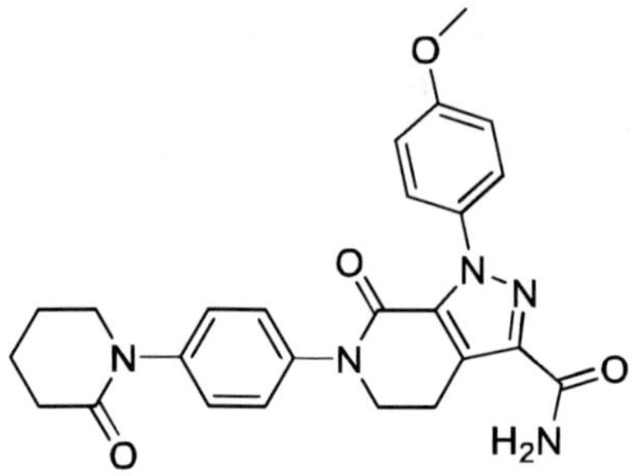

229

Apixaban (**229**), 1-(4-methoxyphenyl)-7-oxo-6-(4-(2-oxopiperidin-1-yl)phenyl)-4,5,6,7-tetrahydro-1*H*-pyrazolo[3,4-c]pyridine-3-carboxamide, is an anticoagulant used to prevent blood clot and reduction of stroke in patients who suffer from atrial fibrillation (irregular heartbeat) [161]. The drug is also used after surgery (e.g., knee or hip replacements) to prevent blood clotting. Apixaban is a direct inhibitor of blood coagulation factor Xa.

The structure of apixaban consists of a novel bicyclic pyrazole core. Metabolism and elimination of apixaban from the body were investigated in 10 healthy male subjects [162]. This study observed mostly unchanged apixaban in the samples collected from urine, blood, and faeces. *o*-demethyl apixaban sulphate was identified as the major metabolite in the blood plasma [163].

Apixaban metabolized by CYP enzymes (Scheme 47). CYP3A4 enzyme O-dealkylates the apixaban, which further undergoes phase II sulphate conjugation by SULT1A1 to give sulphated product. A hydroxylated metabolite is also formed from CYP3A4 metabolism [164].

The synthesis of apixaban **229** started by a reaction between **233** and **234** with pyridine in methylene dichloride (MDC) followed by deprotection under acidic medium afforded **235**. Compound **235** reacted with phenyl hydrazine derivative **236** in aqueous isopropyl alcohol (IPA) to obtain pyrazole compound **237**. Further, **237** undergoes ammonolysis to yield apixaban (**229**) (Scheme 48).

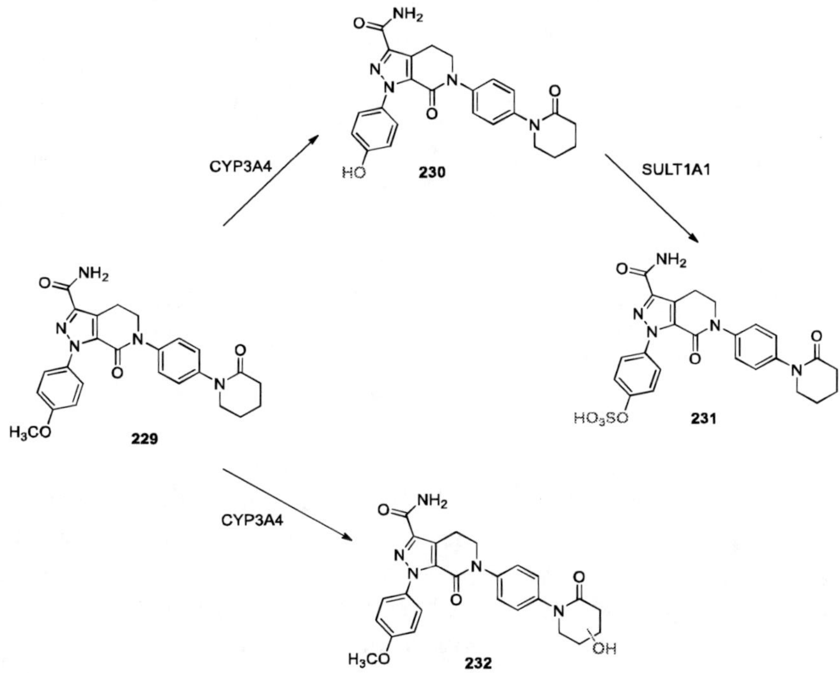

Scheme 47. Metabolic pathway of apixaban (**229**).

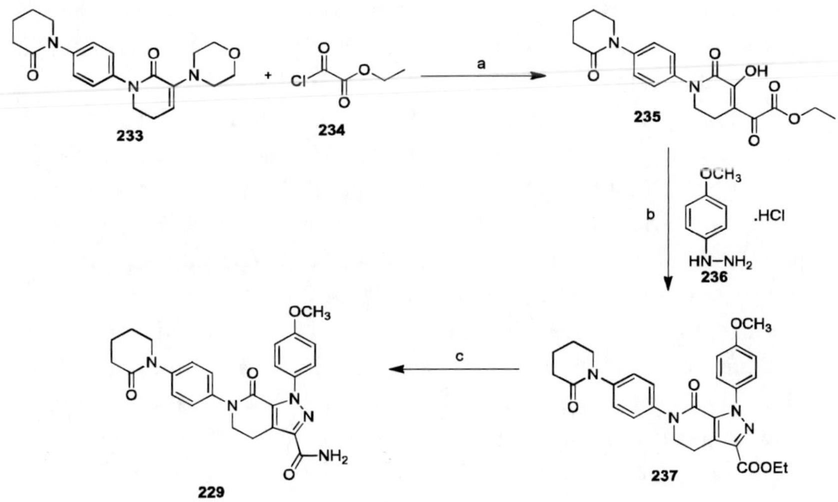

Reagents and conditions: a) i. Pyridine, MDC, ii. 10 % HCl; b) Aq. IPA ; c) ammonia

Scheme 48. Synthesis of apixaban (**229**).

Vericiguat

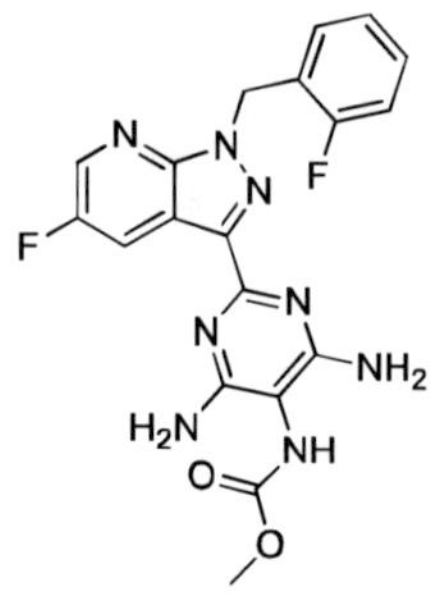

238

Vericiguat (**238**), methyl (4,6-diamino-2-(5-fluoro-1-(2-fluorobenzyl)-1*H*-pyrazolo[3,4-*b*]pyridin-3-yl)pyrimidin-5-yl)carbamate, is used to reduce the risk of heart failure in the event of acute decompensation. It is used to stimulate guanylate cyclase by oral administration. It is well known that the cardiovascular system can be easily regulated by nitric oxide-sGC-cyclic guanosine monophosphate (cGMP). Vericiguat mainly targeting the cyclic guanosine monophosphate (cGMP) pathway by stimulating sGC leads to the relaxation of smooth muscle cells as well as reduced hypertrophy, inflammation, and fibrosis. Patients with vericiguat demonstrated a significant reduction in a composite endpoint of cardiovascular (CV) death or first heart failure hospitalization (HFH) [165, 166, 167, 168].

Vericiguat metabolized in humans via different metabolizing enzymes, i.e., CYP450 enzyme family, as it can directly undergo phase II metabolism via glucuronide conjugation. Phase I metabolism via dealkylation and hydroxylation to produce active metabolite followed by glucuronide conjugation to produce phase II inactive metabolite (Scheme 49) [169].

The vericiguat (**238)** was synthesized (Scheme 50) by condensing carboxylate **242** with aldehyde **243** to obtain **244**. Compound **244** was treated with sodium methoxide under the basic condition to afford compound **245**. Compound **246** was obtained by a reaction between **245** and $POCl_3$ in MeCN, followed by using sodium methoxide under basic condition yield compound **247**. Compound **247** was reacted with **248** in presence of triethylamine in DMF to obtain **249** and subsequent reduction using hydrogen and Pd/C in DMF forms compound **250**. Compound **250** was converted to vericiguat (**238)** in the presence of triethylamine under basic conditions [170].

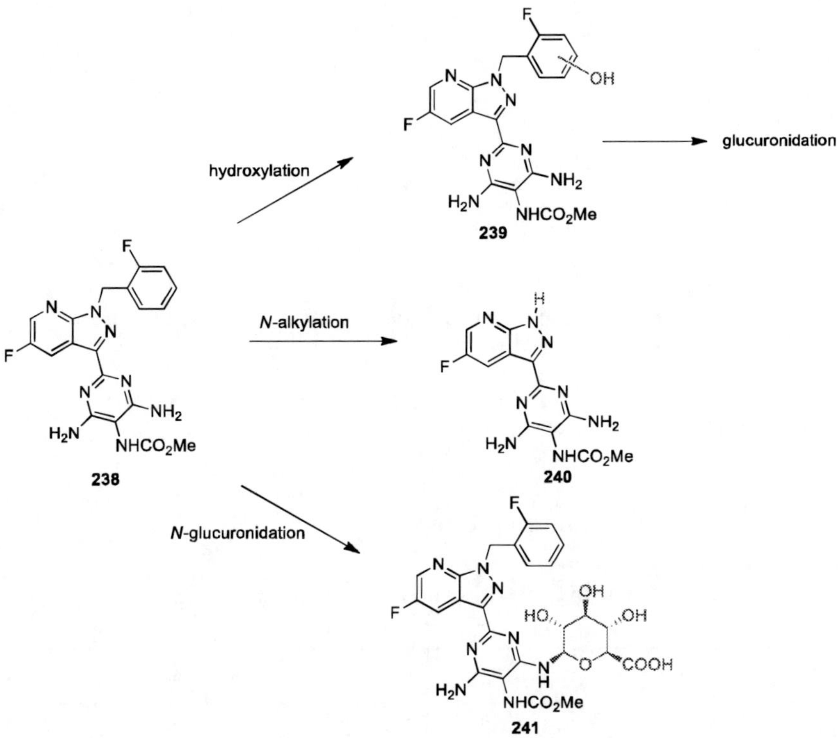

Scheme 49. Metabolic pathway of vericiguat (**238**).

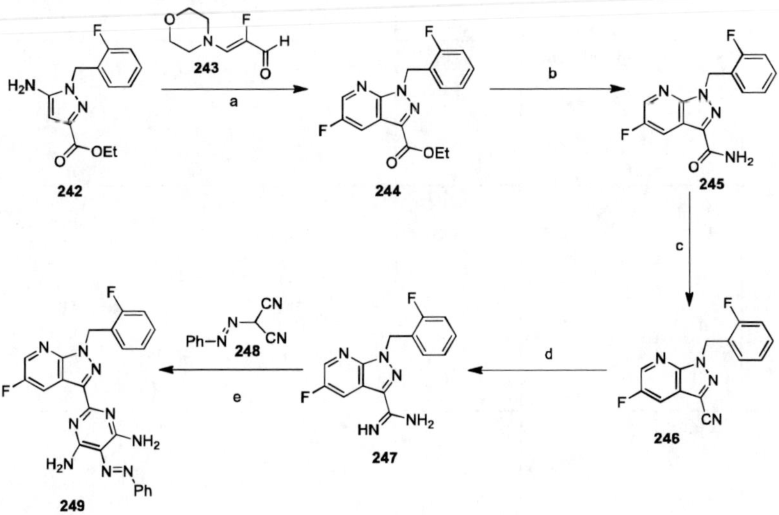

Scheme 50. (Continued).

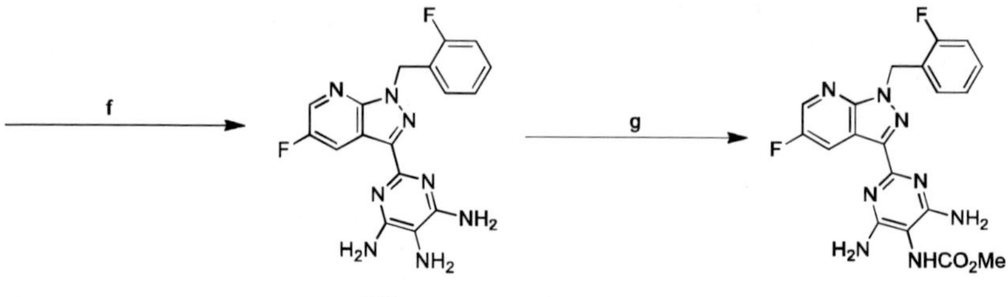

250 238

Reagents and conditions: a) MsOH, LiCl,EtOH; b) Formamide, NaOMe, EtOH-MeOH; c) sulfolane-MeCN, $POCl_3$; d) NaOMe, NH_4Cl, MeOH; e) Et_3N, DMF; f) H_2, Pd/C, DMF; g) $ClCO_2Me$, 2-PrOH, Et_3N, MeOH

Scheme 50. Synthesis of vericiguat (**238**).

Pyridazine Ring Containing Drugs

Table 2. Pyridazine ring containing drugs

Sr. No.	Drug	Class	Structure
39.	Minaprine	MAOI	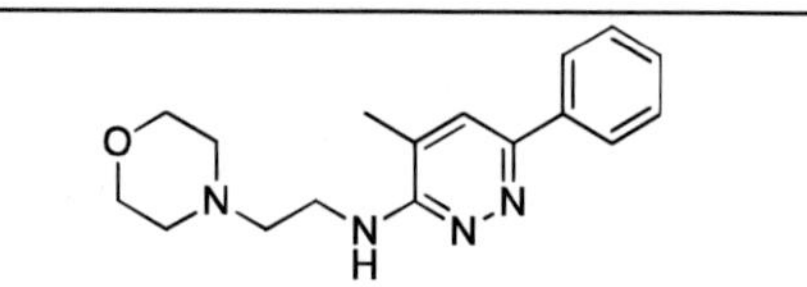
40.	Sulfachloropyridazine	Antibiotic	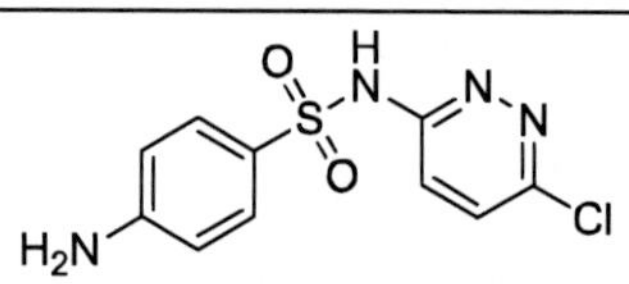
41.	Relugulix	Anticancer	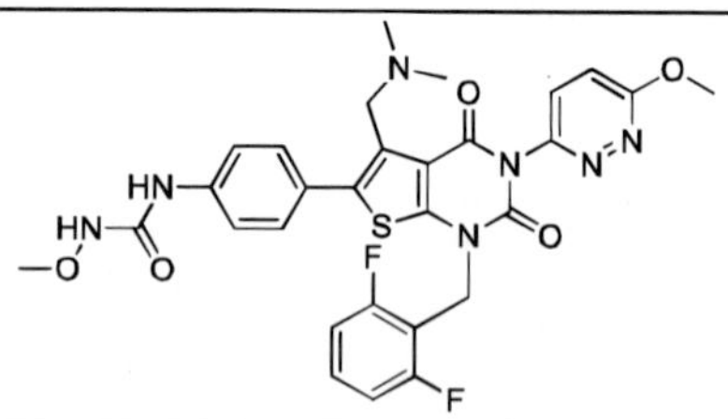
42.	Tepotinib	Anticancer	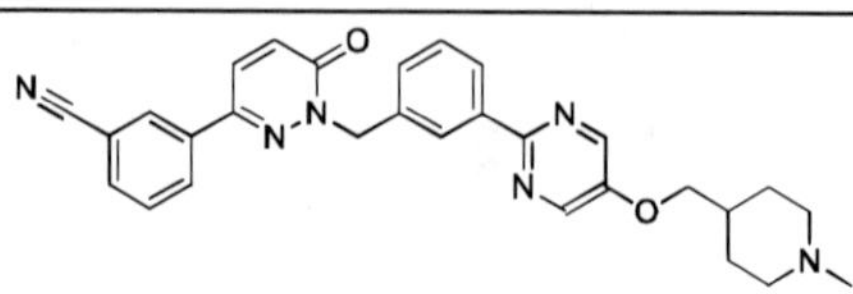

Minaprine

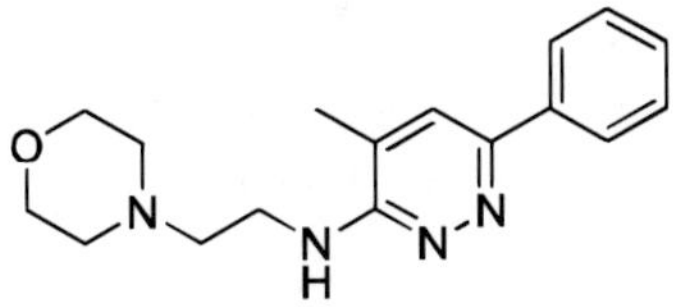

251

Minaprine, 4-methyl-*N*-(2-morpholinoethyl)-6-phenylpyridazin-3-amine, is an antidepressant drug. Its structure is unrelated to the other antidepressant drugs. Minaprine acts via inhibition of MAO type A enzyme. Repeated administration of minaprine can down-regulate the beta-adrenergic functions [171]. It was used in France but later withdrawn from the market because of its side effect of convulsions [172].

The metabolic pathway of minaprine is shown in Scheme 51. minaprine is metabolized in liver microsomal enzyme by CYP540 enzyme into ring hydroxylation **252** and *N*-oxide metabolite **253** [173].

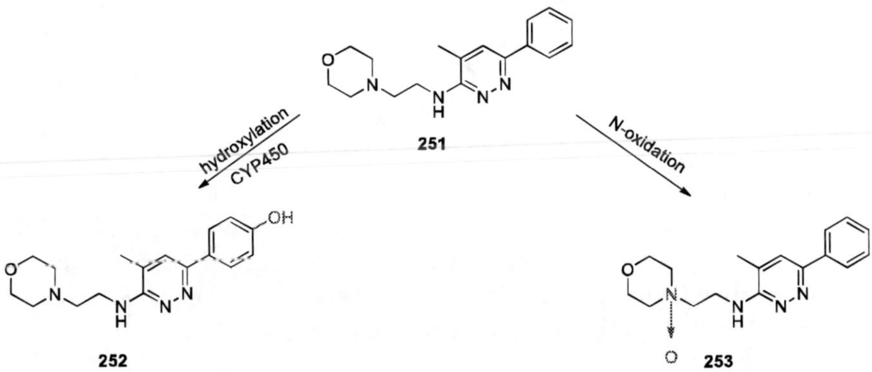

Scheme 51. Metabolic pathway of minaprine (**251**).

Synthesis of minaprine (Scheme 52) involves reaction between acetophenone **254** and pyruvic acid **255** to form α-methyl-β-benzoyl lactic acid **256**. The cyclocondensation of the γ-keto acid of 256 with hydrazine leads to the hydroxydihydropyridazinone **257**. The tertiary alcoholic group of **257** was then easily dehydrated in an acidic medium, resulting in the pyridazinones **258**. Compound **258** converted the pyridazinones into pyridazine chloride **259** by treatment with phosphorus oxychloride. Pyradazine chloride **259** gets condensed with amine **260** to form minaprine (**251**) [174].

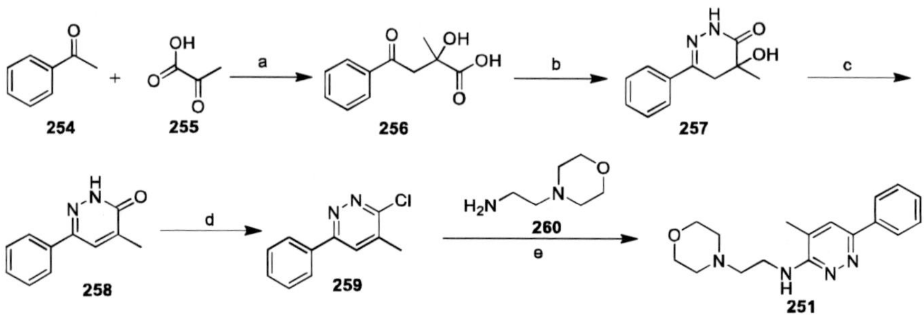

Reagents and conditions: a) KOH ; b) $NH_2NH_2.H_2O$; c) acidic/ H^+ ; d) $POCl_3$; e) NH_4Cl

Scheme 52. Synthesis of minaprine (**251**).

Sulfachloropyridazine

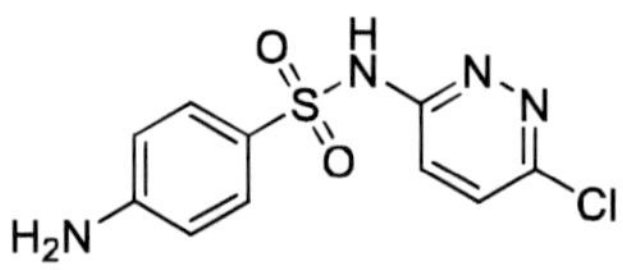

261

Sulfachloropyridazine (261), 4-amino-N-(6-chloropyridazin-3-yl) benzene-sulfonamide is a sulphonamide class of antibiotic drug used in poultry farming. It is a broad-spectrum sulfonamide used against gram-positive and gram-negative aerobic bacteria, effective against chlamydia, and it is considered a common pollutant in surface and ground water [175].

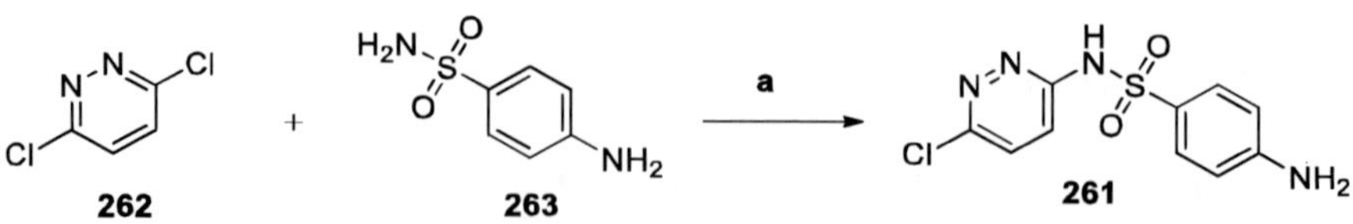

Reagents and conditions: a) $K_2CO_3/NaCl$

Scheme 53. Synthesis of sulfachloropyridazine (**261**).

Synthesis of sulfachloropyridazine is shown in Scheme 53. Dichloropyridazine **262** reacted with aminosulfonamide **263** forms sulfachloropyridazine (**261)** in the presence of K_2CO_3/NaCl [176].

Relugolix

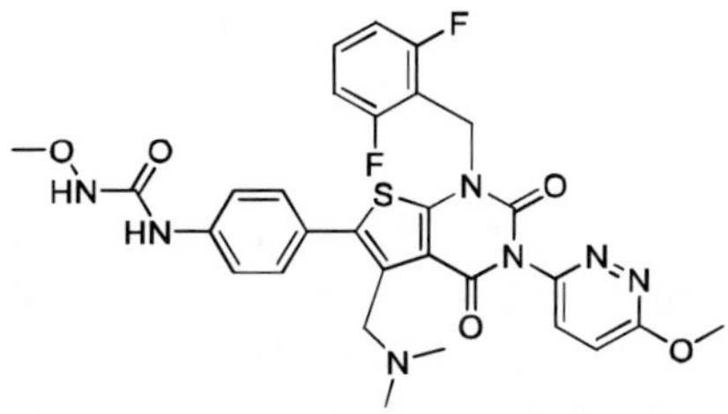

264

Relugolix (**264**), 1-(4-(1-(2,6-difluorobenzyl)-5-((dimethylamino)methyl)-3-(6-methoxypyridazin-3-yl)-2,4-dioxo-1,2,3,4-tetrahydrothieno[2,3-*d*]pyrimidin-6-yl)phenyl)-3-methoxyurea, is an effective gonadotropin-releasing hormone (GnRH) receptor antagonist drug approved in 2020 by US FDA. It treats bleeding during menstruation and estrogen-associated pain in uterine fibroids and endometriosis. It is the first oral antagonist used for several hormone-responsive conditions for adult patients with advanced prostate cancer [177].

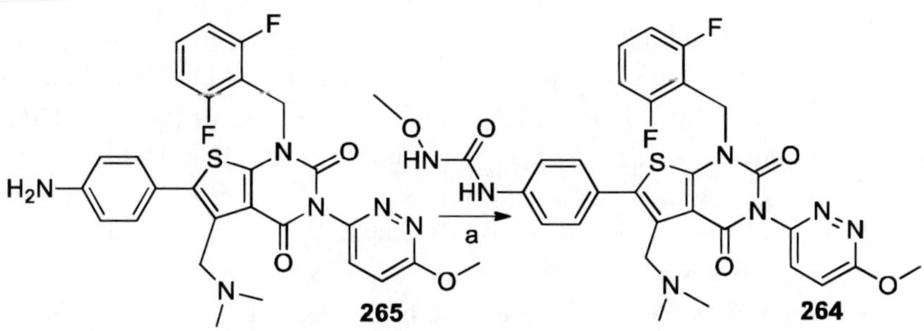

Reagents and conditions: a) CDI, MeCN, H_2O, $MeONH_2$. HCl, Et_3N, RT

Scheme 54. Synthesis of relugolix (**264**).

The synthesis of relugolix is given in Scheme 54. Compound **265** was added to a mixture of CDI, triethylamine, and methoxyamine hydrochloride in acetonitrile reaction mixture stirred for 16 hr to yield relugolix **264** [178].

Tepotinib

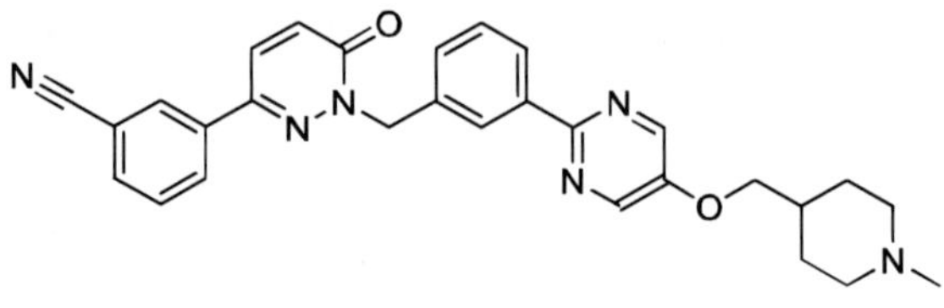

266

Tepotinib (**266**), 3-(1-(3-(5-((1-methylpiperidin-4-yl)methoxy)pyrimidin-2-yl)benzyl)-6-oxo-1,6-dihydropyridazin-3-yl)benzo nitrile is pyridazinone derivative; it is a potent mesenchymal-epithelial transition (MET) inhibitor used as an anti-cancer agent. USFDA approved it in 2021. It also reverses ATP binding cassette sub-family G member -2 (ABCG2) -mediated multidrug resistance *in-vitro* and *in-vivo* by directly acting on ABCG2 and reversibly inhibiting ABCG2 drug reflux activity [179].

Hamim *et al.* studied the penetration of tepotinib into the brain and its efficacy in orthotopic patient-derived models. This study concluded that tepotinib penetrates the brain to achieve intracranial target inhibition and potent activity against MET-driven non-small cell lung cancer brain metastases [180].

Jing *et al.* studied the effect of tepotinib on melanoma. Tepotinib inhibits melanoma cell proliferation, invasion, and migration while inducing apoptosis by blocking the MET and the phosphatidylinositol 3-kinase/protein kinase B (PI3K/AKT) signalling pathways [181].

Tepotinib is metabolized primarily by oxidation (metabolite **267** and **268**), *N*-demethylation (metabolite **270**), and hydroxylation (metabolite **267**) via CYP3A4 and CYP2C8. It also undergoes phase II transformation by glucuronide conjugation (Scheme 55) [182].

The synthesis of tepotinib is shown in Scheme 56. 3-(5-Hydroxy-pyrimidin-2-yl)-benzoic acid methyl ester **271** reacted with *N*-Boc-piperidinemethanol **272** in the presence of triphenyl-phosphine in THF form intermediate **273,** it was further subjected to DIBAL-H in THF form intermediate **274**. intermediate **274** condensed with substituted pyridazinone **275** to form tepotinib (**266**) [183].

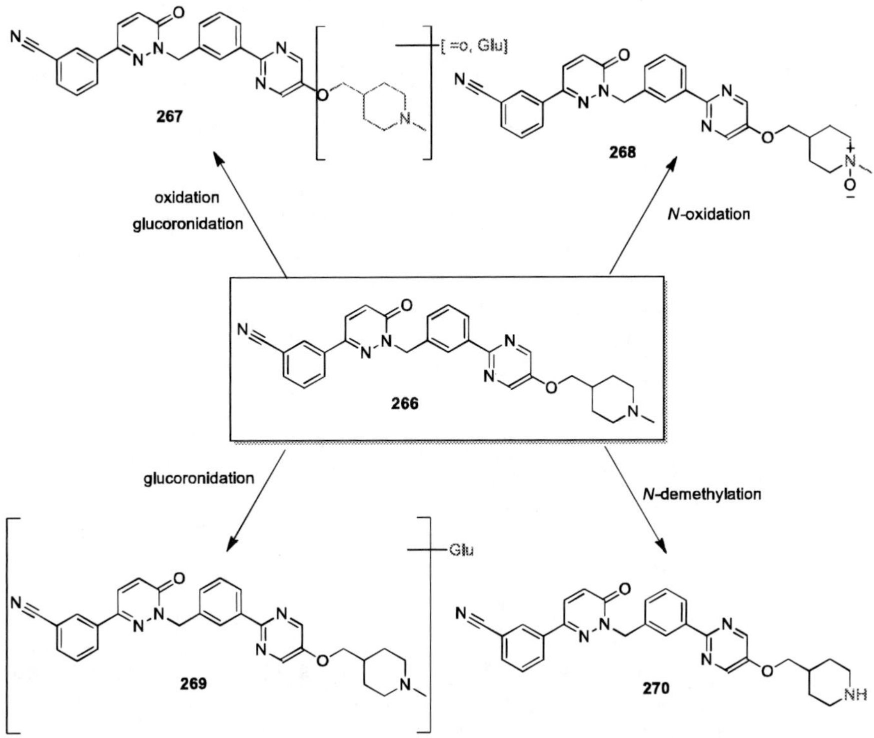

Scheme 55. Metabolic pathway of tepotinib (**266**).

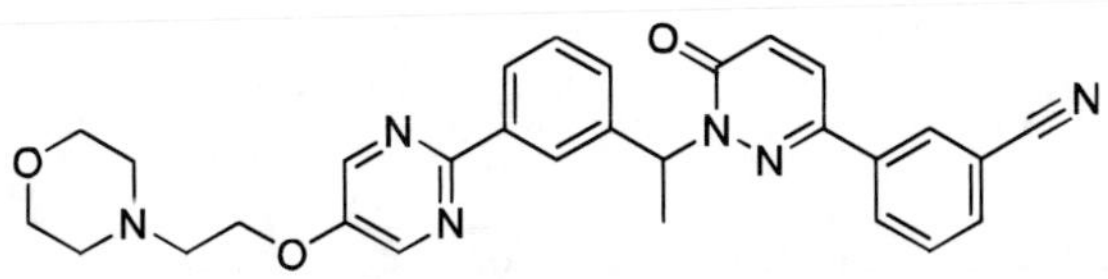

276

Zhang *et al.* synthesized and evaluated new series of tepotinib derivatives for the activity of antiproliferative agents. Of all the synthesized compounds, compound **276** exhibits potent antiproliferative activity with IC_{50} value of 0.026 μM against hepatic carcinoma 97H cells. It offers less toxicity to normal cells. Compound **276** will act via induced G1 phase arrest [183].

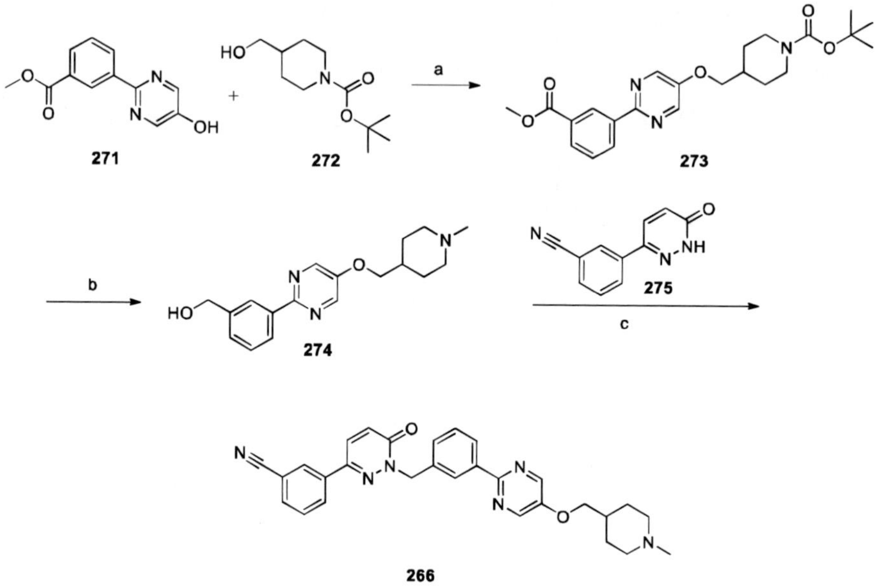

Reagents and conditions: a) PPh_3/DIAD, THF ; b) DIBAL-H, THF ; c) PPh_3/DIAD, THF

Scheme 56. Synthesis of tepotinib (**266**).

Miscellaneous Drugs

Table 3. Miscellaneous drugs

Sr. No.	Drug	Class	Structure
43.	Olaparib	Anticancer (PARP inhibitors)	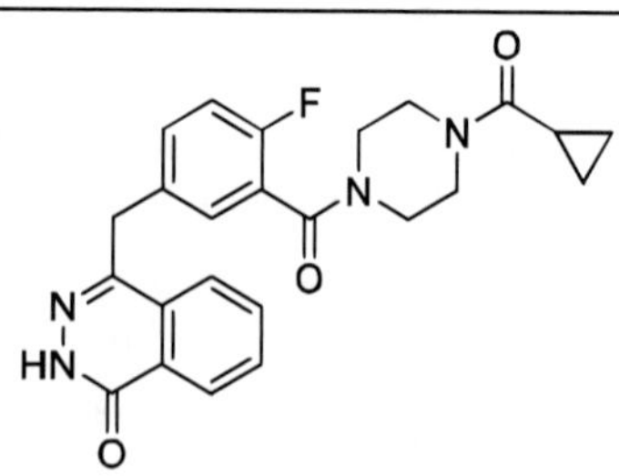

Sr. No.	Drug	Class	Structure
44.	Talazoparib	Anticancer (PARP inhibitors)	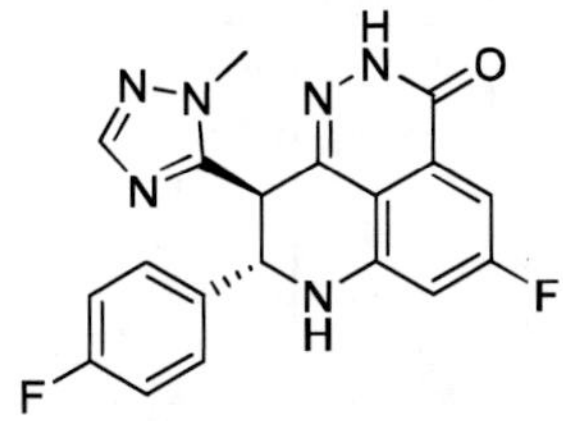
45.	Ponatinib	Anticancer	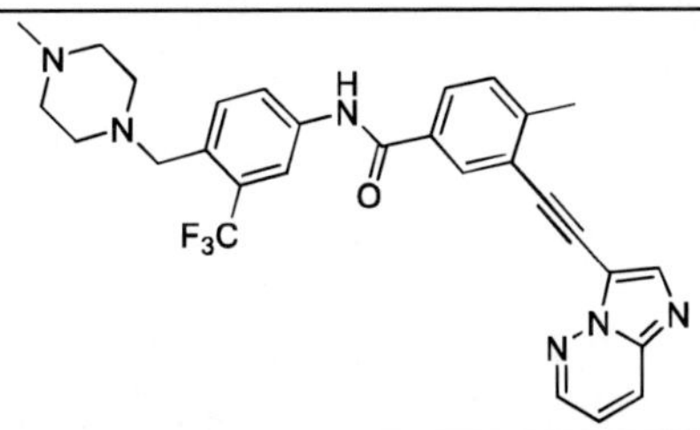
46.	Tucatinib	Anticancer	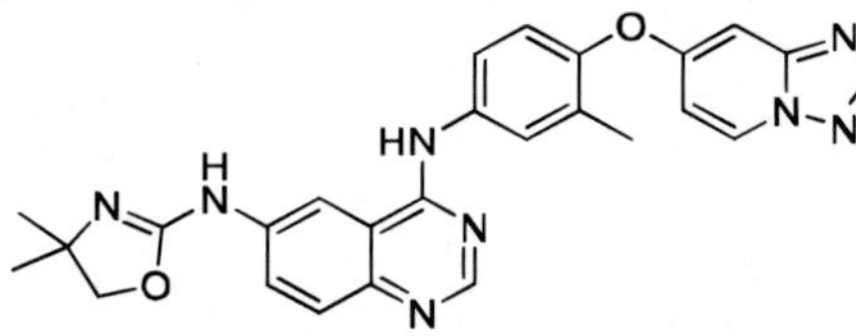
47.	Capmatinib	Anticancer	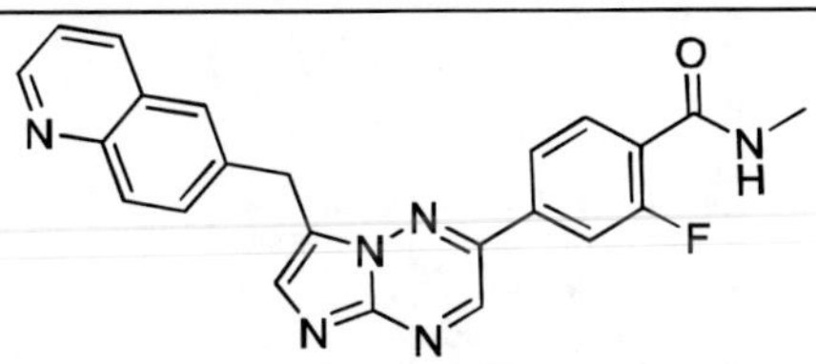
48.	Edaravone	CNS	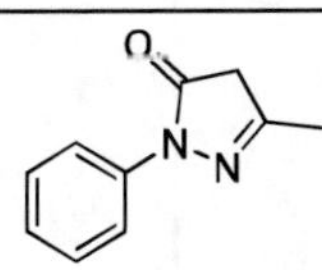
49.	Suvorexant	CNS	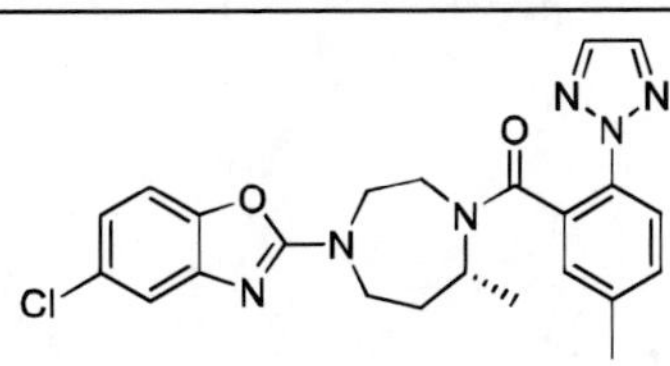

Table 3. (Continued)

50.	Risdiplam	CNS	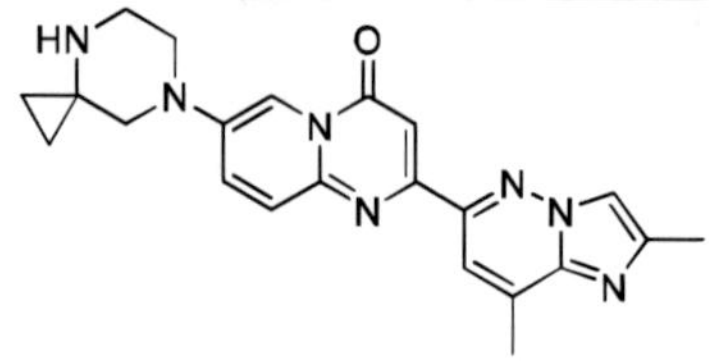
51.	Fostemsavir	Anti-HIV	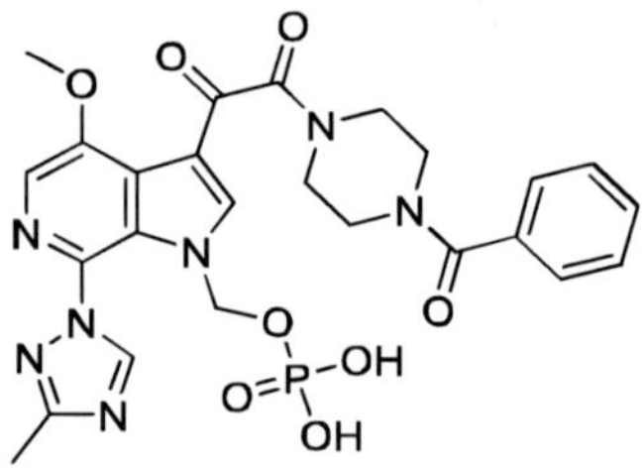
52.	Remdesivir	CoVID-19 treatment	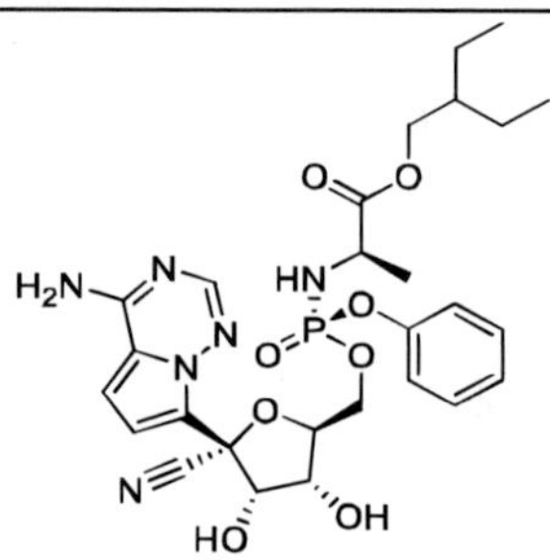
53.	Ibrexafungerp (Brexafemme®) is a triterpenoid	Antifungal	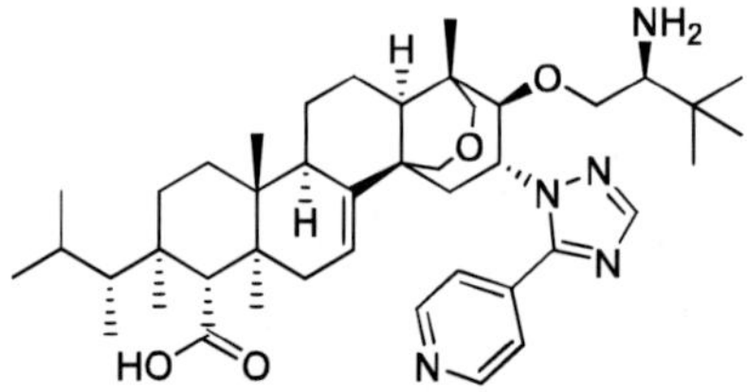

Olaparib

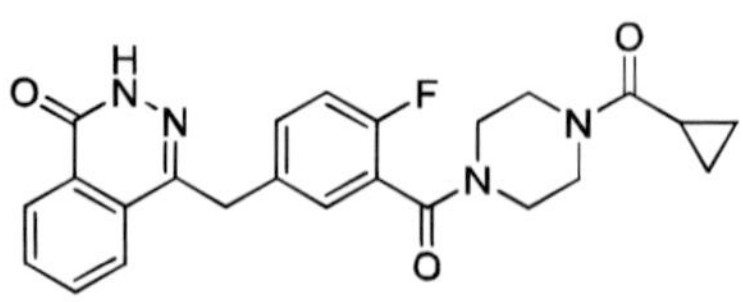

277

Olaparib (**277**), 4-(3-(4-(cyclopropanecarbonyl) piperazine-1-carbonyl)-4-fluorobenzyl)phthalazin-1(2*H*)-one is an anti-cancer drug or PARP inhibitor and FDA-approved this drug developed by AstraZeneca for the treatment of cancer (ovarian, breast, prostate, and pancreatic cancer). Olaparib is mainly designed for breast cancer genes (BRCA), and mutated ovarian cancer. The adverse effect of olaparib is vomiting, nausea, anemia, and fatigue, whereas severe effects include leukemia and myelodysplastic syndrome [184, 185].

Wang *et al.* predicted the *in-vitro* metabolism of olaparib (Scheme 57) in liver microsomes by LC-MS. Olaparib consists of piperazine nucleus, which easily gets metabolized by oxidation to metabolites **279**, **280**, **281**, **282**, and **283**. Olaparib undergoes amide hydrolysis and forms metabolite **278** [186].

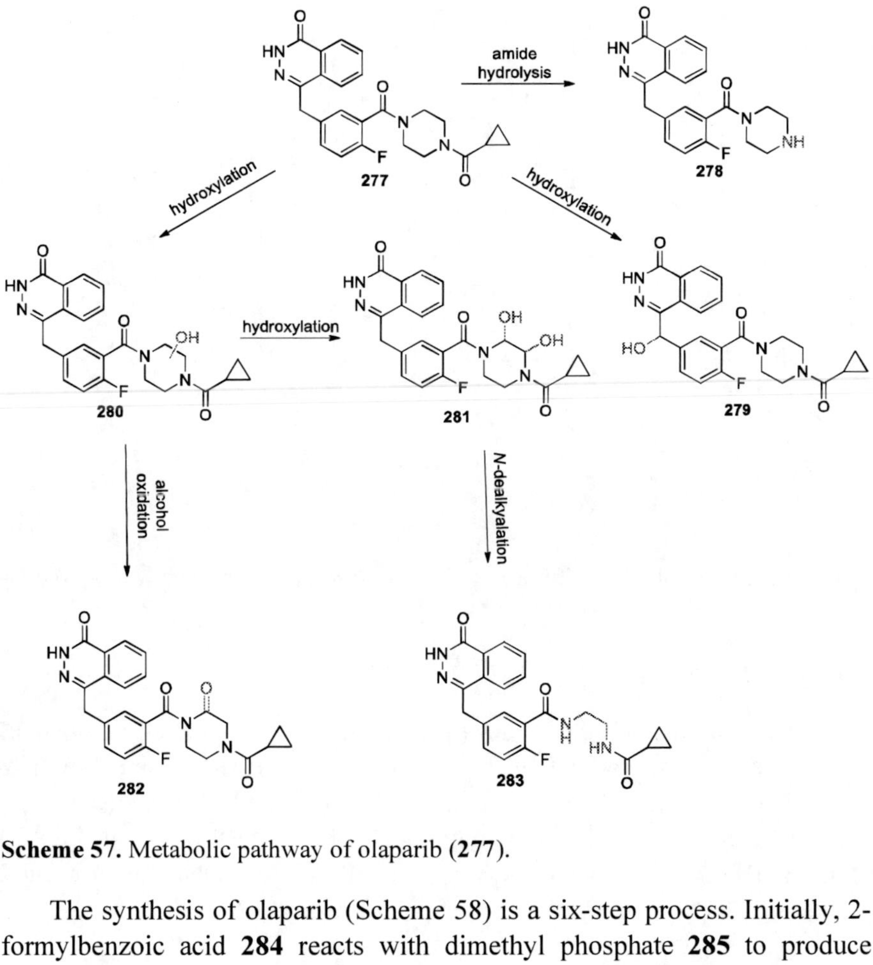

Scheme 57. Metabolic pathway of olaparib (**277**).

The synthesis of olaparib (Scheme 58) is a six-step process. Initially, 2-formylbenzoic acid **284** reacts with dimethyl phosphate **285** to produce phosphonate **286**. The phosphonate reacts with aldehyde **287** by Horner-Wadsworth-Emmons reaction to produce olefin **288**. Olefin **288** was hydrolyzed under alkaline conditions and reacted with hydrazine hydrate to form an intermediate **289** that reacted with *N*-Boc-piperazine **290** and 2-(1*H*-benzotriazol-1-yl)-1,1,3,3-tetramethyluronium hexafluorophosphate to form amide **291**. Upon deprotection of the Boc group, it was reacted with cyclopropane carboxyl chloride **293** to produce olaparib (**277**) [187].

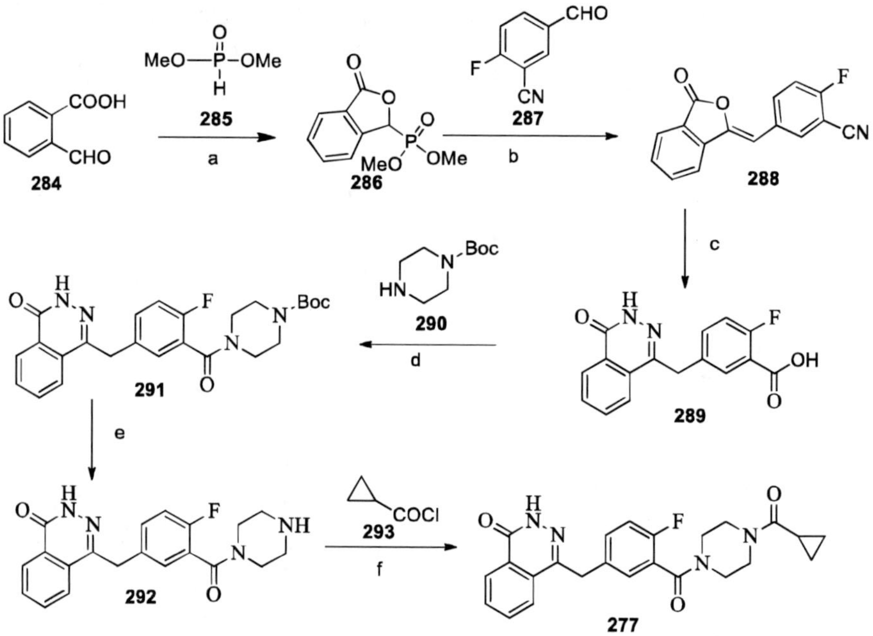

Reagents and conditions: a) NaOMe, methanol; b) Et_3N; c) NaOH, H_2O, 90 °C, NH_2NH_2, H_2O; d) HBTU,DMTA; e) HCl/EtOH; f) Et_3N

Scheme 58. Synthesis of olaparib (**277**).

Olaparib has multiple modes of action. A study suggested that olaparib improved chromatin binding and traps PARP (PARP-1 and PARP-2). Stabilization of the PARP-DNA complex leads to cellular toxicity. Usually, olaparib inhibits PARylation by binding NAD^+ to PARP (PARP-1, PARP-2, and PARP-3). Moreover, olaparib traps PARP on DNA, interfering in repairing DNA and thus leading to cellular toxicity [188, 189].

So far, significant research has been done to improve the anti-cancer properties of olaparib [190, 191]. The advancement in olaparib raised a question regarding the role of PARP treatment, as a study shows that some patients experience a partial response/resistance [192]. Olaparib, combined with other targeted therapy, might overcome issues related to PARP resistance. With the help of chemical modification/combined treatment of olaparib, we can easily change/improve the compound's anti-cancer activity [193].

Talazoparib

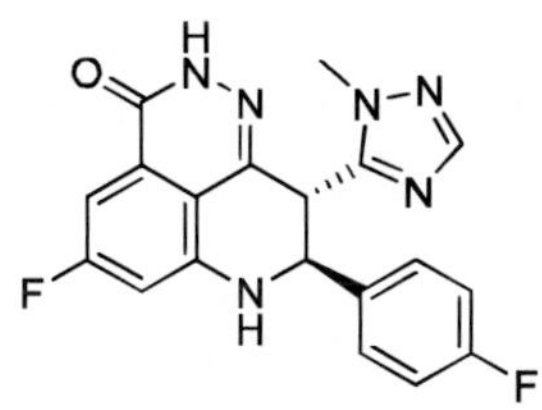

294

Talazoparib (**294**), (8*S*,9*R*)-5-fluoro-8-(4-fluorophenyl)-9-(1-methyl-1*H*-1,2, 4-triazol-5-yl)-2,7,8,9-tetrahydro-3*H*-pyrido[4,3,2-*de*]phthalazin-3-one, developed by Pfizer, is a PARP inhibitor that plays the main role in repairing DNA strands. Talazoparib is an FDA-approved drug for the treatment of BRCA mutated and human epidermal growth factor receptor 2 (HER2) negative breast cancer, which is responsible for the growth of solid tumors and metastatic castration-resistant (CR) prostate cancer (CRPC). The mode of action of talazoparib comprises the inhibition of PARP (PARP-1 and PARP-2) enzymes that play essential roles in detecting and repairing DNA damage and PARP trapping. The PARP protein bind with a PARP inhibitor, and DNA prevents replication, transcription, and repair of DNA, subsequently, cell death [194, 195].

The talazoparib is an effective PARP inhibitor, including PARP-1 and PARP-2, and its binding affinity is approximately the same for both PARP isoforms [196, 197]. Additionally, talazoparib is three times lower effective as a PARP inhibitor compared with other PARP inhibitors like olaparib, veliparib, and rucaparib. It is essential to mention that talazoparib did not show any effects like inhibition of receptors; even at a dose 1 mmol/l talazoparib does not show the effect on poly(ADP-ribose)glycohydrolase (PARG) that is related to PARP, thereby no inhibition of PARP. Talazoparib shows extraordinary anti-cancer effects in combination with a metabolite of irinotecan, temozolomide, and platinum agents [198, 199, 200]. Numerous studies show the improvement of the anti-cancer activity of talazoparib by designing, with the help of nanoformulation and synthetic processes. For example, Baldwin *et al.* synthesized nano-talazoparib and evaluated it based on size, surface charge, and loading. The data suggested that the controlled release of nano-talazoparib formulation efficiently delivers PARP inhibitor therapy for controlling the management of cancer cells. Li *et al.* designed and synthesized a photoactive form of PARP inhibitor, talazoparib. A covalent bond attached the o-nitrobenzyl group to the pharmacophore of talazoparib to provide a photoactive switch. The data suggested the inhibitory activity increased with prolonged irradiation. The photoactive drugs-based strategy offers new insight or biological tools for therapeutic applications.

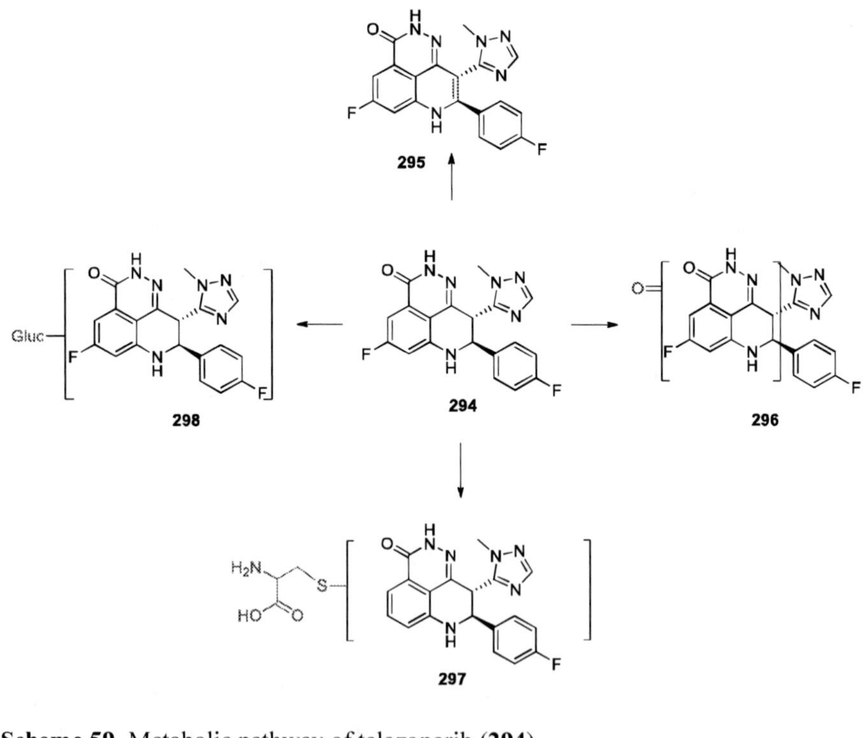

Scheme 59. Metabolic pathway of talazoparib (**294**).

Moreover, stability, cytotoxicity, bioavailability, and animal and cellular studies will be required [201]. Zhou *et al.* synthesized radio [^{18}F] talazoparib and its derivative and evaluated its activity in *in-vitro* and *in-vivo*. The data suggested that the PARP radiotracer has high affinity and selectivity with suitable pharmacological properties that might be served as biomarkers for cancer progression [202]. The literature mentioned above indicated that the controlled release, nanoformulation, photoactive drugs, and radiosynthesis of the drugs might become a next-generation system that enhances the activity of the talazoparib.

Talazoparib gets metabolized by mono-oxidation to form **296**. Dehydrogenation leads to the formation of **295**, cysteine conjugation of mono-desfluoro-talazoparib to form **297**, and glucuronide conjugation of talazoparib to form **298** (Scheme 59) [203].

Scheme 60 shows a multi-step procedure for the radiosynthesis of [^{18}F] talazoparib. The direct radiosynthesis of [^{18}F] talazoparib required significant efforts; thereby, [^{18}F] talazoparib was synthesized from ester **299** and [^{18}F]4-fluorobenzaldehyde **300** to form intermediate **301**, which was treated with hydrazine.

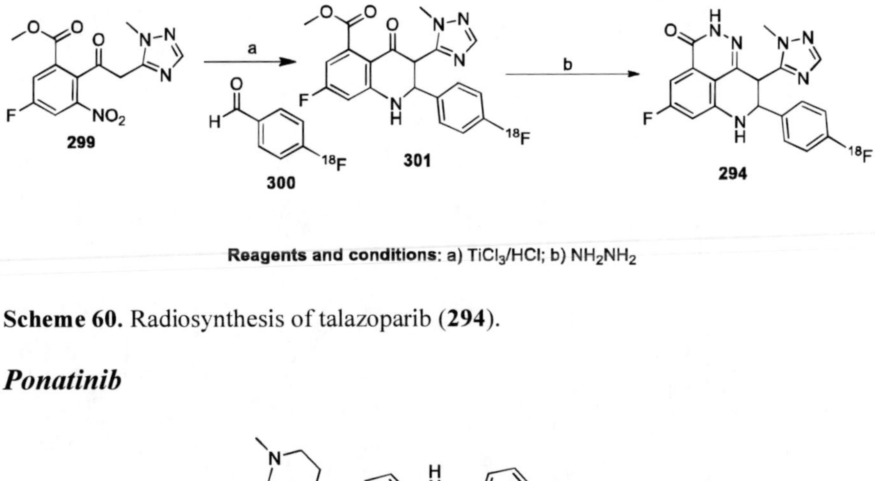

Scheme 60. Radiosynthesis of talazoparib (**294**).

Ponatinib

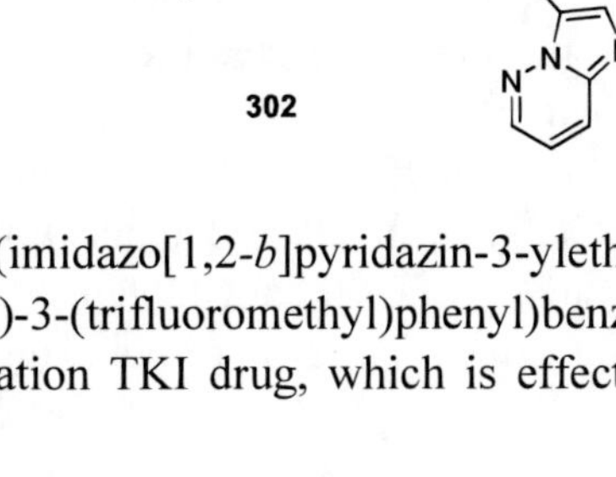

302

Ponatinib (**302**), 3-(imidazo[1,2-*b*]pyridazin-3-ylethynyl)-*N*-(4-((4-methyl-piperidin-1-yl)methyl)-3-(trifluoromethyl)phenyl)benzamide is an FDA-approved third-generation TKI drug, which is effective against BCR-ABL

T315I mutation, thereby efficiently treating CML patients. The results are much better if the treatment of CML is initiated at the chronic phase of CML disease [204, 205, 206].

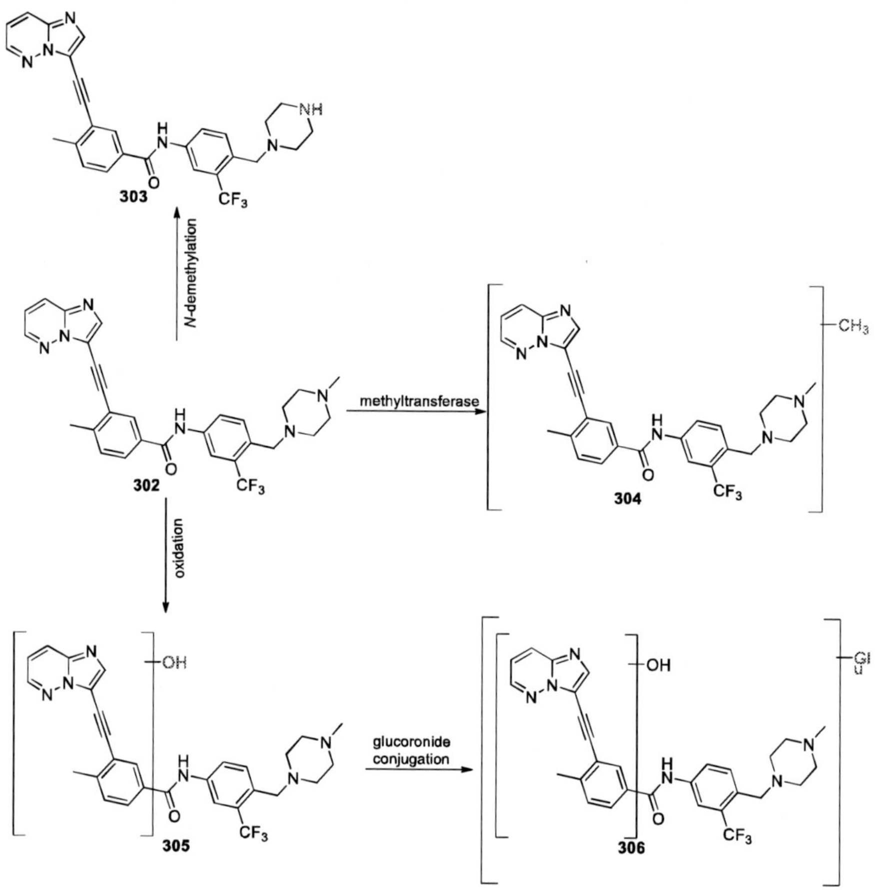

Scheme 61. Metabolic pathway of ponatinib (302).

Ponatinib structure contains a C-C triple bond targeted to BCR-ABL T315I point mutation. The study suggested the anti- BCR-ABL efficiency of the ponatinib drug. Ponatinib was tested against Ba/Fe cell lines expressing BCR-ABL with and without mutation with IC_{50} values ranging (from 0.37-2 nM). The cell proliferation assay indicates that ponatinib shows a considerable decrement in the cell viability of CML cells. Interestingly, approximately 500 times lower IC_{50} values than normal cells attributed to potential ability as anti-

BCR-ABL. The *in-vivo* study suggested that the ponatinib drug increases the survival rate but is mainly dose-dependent [207].

The persistent use of ponatinib would lead to the development of resistance. Few reports are available that focus on the mechanism of ponatinib resistance. Lu *et al.* focus on a different aspect of the mechanism of ponatinib resistance. Ponatinib resistance was developed mainly through longer-term exposure, and the use of higher doses via AXL (tyrosine kinase receptor) overexpression attributed to AXL was the main driver for the development of resistance [208]. Zabriskie *et al.* found around 12-key positions in BCR-ABL mutants that vary in resistance to different drugs, namely, nilotinib, dasatinib, imatinib, and ponatinib [209].

CYP3A4 and methyl transferase enzyme metabolize ponatinib into compounds **303**, **304**, and **305**. Additionally, compound **305** undergoes phase II metabolism by glucuronide conjugation (Scheme 61) [210].

Tucatinib

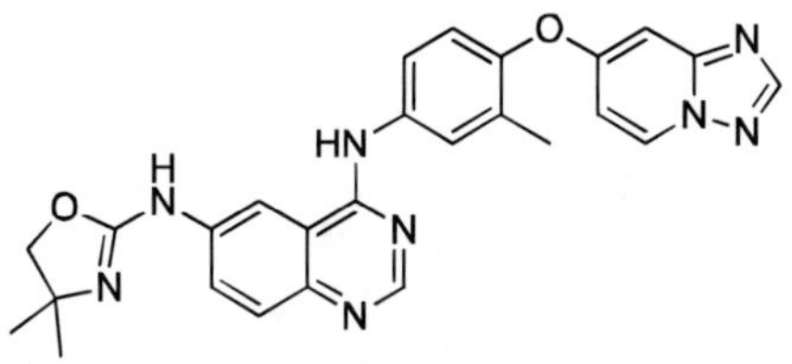

307

Tucatinib **(307)**, *N*4-(4-([1,2,4]triazolo[1,5-*a*]pyridin-7-yloxy)-3-methyl-phenyl)-*N*6-(4,4-dimethyl-4,5-dihydrooxazol-2-yl)quinazoline-4,6-diamine is an anti-cancer drug used for the treatment of breast cancer. Tucatinib is a highly selective tyrosine kinase inhibitor (TKI) and ATP-competitive for human epidermal growth factor receptor-2 (HER2). The cellular assay suggested that tucatinib is 100 times more specific for HER2, whereas the *in-vitro* model determined that tucatinib effectively blocks the phosphorylation and overexpression of HER2. Moreover, tucatinib inhibits HER2 in p95/p110 mutation, which is known to cause trastuzumab resistance. Tucatinib combined with chemotherapeutic drugs shows a synergetic effect [211]. Analogs of tucatinib (lapatinib, neratinib, tucatinib, and erlotinib) function as dual inhibitors for EGFR and HER2. HER2 is a part of epidermal growth factor receptors, including (EGFR)/HER1, HER2, HER3, and HER4. The functions of these receptors are to control cell growth, survival differentiation, and migration [212]. The growth factor receptors, especially HER2, produce a protein (new protein at the outer surface of breast cells), which is mainly responsible for the robust progression of cancer cells. Usually, the HER2 receptor controls the development of breast cells. HER2 can produce multiple copies of itself that cause gene amplification; thereby, several copies of HER2 proteins/receptors lead to the over-expression of the HER2 gene, subsequently leading to the uncontrolled growth of breast cells [213].

Tucatinib prevents the overexpression of the HER2 gene *in-vivo* and *in-vitro*. Tucatinib prevents the phosphorylation of HER2 and HER3 proteins, inhibiting the cells downstream AKT and MAPK signaling. TUKYSA organization demonstrates anti-cancer ability in HER2-expressing cancer cells. The combination of tucatinib with trastuzumab demonstrates higher anti-cancer ability *in-vivo* and *in-vitro* compared to either drug alone [214].

Human liver microsomal enzymes metabolize tucatinib. Major metabolite **310** is formed via CYP2C8. CYP3A4/5 is involved in forming other metabolites (Scheme 62) [215].

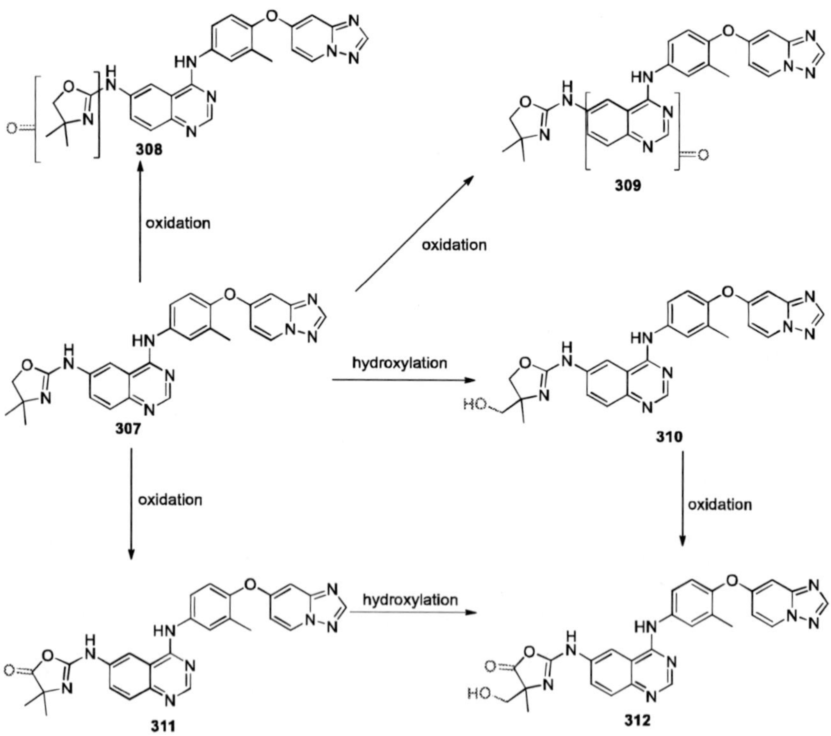

Scheme 62. Metabolic pathway of tucatinib (307).

Synthesis of tucatinib shown in Scheme 63. Pyridine derivative **314** coupled with nitrophenol **313** to produce ether **315**, reacted with ammonium hydroxide to form hydroxylamine **316**, which cyclized to form compound **317**. The nitro group of **317** was reduced to amine **318**. Compound **318** condensed with *N,N*-dimethylphenylformimidamide **319** to obtain the nitroquinazolinamine **320**. The nitro group of **320** was reduced to amine **321**, which further coupled with oxazoline **322** to furnish tucatinib (**307**) [216].

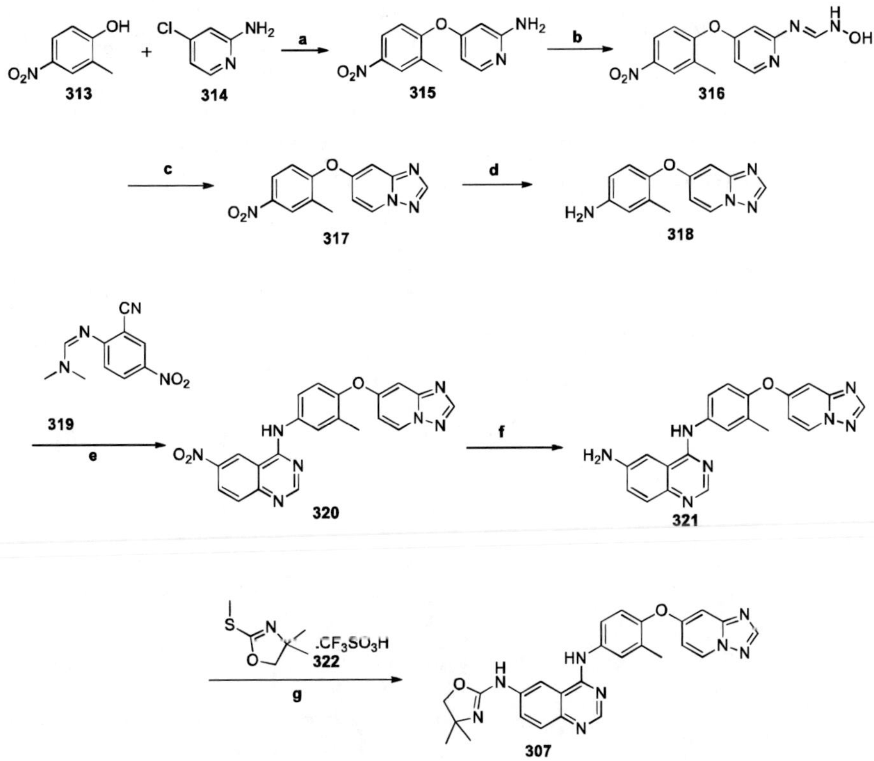

Reagents and conditions: a) DIPEA, NMP,150°C, 72 h, 64%; b) i) DMF-DMA, EtOH, 75°C,3 h; ii) NH_2OH.HCl, 50°C, 3h, 81%; c) TFAA, THF, 10-25°C, 68%; d) H_2,Pd-C, THF, 40°C, 15 h, 93%; e) AcOH, 85°C, 2 h, 83%; f) H_2,Pd-C, rt, 5 h, 92%; g) DMF, Cs_2CO_3, 125°C, 20 h,76%

Scheme 63. Synthesis of tucatinib (**307**).

Capmatinib

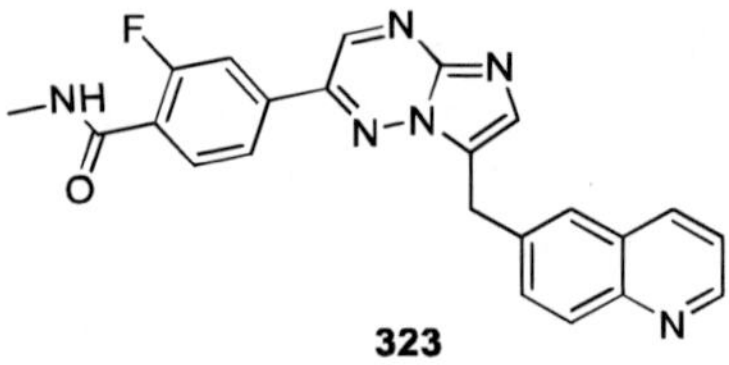

323

Capmatinib **(323)**, 2-fluoro-*N*-methyl-4-(7-(quinolin-6-ylmethyl)imidazo-[1,2-*b*][1,2,4]-triazin-2-yl)benzamide is an anti-cancer drug used for treating lung cancer. Capmatinib is targeted at the c-MET oncogenic driver, a tyrosine kinase receptor. MET is typically expressed in epithelial cells and activates signaling cascades involving organ regeneration and tissue repair [217].

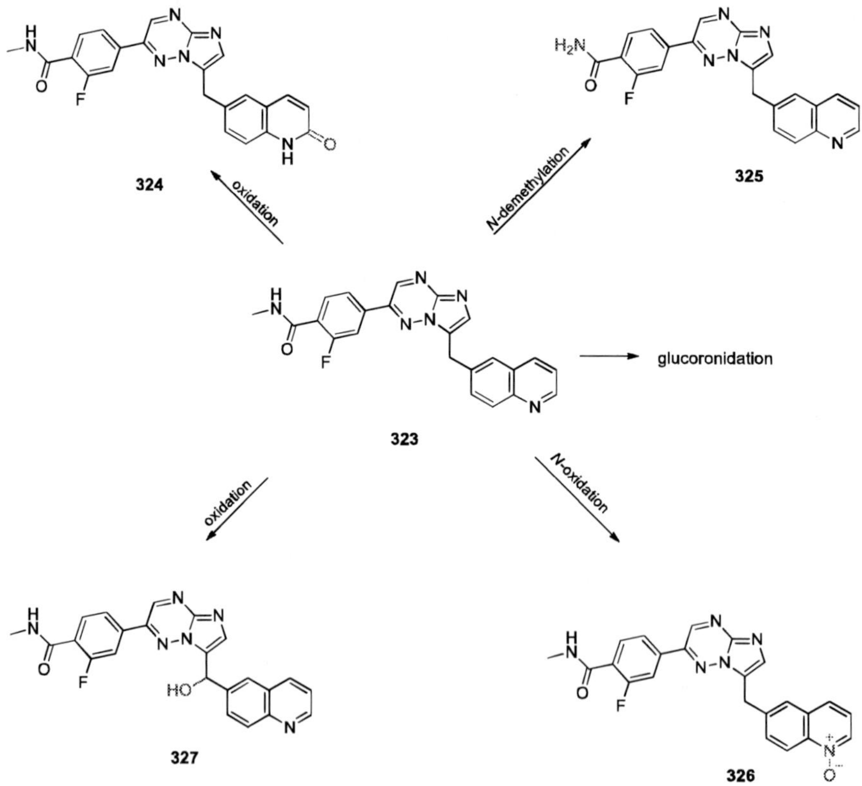

Scheme 64. Metabolic pathway of capmatinib **(323)**.

Capmatinib selectively acts as a MET inhibitor, blocking MET phosphorylation and the activation of the downstream effectors in the MET-dependent tumor cell lines. Capmatinib demonstrated opposite effects of MET activation on the EGFR and HER-3 pathways that reestablish sensitivity to the EGFR-TKI erlotinib in *EGFR*-mutant NSCLC cell lines with acquired EGFR-TKI resistance [218].

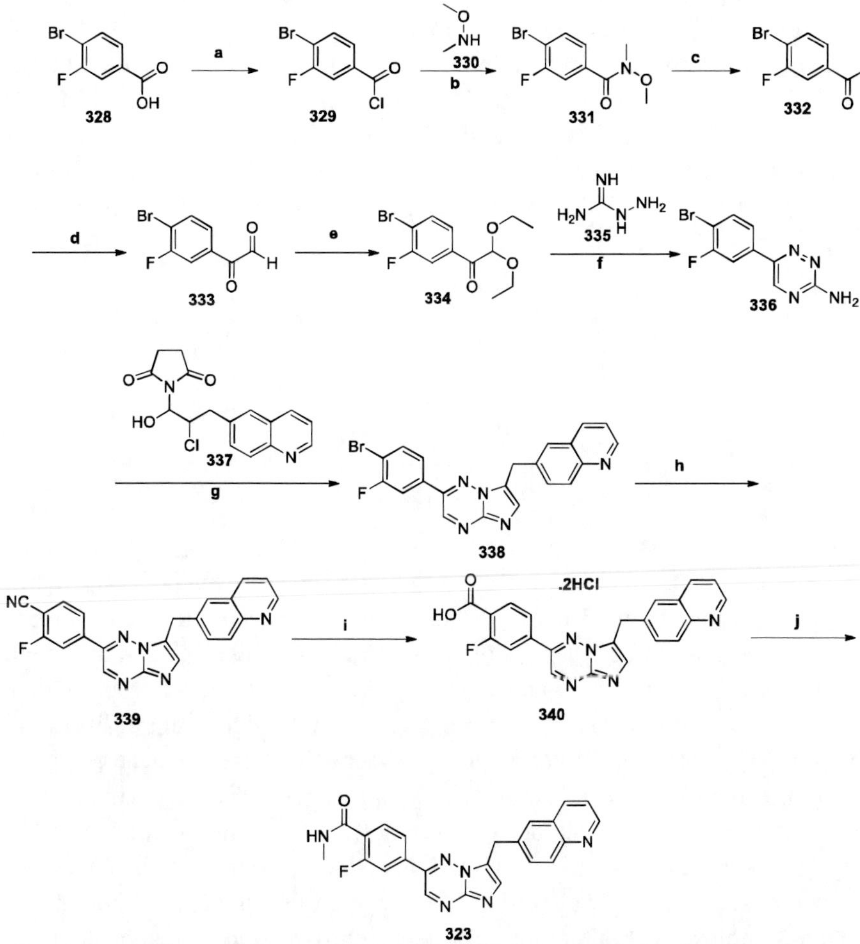

Reagents and conditions: a) oxalyl chloride, DCM, DMF,rt; b) TEA, 96%; c) MeMgCl, THF, 0°C, 98%; d) 48% HBr/DMSO ,60°C, 100%; e) $HC(OEt)_3$, TsOH, 110°C, 92%; f) KOH, H_2O, EtOH, 75°C, 53%; g) 1-butanol, 110°C, 50%; h) $Zn/Zn(CN)_2$, $Pd_2(dppf)_2Cl_2$, 110°C, 73%; i) conc. HCl, H_2O, 100°C, 93%; j) $MeNH_2$, ByBOP, TEA, DMF, 85%

Scheme 65. Synthesis of capmatinib **(323)**.

CYP3A is the major P450 enzyme subfamily involved in the hepatic microsomal metabolism of capmatinib. Capmatinib will undergo oxidation, N-dealkylation, and *N*-oxidation. Further, it gets metabolized by glucuronide conjugation (Scheme 64) [219].

The synthesis of capmatinib is shown in Scheme 65. Chlorination of 4-bromo-3-fluorobenzoic acid **(328)** gives acid chloride **329**, which was substituted with *N*, O-dimethylhydroxylamine **(330)** to form hydroxamic acid derivative **331** in two steps. Compound **331** was methylated by Grignard reagent to obtain acetophenone **332**, which was oxidated to α-keto aldehyde **333**, followed by acetalization of the aldehyde group to produce **334**. The cycloaddition reaction of **334** and **335** forms **336** under a basic condition, which is reacted with **337** to form fused imidazole **338**. The addition of cyanide group yields **339**, which undergoes hydrolysis and amidation to form capmatinib **(323)** [220].

Edaravone

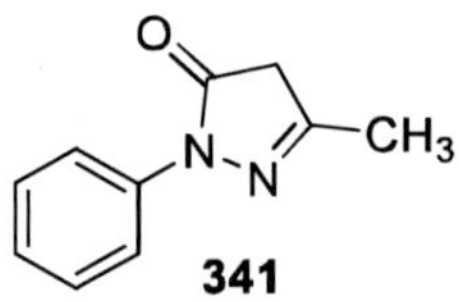

341

Edaravone (**341**) is a neuroprotective agent (FDA-approved drugs) (3-methyl-1-phenyl-2-pyrazoline-5-one) that delays the progression of ALS. Moreover, edaravone is an effective pyrazolone free-radical scavenger and antioxidant that decreases oxidative stress in ALS [221, 222]. Amyotrophic lateral sclerosis (ALS)/Lou Gehrig's disease is a lethal, degenerative disease that mainly distresses the motor neurons (especially those connected with the spinal cord and brain) and subsequently causes paralysis and death [223].

Edaravone and its analog were synthesized through the condensation of hydrazine and esters. The 2-pyrazoline-5-one was used as a template to generate analogs by (1) alteration of the phenyl group, (2) substituents on 1-phenyl moiety, (3) alteration of methyl group, and (4) substituents on methylene moiety. Earlier studies suggested that edaravone has a lipophilic substituent that effectively inhibits lipid peroxidation. The mode of action of edaravone drugs is unclear; reactive oxygen species (ROS), glutamate excitotoxicity, and oxidative stress are considered important factors in ALS. Usually, edaravone acts as a ROS scavenger that inhibits peroxyl radical

(LOO•). Edaravone scavenges H_2O_2 and protects cells from oxidative stress via upregulation of peroxiredoxin-2, inhibiting apoptosis, and down-regulation of protein disulphide isomerase. Edaravone reacts with peroxynitrite anion ($ONOO^-$) and traps •OH radical. The trapping of •OH radical inhibits the OH-dependent lipid peroxidation/tyrosine nitration encouraged by $ONOO^-$ anion [224, 225, 226].

Numerous studies suggested that the antioxidant radical scavenging activity suggested inhibition of both enzymatic and non-enzymatic lipid peroxidation [227], no effects on the production of superoxide [228], cellular damage of endothelial cells [229], and formation of dihydroxybenzoic acid (DHBA) from salicylate [230]. It is important to mention here as its plausible mode of action makes promising candidates not only for the treatment of ALS but also for the treatment of other ROS-related diseases. Moreover, other analogs did not show comparable efficacy with edaravone.

The glucuronide conjugation of edaravone involves multiple uridine diphosphate glucuronosyltransferase (UGT) isoforms (UGT1A6, UGT1A9, UGT2B7, and UGT2B17) in the liver and kidney. In human plasma, edaravone is mainly detected as the sulphate conjugate, which is presumed to be formed by sulfotransferases [231].

The synthesis of edaravone is given in **Scheme 66**. Ethyl acetoacetate **343** reacted with phenylhydrazine **342** in ethanol under reflux, giving edaravone (**341**) [232].

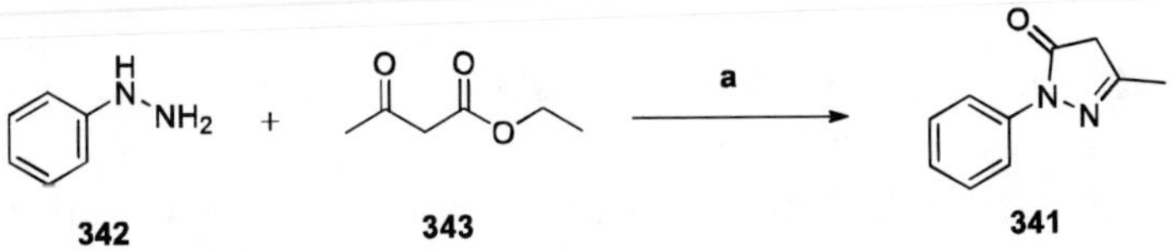

Reagents and conditions: EtOH, reflux

Scheme 66. Synthesis of edaravone (**341**).

Suvorexant

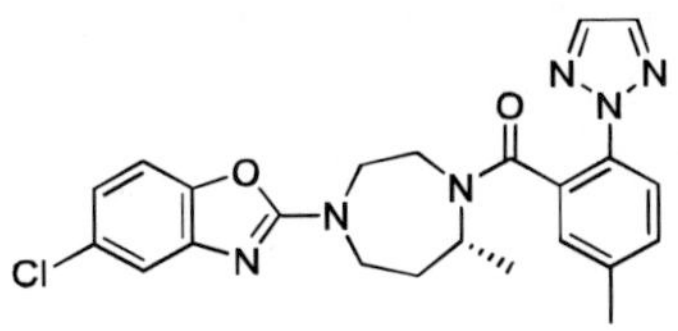

344

Insomnia leads to a lack of focus, anxiety, depression, and lower energy. Suvorexant (MK-4305), (*R*)-(4-(5-chlorobenzo[*d*]oxazol-2-yl)-7-methyl-1,4-diazepan-1-yl)(5-methyl-2-(2*H*-1,2,3-triazol-2-yl)phenyl)methanone is a novel treatment that selectively and effectively binds the neuropeptide orexin receptor antagonists A (OX1R) and B (OX2R). The primary mode of action of suvorexant is blocking OX1R and OX2R. The blocking of OX1R and OX2R decreases the awakenings and improves consolidated sleep. The 7-methyl on the diazepine offers additional efficacy when joined with chlorobenzoxazole. The suvorexant shows effectiveness in the dose-dependent manner.

So far, numerous clinical trials have been done for the efficacy of the suvorexant against insomnia. Michelson *et al.* used Suvorexant for safety and efficacy over three months and assessed after one year of treatment in insomnia patients. The study suggested that suvorexant drugs are effective and safer for use over one year [233].

Herring *et al.* use suvorexant as an antagonist of OX1R and OX2R to treat insomnia. The data suggested that the suvorexant 40/30 mg shows an extraordinary response in all patients [234]. Kuriyama *et al.* reviewed four randomized suvorexant trials for insomnia treatment. Moreover, the suvorexant's adverse effects are tiredness, somnolence, and abnormal dreams [235]. A longer-term study of Suvorexant drugs is required to improve insomnia treatment and understand their adverse effect.

Suvorexant metabolized by CYP540 enzyme. Hydroxylation of the aliphatic chain gives metabolite **345**, **350**. Ring oxidation gives metabolites **346**, **348**, **and 349**. All these metabolites further undergo glucuronide conjugation (Scheme 67) [236].

The core structure of diazepane (**357**) was prepared and purified by chiral HPLC. The synthesis started with a reaction of unsaturated ketone **351** and Boc-protected amine **352** to form compound **353**, which was cyclized to generate diazepane **354**. Upon Boc protection of **354** it produced **355**, which was resolved using chiral HPLC column chromatography to produce pure isomer **356**. After Boc removal, the amine **357** was coupled with acid **358** to form intermediate **359**, which upon hydrogenolysis of the Cbz group, formed amine **360**. Amine **360** was treated with chloride **361** in the presence of potassium carbonate to furnish suvorexant (**344**) (Scheme 68) [237].

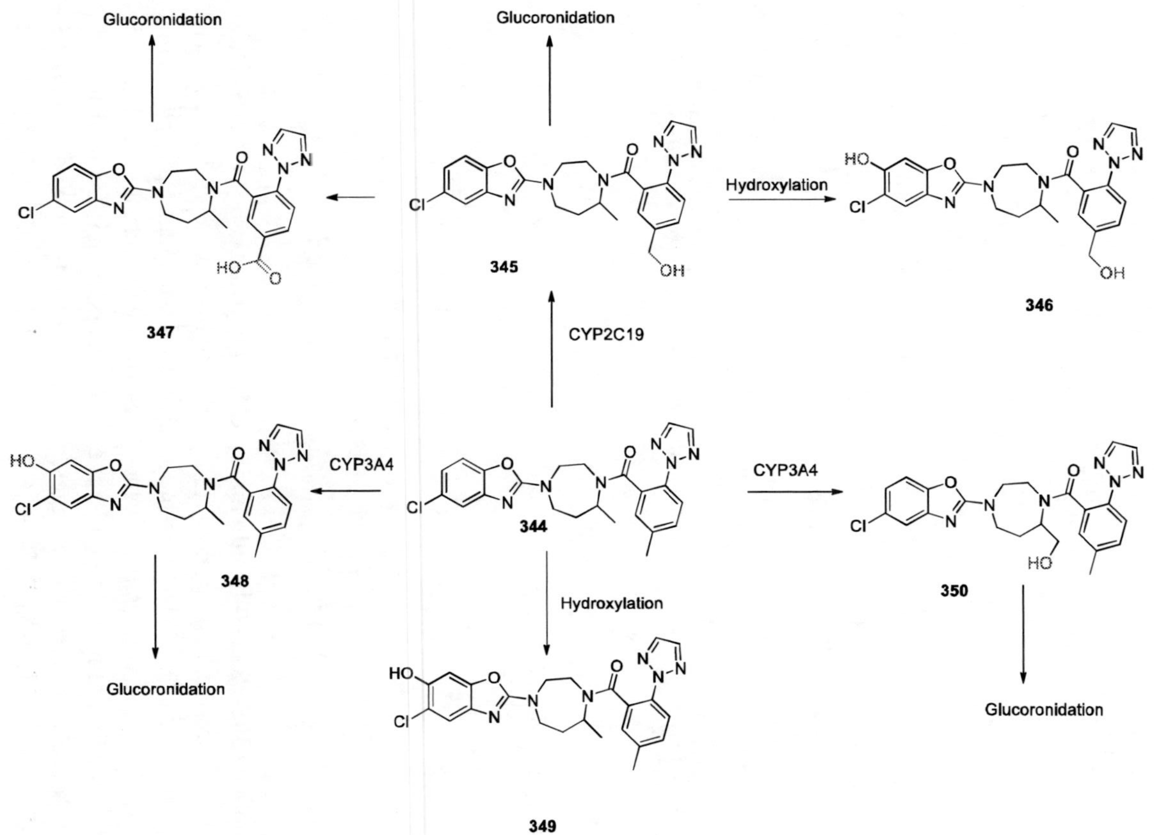

Scheme 67. Metabolic pathway of suvorexant (**344**).

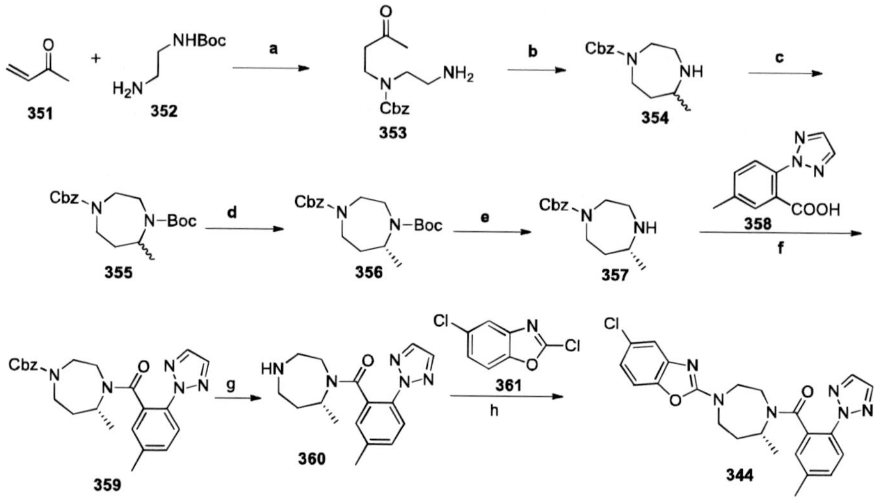

Reagents and conditions: a) Et_2O, Cbz-Cl, TEA; b) i) HCl, EtOAc, ii) $Na(OAc)_3BH$, HOAc, CH_2Cl_2; c) Boc_2O; d) seperation by chiral HPLC; e) HCl, EtOAc; f) EDC, HOAt, NMM, DMF, 87%; g) $Pd(OH)_2$, H_2, MeOH/EtOAc; h) K_2CO_3, DMF, 75%

Scheme 68. Synthesis of suvorexant (**344**).

Risdiplam

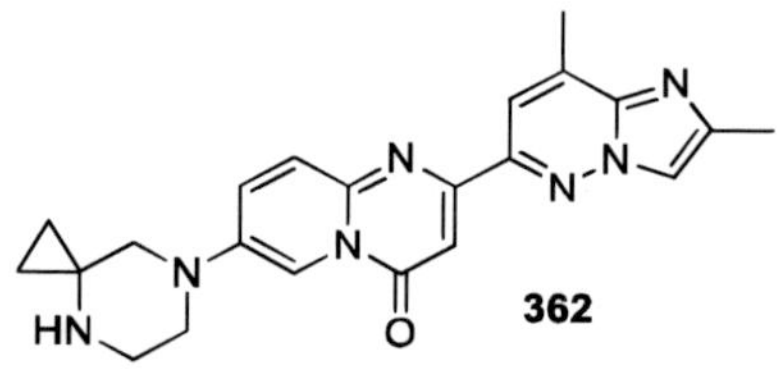

362

Risdiplam (**362**), 2-(2,8-dimethylimidazo[1,2-*b*]pyridazin-6-yl)-7-(4,7-diaza-spiro[2.5]octan-7-yl)-4H-pyrido[1,2-*a*]pyrimidin-4-one, is a FDA approved drug that can alter SMN2 pre-messenger-RNA splicing. Spinal muscular atrophy (SMA) is an unusual, advanced neuromuscular disease that causes by the deletion of the motor neuron 1 gene (SMN1), resulting in inadequate levels of SMN proteins. The study suggested that risdiplam can bind two sites in SMN2 pre-messenger-RNA (1) splicing enhancer 2 in exon 7 and (2) 5' splice site in intron 7. Risdiplam stimulates the addition of exon 7, enhancing the expression of SMN2 pre-messenger-RNA and SMN protein.

Moreover, Risdiplam was evaluated in two mouse SMA models, (1) C/C-allele mice (insignificant SMA with normal life span with reduced body weight and weakening of muscles) treated for 10 days, and (2) SMAΔ7 (a severe form of SMA) treated for 3-23 days postnatal than the oral dose. The data suggested that the SMN protein increased in both groups. Furthermore, treated mice show improvements in neuromuscular architecture and motor neurons in a dose-dependent manner. Additionally, Risdiplam distributes to the central nervous system (CNS) [238, 239].

Several clinical trials, such as placebo-controlled phase-1 [240], FIREFISH trial [241], SUNFISH trial [242], JEWELFISH trials [243], and RAINBOWFISH trial [244], have been done so far for the safety and evaluation of risdiplam. These clinical trials suggested that the risdiplam was effective against SMA. It is important to mention that risdiplam treatment does not show any retinal and renal toxicity in humans.

Risdiplam is metabolized into human liver microsomes to *N*-hydroxylated metabolite **363** as the primary metabolite (**Scheme 69**) [245].

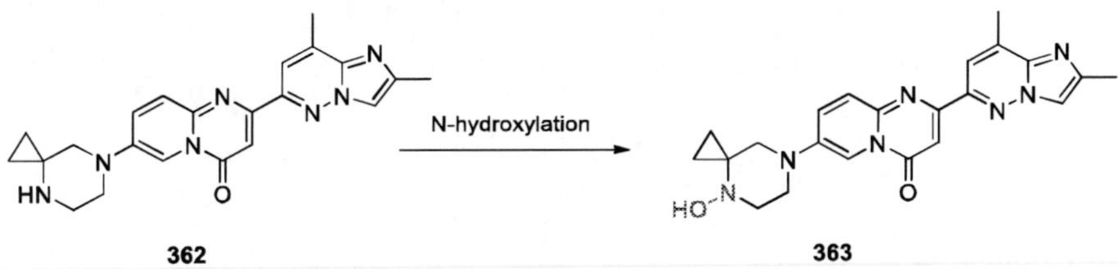

Scheme 69. Metabolic pathway of risdiplam (**362**).

Substituted pyridazine **364** undergoes nucleophilic substitution with aqueous ammonia, giving a mixture of regioisomers. A subsequent *N*-alkylation with 1-bromo-2,2-dimethoxypropane followed by an intramolecular cyclization gave the imidazopyridazine **365** (both regioisomers were separated), converted into boronic ester **366**. In another step, a condensation between the 5-fluoropyridin-2-amine **367** and dimethyl malonate, followed by chlorination with $POCl_3$, gives intermediate **368**. A Suzuki cross-coupling between boronic ester **366** and intermediate **368** forms coupled product **369,** which upon aromatic nucleophilic substitution with 4,7-diazaspiro[2,5]octane gave risdiplam (**362**) (Scheme 70) [245].

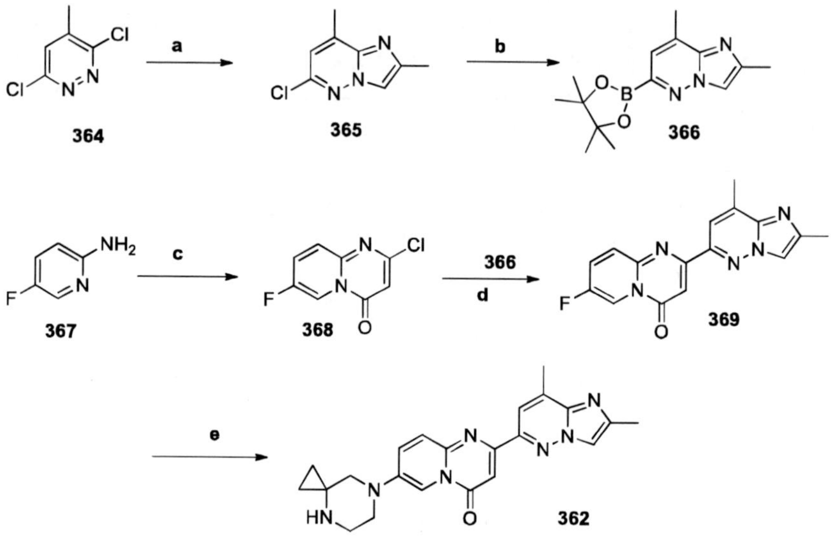

Reactions and Conditions: a) i) aqueous ammonia, 110 °C, 48 h; ii) 1-bromo-2,2-dimethoxypropane (1.2 equiv), PPTS (0.072 equiv), 2-propanol, 105 °C,12 h, 21% over 2 steps; b) 4,4,4',4',5,5,5',5'-octamethyl-2,2'-bi(1,3,2-dioxaborolane (1.0 equiv), KOAc (2.0 equiv), dioxane, 110 °C; c) (i) dimethyl malonate (5 equiv), 230 °C, 1.5 h; (ii) $POCl_3$, i-Pr_2NEt (1.0 equiv), 110 °C, 15 h, 50% over 2 steps; d) $Pd(PPh_3)_4$ (0.05 equiv), aqueous K_2CO_3 (2M, 3.0 equiv), CH_3CN, 100 °C, 6.0 h, 60%; e) 4,7-diazaspiro[2,5]octane dihydrochloride (3.0 equiv), i-Pr_2NEt (4.0 equiv), DMSO, 130 °C, 48 h, 18%.

Scheme 70. Synthesis of risdiplam (**362**).

Fostemsavir

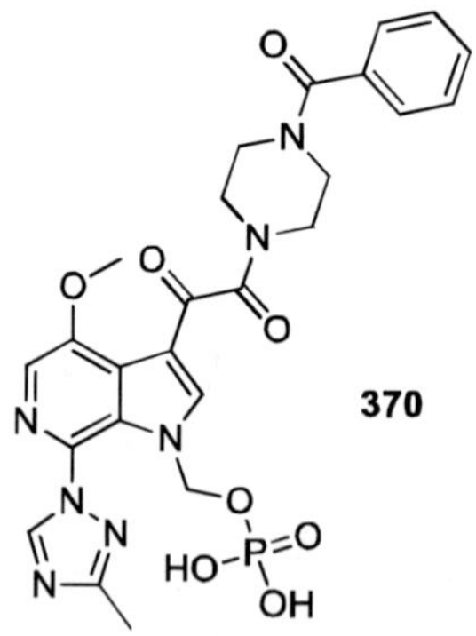

370

Fostemsavir (**370**), (3-(2-(4-benzoylpiperazin-1-yl)-2-oxoacetyl)-4-methoxy-7-(3-methyl-1*H*-1,2,4-triazol-1-yl)-1*H*-pyrrolo[2,3-*c*]pyridin-1-yl)methyl dihydrogen phosphate is a prodrug developed for HIV-1 patients. Fostemsavir binds directly to the glycoprotein 120 (gp120) adjacent to the binding site of the CD4+. The newer mode of action of fostemsavir binds gp120, which prohibits the conformational changes necessary for interacting between CD4 cells and virus surface receptors. Therefore, neither attachments nor subsequent entry was found to the immune and host T cells [246].

Numerous clinical trials have been so far attempted to ascertain the efficacy of the fostemsavir against HIV-1 patients, which shows a considerable decrease in the level of HIV-1 RNA [247], RNA load of HIV-1 is decreased around less than 50 copy/mL [248, 249, 250] These studies suggested that fostemsavir has effectively reduced the level of HIV-1 RNA and assured a safety profile without any significant effects on metabolic parameters. These properties of the fostemsavir might benefit a modified strategy for suppressing viral load even in the early treatment of HIV-1. More clinical data or studies are required to understand the potential use of Fostemsavir in HIV-1 patients.

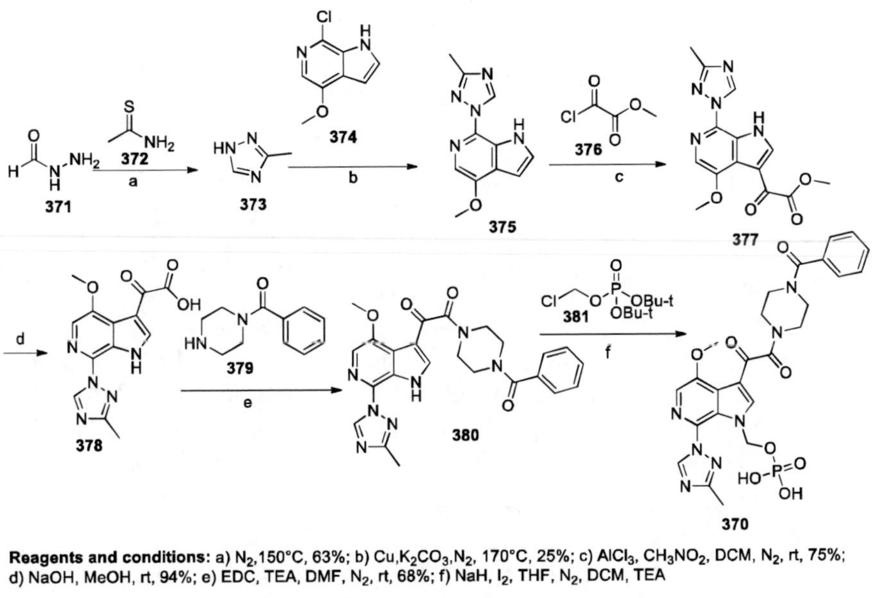

Reagents and conditions: a) N_2,150°C, 63%; b) Cu,K_2CO_3,N_2, 170°C, 25%; c) $AlCl_3$, CH_3NO_2, DCM, N_2, rt, 75%; d) NaOH, MeOH, rt, 94%; e) EDC, TEA, DMF, N_2, rt, 68%; f) NaH, I_2, THF, N_2, DCM, TEA

Scheme 71. Synthesis of fostemsavir (**370**).

The synthetic approach of fostemsavir involved a linear six-step synthetic route, as shown in Scheme 71. Cycloaddition reaction between formo-hydrazide **371** and ethane thioamide **372** provides triazole **373**. Copper-catalyzed C-N coupling reaction between triazole **373** and azaindole **374** generated the intermediate **375**. Friedel-Crafts reaction between compound **375** and methyl 2-chloro-2-oxoacetate **376** furnishes compound **377**, which upon hydrolysis of ester yield α-oxo acid **378**. Compound **378** reacted with phenyl (piperazin-1-yl)methanone **379** in the presence of EDC, producing amide **380** in 68% yield. Substitution reaction of **380** with **381**, followed by trifluoroacetic acid (TFA), promoted deprotection yield fostemsavir (**370**) [251].

Remdesivir

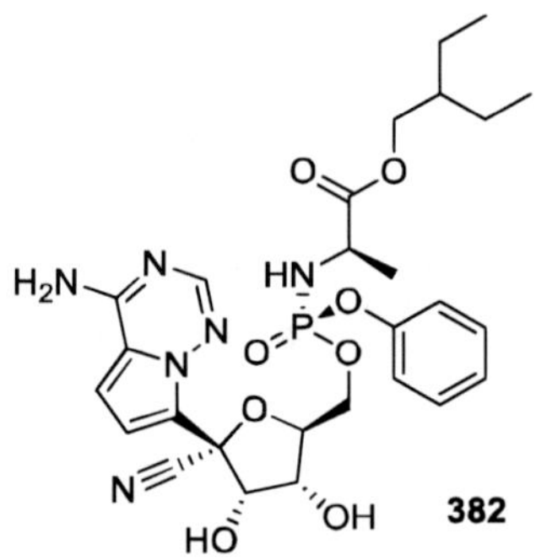

382

Gilead Sciences developed Remdesivir, 2-ethylbutyl ((*R*)-(((2*S*,3*R*,4*S*,5*S*)-5-(4-aminopyrrolo[2,1-*f*][1,2,4]triazin-7-yl)-5-cyano-3,4-dihydroxytetrahydro-furan-2-yl)methoxy) (phenoxy)phosphoryl)-*D*-alaninate. Modified nucleosides and prodrugs are highly permeable and metabolized to release phosphorylated nucleosides within the cells. The 1-CN altered adenosine C-nucleoside knockout with prodrugs (modified nucleosidase) to produce remdesivir that has high potential against the suppression of COVID-19 [252]. The prodrug remdesivir is metabolized into alanine metabolite inside a cell. The alanine metabolite further ensued into the monophosphate derivative, subsequently forming nucleoside triphosphate that inhibits RNA-dependent RNA-polymerase.

The Coronavirus 2019 (COVID-19) is a viral disease caused by severe acute respiratory syndrome coronavirus 2 (SARS-CoV-2) [253, 254, 255]. With the occurrence of the COVID-19 pandemic, few clinical trials have been done to ascertain the safety and efficacy of remdesivir. The clinical trials suggested that the remdesivir dose of 200 mg on the first day and later 100 mg up to 9 days effectively normalized the fever, oxygen level, cough, and respiratory rate. Overall, it suppresses the COVID-19 viral load [256, 257].

Moreover, adverse effects of remdesivir, like anemia, hypoglycemia, pyrexia, constipation, increased bilirubin, and hypoalbuminemia were observed. Additionally, severe adverse effects, mainly septic shock, kidney injury, and dysfunction of multiple organs, were reported [258].

Coupling of 2-ethylbutyl-*L*-alanine chloride **383** with $PO(OPh)Cl_2$ **384** in Et_3N and DCM yielded the phosphoramidoyl chloridate **385**, which was treated with *p*-nitrophenol **386** to yield enantiomers **387** which upon chiral column chromatography purification provide intermediate **388**. Coupling of **388** with the D-ribose-derived alcohol **389** in the presence of $MgCl_2$ and diisopropylethylamine forms (*S*)-P-phosphoamidate **390,** which underwent acidic cleavage of the acetal group to produce remdesivir (**382**) (Scheme 72) [259].

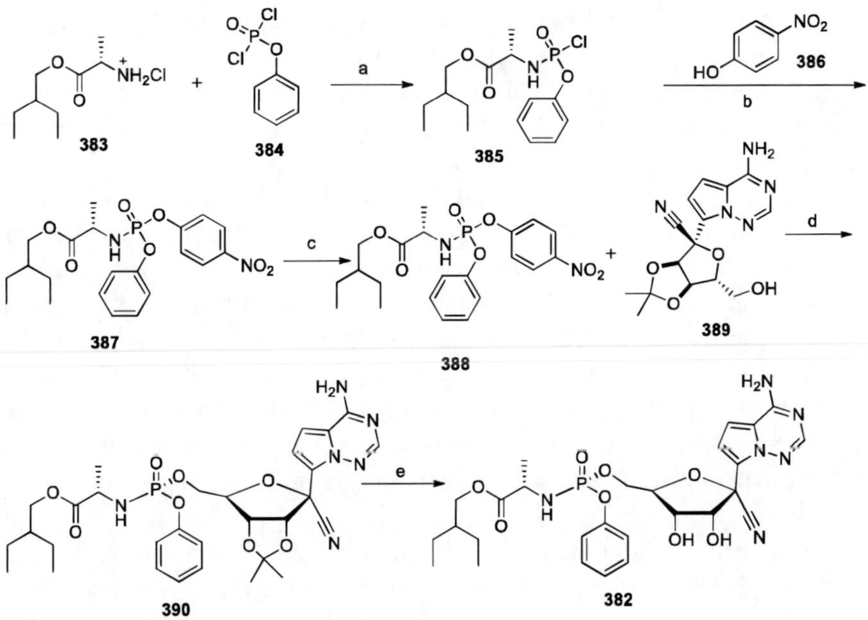

Reagents and conditions: a) Et_3N, CH_2Cl_2, -78°C; b) Et_3N, 0°C,80%; c) chiral seed for recrystallization, 39%; d) $MgCl_2$, (*i*- Pr)$_2$NEt, ACN, 50°C, 70%; e) 37% HCl, THF, RT, 69%

Scheme 72. Synthesis of remdesivir (**382**).

Ibrexafungerp (Brexafemmetm)

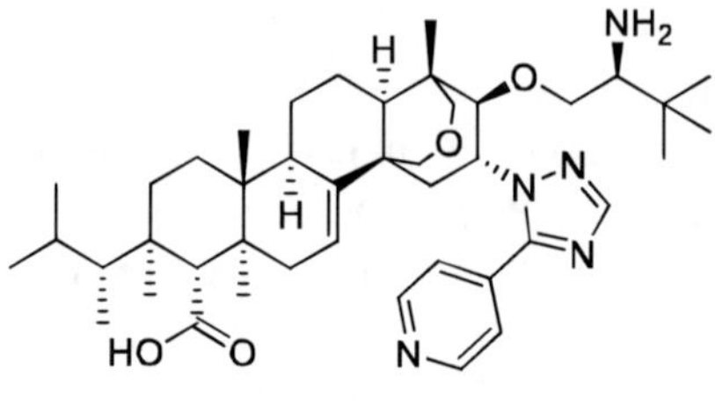

391

Ibrexafungerp, (1*S*,2*R*,3*R*,4a*R*,6a*S*,7*R*,8*R*,10a*R*,10b*R*,12a*R*)-2-((*S*)-2-amino-3,3-dimethylbutoxy)-1,6a,8,10a-tetramethyl-8-((*R*)-3-methylbutan-2-yl)-3-(5-(pyridin-4-yl)-1*H*-1,2,4-triazol-1-yl)-1,3,4,6,6a,7,8,9,10,10a,10b,11,12,12a-tetradecahydro-2*H*-1,4a-(methanooxymethano)chrysene-7-carboxylic acid is a triterpenoid, an antifungal drug mainly used for the treatment of vulvovaginal candidiasis. This drug can actively function against the most prevalent fungus pathogens, candida and aspergillus.

Additionally, there are several advantages, fast tissue penetration, low cytotoxicity, minimal risk of drug-to-drug interaction, high activity at low pH, etc. Ibrexafungerp mainly inhibits the biosynthesis of β-(1,3)-D-glucan, a chemical in most fungi cell walls. These drugs mainly reduce an enzyme, β-(1,3)-D-glucan synthase, found in most eukaryotes, and reduce their cytotoxic effect. Before ibrexafungerp, echinocandins were the broad-spectrum antifungal drug used to treat several fungal infections by inhibiting glucan synthase, but parenteral administration limits its applicability. The working mechanism of ibrexafungerp is like echinocandins. However, its active site is not precisely similar.

Both drugs inhibit the activity of the β-(1,3) D-glucan synthase enzyme complex. However, the binding sites of ibrexafungerp and echinocandin are different, thereby, have limited cross-resistance due to the partial overlapping. The resistance to echinocandins is mainly due to the FKS gene mutation (mainly two areas hot spots 1 and 2) that reduces the susceptibility to echinocandins. Ibrexafungerp has potent activity against FKS gene mutation. It can alter gene mutations in the FKS1 (*F625del*) and FKS2 genes (F659del) with two additional mutations at the different areas designated as W715L and A1390D, outside the hotspot two regions in the FKS2 gene, resulting in a high fold increase in the activity of ibrexafungerp [260].

Conclusion

Heterocyclic aromatic compounds containing two nitrogen atoms adjacent to each other, as in pyrazole, pyridazine, etc. containing drugs immensely contribute towards medications coverage, which navigates almost all therapeutic areas which require the attention of the medical fraternity [3]. More than 65 drugs from this category immensely contribute to the treatment options available for various sufferings of human beings. Pyrazole alone supplies more than 50 drugs from its category, widely applicable in different therapeutic segments, which connotes its significance in the drug design aspects.

In addition to the currently available medications from these classes, newer research analogs containing these heterocyclic features (such as pyrazole, pyridazine, and miscellaneous) are being tested and added uninterruptedly to the pool as potential drug candidates. This continuous addition to the existing list propels its marching towards a century of drugs from these heterocycles expeditiously. The popularity of drug-like features, easy, convenient, and established chemistry, and this review related to heterocyclic moieties derived from hydrazine would provide directions for the future drug design and development processes to enrich the remedial science for various human diseases and disorders, including terminal, rare/orphan, neglected and third world countries human ailments.

References

[1] Schmidt, A., and Dreger, A. (2011). "Recent advances in the chemistry of pyrazoles. Properties, biological activities, and syntheses." *Current Organic Chemistry*, 15, 9, 1423-1463.

[2] Karrouchi, K., Radi, S., Ramli, Y., Taoufik, J., Mabkhot, Y. N., Al-Aizari, F. A., and Ansar, M. (2018). "Synthesis and pharmacological activities of pyrazole derivatives: A review." *Molecules*, 23, 1, 134.

[3] Khan, M. F., Alam, M. M., Verma, G., Akhtar, W., Akhter, M., and Shaquiquzzaman, M. (2016). "The therapeutic voyage of pyrazole and its analogs: a review." *European Journal of Medicinal Chemistry*, 120, 170-201.

[4] Richter, G., Liao, V. W. Y., Ahring, P. K., and Chebib, M. (2020). "The Z-Drugs Zolpidem, Zaleplon, and Eszopiclone Have Varying Actions on Human GABAA Receptors Containing γ1, γ2, and γ3 Subunits." *Front. Neurosciences.*, 14, 599812.

[5] Padmanathan, T. (1998). "Process improvement in the synthesis of *N*-3-(3-cyano-pyrazolo[1,5-*a*]pyrimidin-7-yl) phenyl-*N*-ethylacetamide." U.S. Patent 5,714,607, issued February 3.

[6] Weitzel, K. W., Wickman, J. M., Augustin, S. G., and Strom, J. G. (2000). "Zaleplon: a pyrazolopyrimidine sedative-hypnotic agent for the treatment of insomnia." *Clinical Therapeutics*, 22(11), 1254–1267.

[7] Klimm, H. D., Dreyfus, J. F., and Delmotte, M. (1987). "Zopiclone Versus Nitrazepam: A Double-Blind Comparative Study of Efficacy and Tolerance in Elderly Patients with Chronic Insomnia." *Sleep*, 10(suppl_1), 73–78.

[8] Renwick, A. B., Mistry, H., Ball, S. E., Walters, D. G., Kao, J., and Lake, B. G. (1998). "Metabolism of Zaleplon by human hepatic microsomal cytochrome P450 isoforms." *Xenobiotica*, 28(4), 337–348.

[9] https://patents.google.com/patent/US5714607/enIt.

[10] Bharathi, K., Prabahar, K. J., Prasad, C. S., Kumar, M. S., Magesh, S., Handa, V. K., Dandala, R., and Naidu, A. (2007). "Impurity profile study of Zaleplon." *Journal of Pharmaceutical and Biomedical Analysis*, 44, 1, 101-109.

[11] Radl, S., Blahovcova, M., and Tkadlecova, M. (2010). "Synthetic studies connected with the preparation of *N*-[3-(3-cyanopyrazolo [1,5-*a*]pyrimidin-5-yl) phenyl]-*N*-ethylacetamide, a Zaleplon regioisomer." *Heterocycles*, 80, 2, 1359-1379.

[12] Baraldi, P. G., Fruttarolo, F., Tabrizi, M. A., Romagnoli, R., Preti, D., Ongini, E., El-Kashef, H., Carrión, M. D., and Borea, P. A. (2007). "Synthesis of a new series of pyrazolo [1, 5-a] pyrimidines structurally related to Zaleplon." *Journal of Heterocyclic Chemistry*, 44, 2, 355-361.

[13] Behbehani, H., Ibrahim, H. M., and Makhseed, S. (2010). "Synthesis of 7-substituted pyrazolo [1, 5-apyrimidine-3-carboxamides as potential non benzodiazepine hypnotics." *ARKIVOC*, 4, 267-282.

[14] Lankford, A. (2007) "Indiplon in the treatment of sleep disorders." *Neuropsychiatric Disease and Treatment*, 3, 6, 765-773.

[15] Burgos, C. F., Sanchéz, C., Sepúlveda, C., Fuentes, E., Palomo, I., and Alarcón, M. (2019) "Anti-aggregation effect on platelets of Indiplon a hypnotic sedative non-benzodiazepine drug." *Biomedicine & Pharmacotherapy*, 111, 378-385.

[16] Shimizu, M., Fukami, T., Ito, Y., Kurokawa, T., Kariya, M., Nakajima, M., and Yokoi, T. (2014). *Indiplon Is Hydrolyzed by Arylacetamide Deacetylase in Human Liver. Drug Metabolism and Disposition*, 42, 4, 751–758.

[17] Koilpillai, J. P., Kale, S. A., Kelkar, L. M., Zope, S. S., and Khan, M. A. (2012). "*Processes for the Preparation of Indiplon and Intermediates Thereof.*" U.S. Patent Application 13/260,247, filed February 2.

[18] Atack, J. R. (2005). "The benzodiazepine binding site of GABAA receptors as a target for the development of novel anxiolytics." *Expert Opinion on Investigational Drugs*, 14, 5, 601-618.

[19] Lippa, A., Czobor, P., Stark, J., Beer, B., Kostakis, E., Gravielle, M., Bandyopadhyay, S., Russek, S. J., Gibbs, T. T., Farb, D. H., and Skolnick, P. (2005). "Selective anxiolysis produced by Ocinaplon, a GABAA receptor modulator." *Proceedings of the National Academy of Sciences*, 102, 20, 7380-7385.

[20] Helton, D. Reed, and E. Pfadenhauer. (2019). "*Treatment of organophosphate exposure with ocinaplon.*" U.S. Patent 10,478,438, issued November 19.

[21] Basile A. S., Kostakis, E., Downing, S., Farb, D. H., Berezhnoy, D., Gravielle, M. C., Skolnick, P., and Gibbs, T. (2008). "Pharmacological Properties of DOV 315,090, an Ocinaplon Metabolite." 8, 11, 1-10.

[22] Baumann, M., and Baxendale, I. R. (2013). "An overview of the synthetic routes to the best selling drugs containing 6-membered heterocycles." *Beilstein Journal of Organic Chemistry*, 9, 1, 2265-2319.

[23] Baizman, E. R., Ezrin, A. M., Ferrari, R. A., and Luttinger, D. (1987). "Pharmacologic profile of Fezolamine fumarate: a nontricyclic antidepressant in animal models." *Journal of Pharmacology and Experimental Therapeutics*, 243, 1, 40-54.

[24] Zisook, S., Mendels, J., Janowsky, D., Feighner, J., John, L., and Fritz, A. (1987). "Efficacy and safety of Fezolamine in depressed patients." *Neuropsychobiology*, 17, 3, 133-138.

[25] McCoy, L., Skee, D., and Edelson, J. (1985). "Determination of fezolamine and its desmethyl metabolite in human plasma and urine by high-performance liquid chromatography: Intravenous pharmacokinetics in the beagle hound." *Journal of Chromatography B: Biomedical Sciences and Applications*, 344, 211-220.

[26] Bailey, D. M., Hansen, P. E., Hlavac, A. G., Baizman, E. R., Pearl, J., DeFelice, A. F., and Feigenson, M. E. (1985). "3,4-Diphenyl-1*H*-pyrazole-1-propanamine antidepressants." *Journal of Medicinal Chemistry*, 28, 2, 256-260.

[27] Katz, J. R. (1984) "Effects of Zometapine, a structurally novel antidepressant, in an animal model of depression." *Pharmacology Biochemistry and Behavior*, 21, 4, 487-490.

[28] Kinney, G. G., O'Brien, J. A., Lemaire, W., Burno, M., Bickel, D. J., Clements, M. K., Chen, T. B., Wisnoski, D. D., Lindsley, C. W., Tiller, P. R., Smith, S., Jacobson, M. A., Sur, C., Duggan, M. E., Pettibone, D. J., Conn, P. J., and Williams Jr, D. L. (2005). "A novel selective positive allosteric modulator of metabotropic glutamate receptor subtype 5 has *in vivo* activity and antipsychotic-like effects in rat behavioral models." *Journal of Pharmacology and Experimental Therapeutics*, 313, 1, 199-206.

[29] Uslaner, J. M., Parmentier-Batteur, S., Flick, R. B., Surles, N. O., Lam, J. S. H., McNaughton, C. H., Jacobson, M. A., and Hutson, P. H. (2009). "Dose-dependent effect of CDPPB, the mGluR5 positive allosteric modulator, on recognition memory is associated with GluR1 and CREB phosphorylation in the prefrontal cortex and hippocampus." *Neuropharmacology*, 57, 5-6, 531-538.

[30] Shallcross, J., Hámor, P., Bechard, A. R., Romano, M., Knackstedt, L., and Schwendt, M. (2019). "The divergent effects of CDPPB and cannabidiol on fear extinction and anxiety in a predator scent stress model of PTSD in rats." *Frontiers in Behavioral Neuroscience*, 13, Article 91.

[31] Bellozi, P. Q., Gomes, G. F., Machado da Silva, M. C., de Assis Lima, I. V., Álvares Batista, C. R., Carneiro Junior, W. O., Dória, J. G., *et al.* (2019). "A positive allosteric modulator of mGluR5 promotes neuroprotective effects in mouse models of Alzheimer's disease." *Neuropharmacology*, 160, 107785.

[32] Taciak, P. P., Lysenko, N., and Mazurek, A. P. (2018). "Drugs which influence serotonin transporter and serotonergic receptors: pharmacological and clinical properties in the treatment of depression." *Pharmacological Reports*, 70, 1, 37-46.

[33] Dotevall, G., and Groll, E. (1974). "Controlled clinical trial of Mepiprazole in irritable bowel syndrome." *Br Med J*, 4, 5935, 16-18.

[34] Gijsman, H. J., Cohen, A. F., and Van Gerven, J. M. A. (2004). "The application of the principles of clinical drug development to pharmacological challenge tests of the serotonergic system." *Journal of Psychopharmacology*, 18, 1, 7-13.

[35] Lawlor, B. A., Sunderland, T., Mellow, A. M., Hill, J. L., Newhouse, P. A., and Murphy, D. L. (1989). "A preliminary study of the effects of intravenous m-chlorophenylpiperazine, a serotonin agonist, in elderly subjects." *Biological Psychiatry*, 25, 6, 679-686.

[36] Gijsman, H. M., Gerven, J. V., Tieleman, M. C., Schoemaker, R. C., Pieters, M., Ferrari, M. D., Cohen, A. F., and Van Kempen, G. M. J. (1998). "Pharmacokinetic and pharmacodynamic profile of oral and intravenous meta-chlorophenylpiperazine in healthy volunteers." *Journal of Clinical Psychopharmacology*, 18, 4, 289-295.

[37] Hoy, S. M. (2019). "Elexacaftor/ivacaftor/tezacaftor: first approval." *Drugs*, 79, 18, 2001-2007.

[38] David, L. H. (2019). 'Patent Review of Synthetic Routes and Crystalline Forms of the Cftr-Modulator Drugs Ivacaftor, Lumacaftor, Tezacaftor, and Elexacaftor', *Organic Process Research & Development*, 23, 2302-22.

[39] Heijerman, H. G. M., McKone, E. F., Downey, D. G., Van Braeckel, E., Rowe, S. M., Tullis, E., Mall, M. A., Welter, J. J., Ramsey, B. W., McKee, C. M., Marigowda, G., Moskowitz, S. M., Waltz, D., Sosnay, P. R., Simard, C., Ahluwalia, N., Xuan, F., Zhang, Y., Taylor-Cousar, J. L., and McCoy, K. S. (2019). 'Efficacy and Safety of the Elexacaftor Plus Tezacaftor Plus Ivacaftor Combination Regimen in People with Cystic Fibrosis Homozygous for the F508del Mutation: A Double-Blind, Randomised, Phase 3 Trial', *Lancet*, 394, 1940-1948.

[40] Onofrio, L., Bartlett, C., Gunawardena, T. N. A., Ouyang, H., Eckford, P. D. W., Moraes, T. J., Bear, C. E., and Gonska, T. (2021). 'Rescue of Multiple Class Ii Cftr Mutations by Elexacaftor+Tezacaftor+Ivacaftor Mediated in Part by the Dual Activities of Elexacaftor as Both Corrector and Potentiator', *European Respiratory Journal*, 57, 2002774.

[41] Mall, G. M. P. M. A., Dřevínek, P., Lands, L. C., McKone, E. F., Polineni, D., Ramsey, B. W., Taylor-Cousar, J. L., Tullis, E., Vermeulen, F., Marigowda, G., McKee, C. M., Moskowitz, S. M., Nair, N., Savage, J., Simard, C., Tian, S., Waltz, D., Xuan, F., Rowe, S. M., and Jain, R. (2019) 'Elexacaftor–Tezacaftor–Ivacaftor for Cystic Fibrosis with a Single Phe508del Allele', *New England Journal of Medicine*, 381, 1809-1819.

[42] Matthias, G., Costa, S., Linnemann, R. W., Mall, M. A., McKone, E. F., Polineni, D., Quon, B. S., Ringshausen, F. C., Taylor-Cousar, L., Withers, N. J., Moskowitz, S. M., and Daines, C. L. (2020). 'Safety and Efficacy of Elexacaftor/Tezacaftor/ Ivacaftor for 24 Weeks or Longer in People with Cystic Fibrosis and One or More F508del Alleles: Interim Results of an Open-Label Phase 3 Clinical Trial', *American Journal of Respiratory and Critical Care Medicine*, 203, 381-385.

[43] Katz, R. J. (1984). "Effects of Zometapine, a structurally novel antidepressant, in an animal model of depression." *Pharmacology Biochemistry and Behavior*, 21, 4, 487-490.

[44] Kawashima, Y., Lin, S. Y., Ogawa, M., Handa, T., and Takenaka, H. (1985). "Preparations of agglomerated crystals of polymorphic mixtures and a new complex of indomethacin—Epirizole by the spherical crystallization technique." *Journal of Pharmaceutical Sciences*, 74, 11, 1152-1156.

[45] Onoa, G. B., Moreno, V., Font-Bardia, M., Solans, X., Pérez, J. M., and Alonso, C. (1999). "Structural and cytotoxic study of new Pt (II) and Pd (II) complexes with the bi-heterocyclic ligand mEpirizole." *Journal of Inorganic Biochemistry*, 75, 3, 205-212.

[46] Lees, P., and Toutain, P. (2013). "Pharmacokinetics, pharmacodynamics, metabolism, toxicology and residues of Phenylbutazone in humans and horses." *The Veterinary Journal*, 196, 3, 294-303.

[47] Worboys, M., and Toon, E. (2018). "Phenylbutazone (Bute, PBZ, EPZ): One drug across two species." *History and Philosophy of the Life Sciences*, 40, 2, 1-24.

[48] Dodman, N., Blondeau, N., and Marini, A. M. (2010). "Association of Phenylbutazone usage with horses bought for slaughter: A public health risk." *Food and Chemical Toxicology*, 48, 5, 1270-1274.

[49] Borges, R. S., Palheta, I. C., Ota, S. S. B., Morais, R. B., Barros, V. A., Ramos, R. S., Silva, R. C., Costa, J., Silva, C., Campos, J. M., and Santos, C. B. R. (2019). "Toward of safer Phenylbutazone derivatives by exploration of toxicity mechanism." *Molecules*, 24, 1, 143.

[50] Mathews, K. A. (1996). "Nonsteroidal anti-inflammatory analgesics in pain management in dogs and cats." *The Canadian Veterinary Journal*, 37, 9, 539–545.

[51] Singh, A., Hakk, H., and Lupton, S. J. (2018). "Facile synthesis of high specific activity 4-[1-14 C] butyl-1, 2-diphenylpyrazolidine-3, 5-dione (phenylbutazone) using nucleophilic substitution." *Journal of Labelled Compounds and Radiopharmaceuticals*, 61, 4, 386-390.

[52] Raulf, M., and König, W. (1990) "Effects of the nonsteroidal anti-inflammatory compounds Lonazolac Ca, indomethacin and NDGA on inflammatory mediator generation and release from various cells." *Immunopharmacology*, 19, 2, 103-111.

[53] Vinge, E., and Björkman, S. (1986). "Compared effects of the two antiinflammatory drugs indoprofen and Lonazolac on thromboxane generation and on platelet aggregation, related to plasma concentrations after single oral doses." *Acta Pharmacologica et Toxicologica*, 59, 3, 165-174.

[54] Matsuzaki, H., Takeda, N., Yasui, M., Ito, Y., Konishi, K., and Ueda, M. (2020). "Synthesis of Pyrazoles Utilizing the Ambiphilic Reactivity of Hydrazones." *Organic Letters*, 22, 23, 9249-9252.

[55] Ismail, A. H. M., Lehmann, J., El Ella, D.l A. A., Albohy, A., and Abouzid, K. A. M. (2009). "Lonazolac analogues: molecular modeling, synthesis, and *in vivo* anti-inflammatory activity." *Medicinal Chemistry Research*, 18, 9, 725-744.

[56] Harras, M. F., Sabour, R., and Alkamali, O. M. (2019). "Discovery of new non-acidic Lonazolac analogues with COX-2 selectivity as potent anti-inflammatory agents." *MedChemComm*, 10, 10, 1775-1788.

[57] Abdellatif, R. A. K., Fadaly, W. A. A., Elshaier, Y. A. M. M., Ali, W. A. M., and Gehan M. Kamel. (2018). "Non-acidic 1, 3, 4-trisubstituted-pyrazole derivatives as Lonazolac analogs with promising COX-2 selectivity, anti-inflammatory activity and gastric safety profile." *Bioorganic Chemistry*, 77, 568-578.

[58] Tsutomu, K., and Toshitaka, N. (1978). "Effects of 1, 3-diphenyl-5-(2-dimethylaminopropionamide)-pyrazole [Difenamizole] on a conditioned avoidance response." *Neuropharmacology*, 17, 4-5, 249-256.

[59] Nabeshima, T., Yamaguchi, K., and Kameyama, T. (1978). "Effects of Difenamizole on content of catecholamines and metabolites in mouse brain." *The Japanese Journal of Pharmacology*, 28, 4, 642-646.

[60] Secci, D., Bolasco, A., Chimenti, P., and Carradori, S. (2011). "The state of the art of pyrazole derivatives as monoamine oxidase inhibitors and antidepressant/ anticonvulsant agents." *Current Medicinal Chemistry*, 18, 33, 5114-5144.

[61] Brune, K. (1997). "The early history of non-opioid analgesics." *Acute Pain*, 1, 1, 33-40.

[62] Elattar K. M. and Fadda, A. A. (2016). "Chemistry of Antipyrine." *Synthetic Communications*, 46, 19, 1567-1594.

[63] Iguchi, S., Goromaru, T., and Noda, A. (1975). "A new metabolite of aminopyrine (aminophenazone) in man, 4-formylaminoantipyrine." *Chemical and Pharmaceutical Bulletin*, 23, 4, 932-934.

[64] Kampffmeyer, H. G. (1976). "On aminophenazone metabolism in the isolated perfused rat liver." *European Journal of Drug Metabolism and Pharmacokinetics*, 1, 4, 182-187.

[65] Zhao, Z., Dai, X., Li, C., Wang, X., Tian, J., Feng, Y., Xie Cong Ma, J., Nie,, Peinan Fan, M. Qian, X.He, Shaoping Wu, Y. Zhang, and Zheng, X. (2020). "Pyrazolone structural motif in medicinal chemistry: Retrospect and prospect." *European Journal of Medicinal Chemistry*, 186, 111893.

[66] Knorr, L. (1884). "Synthese von Chinolinderivaten [*Synthesis of Quinoline Derivatives*]." *Berichte Der Deutschen Chemischen Gesellschaft*, 17, 1, 540-546.

[67] Krohn, K., and Stenns, C. (1989). "Chemical Synthesis of Phase-I-and Phase-II-Metabolites of Antipyrine." *Archiv der Pharmazie*, 322, 6, 351-354.

[68] Brogden, R. N. (1986). "Pyrazolone derivatives." *Drugs*, 32, 4, 60-70.

[69] Gulmez, S. E., Tulunay, F. C., Beder, S., Kayacan, O., and Karnak, D. (2006). "Does Dipyrone have any effect on respiratory function in COPD patients?." *Respiratory Medicine*, 100, 5, 828-834.

[70] Badian, M., Le Normand, Y., Rupp, W., and Zapf, R. (1984). "There is no interaction between Dipyrone (metamizol) and the anticoagulants, phenprocoumon and ethylbiscoumacetate, in normal Caucasian subjects." *International Journal Of Pharmaceutics*, 18, 1-2, 9-15.

[71] Levy, M., Zylber-Katz, E., and Rosenkranz, B. (1995). "Clinical pharmacokinetics of dipyrone and its metabolites." *Clinical Pharmacokinetics*, 28, 3, 216-234.

[72] Lutz, M. (2019). "Metamizole (dipyrone) and the liver: a review of the literature." *The Journal of Clinical Pharmacology*, 59, 11, 1433-1442.

[73] http://office.oc.chemie.tu-darmstadt.de/DaMocles/files/molecule083.pdf. Accessed on 13 Sept. 2022.

[74] Garcia, S., Canionero, M., and Lopes, G. (2006). "Dipyrone-induced granulocytopenia: a case for awareness." *Pharmacotherapy: The Journal of Human Pharmacology and Drug Therapy*, 26, 3, 440-442.

[75] Bar-Oz, B., Bulkowstein, M., Benyamini, L., Greenberg, R., Soriano, I., Zimmerman, D., Bortnik, O., and Berkovitch, M. (2003). "Use of antibiotic and analgesic drugs during lactation." *Drug Safety*, 26, 13, 925-935.

[76] Abe, M., Kumagi, T., Nakanishi, S., Yamagami, T., Michitaka, K., Abe, K., Okura, I., Yamashita, H., Horiike, N., and Onji, M. (2002). "Drug-induced hepatitis with hepatic granuloma due to saridon." *Journal of Gastroenterology*, 37, 12, 1068-1072.

[77] Allison, M. C. (1992). "Gastrointestinal Damage Associated with Nonsteroidal Anti-inflammatory Drugs-Reply." *New England Journal of Medicine*, 327, 26, 1883-1883.

[78] Volz, M. A. N. F. R. E. D., and HANS-MARTIN Kellner. (1980). "Kinetics and metabolism of pyrazolones (propyphenazone, aminopyrine and dipyrone)." *British Journal of Clinical Pharmacology*, 10, S2, 299S-308S.

[79] Tam, S. S. C., Lee, D. H. S., Wang, E. Y., Munroe, D. G., and Lau, C. Y. (1995). "Tepoxalin, a Novel Dual Inhibitor of the Prostaglandin-H Synthase Cyclooxygenase and Peroxidase Activities." *Journal of Biological Chemistry*, 270, 23, 13948-13955.

[80] McQuerry, J. A., Chen, J., Chang, J. T., and Bild, A. H. (2021). "Tepoxalin increases chemotherapy efficacy in drug-resistant breast cancer cells overexpressing the multidrug transporter gene ABCB1." *Translational oncology*, 14, 10, 101181.

[81] Zhou, L., Pope, B. L., Chourmouzis, E., Fung-Leung, W., and Lau, C. Y. (1996). "Tepoxalin blocks neutrophil migration into cutaneous inflammatory sites by inhibiting Mac-1 and E-selectin expression." *European Journal Of Immunology*, 26, 1, 120-129.

[82] De Boever, S., Neirinckx, E., Baert, K., De Backer, P., and Croubels, S. (2009). "Pharmacokinetics of Tepoxalin and its active metabolite in broiler chickens." *Journal of Veterinary Pharmacology and Therapeutics*, 32, 1, 97-100.

[83] Murray, W. V., and Hadden, S. K. (1992). "A facile synthesis of tepoxalin, 5-(4-chlorophenyl)-*N*-hydroxy-1-(4-methoxyphenyl)-*N*-methyl-1*H*-pyrazole-3-propanamide." *The Journal of Organic Chemistry*, 57, 24, 6662-6663.

[84] Smith, C. J., "Zhang, Y., K. CM, Muhammad, J., Zweifel, B. S., Shaffer, A., Talley, J. J., Masferrer, J. L., Seibert, K., and Isakson, P. C." (1998). *Pharmacological Analysis of Cyclooxygenase-1 in Inflammation. Proc Natl Acad Sci USA*, 95, 13, 313-313.

[85] Lee, E. M., Hong, S. H., Lee, Y. J., Kang, Y. H., Choi, K. C., Choi, S. H., Kim, I. H., and Lim, S. J. (2004). "Liposome-complexed adenoviral gene transfer in cancer cells expressing various levels of coxsackievirus and adenovirus receptor." *Journal of Cancer Research And Clinical Oncology*, 130, 3, 169-177.

[86] Wang, Y., Han, J., Chen, J., and Cao, W. (2015). "An efficient route to 3-trifluoromethylpyrazole via cyclization/1, 5-H shift and its applications in the synthesis of bioactive compounds." *Tetrahedron*, 71, 43, 8256-8262.

[87] Smith, C. J., "Zhang, Y., Koboldt, C. M., Muhammad, J., Zweifel, B. S., Shaffer, A., Talley, J. J., Masferrer, J. L., Seibert, K., and Isakson, P. C. (1998).

Pharmacological Analysis of Cyclooxygenase-1 in Inflammation. Proc Natl Acad Sci USA, 95, 13 313-13.

[88] Clemett, K. L. (2000). Goa. "Celecoxib." *Drugs*, 59, 4, 957-980.

[89] Giorgi, M., Mengozzi, G., Raffaelli, A., and Saba, A. (2011). "Characterization of *in vivo* plasma metabolites of tepoxalin in horses using LC–MS–MS." *Journal of Pharmaceutical and Biomedical Analysis*, 56, 1, 45-53.

[90] Vardanyan, R., and Hruby, V. (2016). *Synthesis of Best-Seller Drugs*. Academic Press.

[91] Jonkers, F. C., Guiot, Y., Rahier, J., and Henquin, J. (2001). "Tolbutamide stimulation of pancreatic β-cells involves both cell recruitment and increase in the individual Ca^{2+} response." *British Journal of Pharmacology*, 133, 4, 575-585.

[92] Monteiro-Steagall, B. P., Steagall, P. V. M., and Lascelles, B. D. X. (2013). "Systematic review of nonsteroidal anti-inflammatory drug-induced adverse effects in dogs." *Journal of Veterinary Internal Medicine*, 27, 5, 1011-1019.

[93] Davis, J. L., Marshall, J. F., Papich, M. G., Blikslager, A. T., and Campbell, N. B. (2011). "The pharmacokinetics and *in vitro* cyclooxygenase selectivity of Deracoxib in horses." *Journal of Veterinary Pharmacology And Therapeutics*, 34, 1, 12-16.

[94] Gong, L., Thorn, C. F., Bertagnolli, M. M., Grosser, T., Altman, R. B., and Klein, T. E. (2012). "Celecoxib pathways: pharmacokinetics and pharmacodynamics." *Pharmacogenetics and Genomics*, 22, 4, 310-318.

[95] Wan, C., Pang, J., Jiang, W., Zhang, X., and Hu, X. (2021). "Copper-Catalyzed Reductive Ring-Cleavage of Isoxazoles: Synthesis of Fluoroalkylated Enaminones and Application for the Preparation of Celecoxib, Deracoxib, and Mavacoxib." *The Journal of Organic Chemistry*, 86, 6, 4557-4566.

[96] Choi, Y. L., Soda, M., Yamashita, Y., Ueno, T., Takashima, J., Nakajima, T., Yatabe, Y., Takeuchi, K., Hamada, T., Haruta, H., Ishikawa, Y., Kimura, T., Mitsudomi, H., Yoshiro Tanio, and Mano, H. (2010). "EML4-ALK mutations in lung cancer that confer resistance to ALK inhibitors." *New England Journal of Medicine*, 363, 18, 1734-1739.

[97] Sun, H. Y., and Ji, F. (2012). "A molecular dynamics investigation on the Crizotinib resistance mechanism of C1156Y mutation in ALK." *Biochemical and biophysical research communications*, 423, 2, 319-324.

[98] Johnson, T. R., Tan, W., Goulet, L., Smith, E. B., Yamazaki, S., Walker, G. S., O'Gorman, M. T., *et al.* (2015). "Metabolism, excretion and pharmacokinetics of [14C] crizotinib following oral administration to healthy subjects." *Xenobiotica*, 45, 1, 45-59.

[99] Qian, J., Yan, P., Che, D., Zhou, Q., and Li, Y. (2014). "A novel approach for the synthesis of Crizotinib through the key chiral alcohol intermediate by asymmetric hydrogenation using highly active Ir-Spiro-PAP catalyst." *Tetrahedron Letters*, 55, 9, 1528-1531.

[100] Frankel, P., Ruel, C., Uche, A., Choy, E., Okuno, S., Somiah, N., and Chow, W. A. (2022). "Pazopanib in Patients with Osteosarcoma Metastatic to the Lung: Phase 2 Study Results and the Lessons for Tumor Measurement." *Journal of Oncology*, 2022, 3691025.

[101] Karschnia, P., Barbiero, F. J., Schwaiblmair, M. H., Kaulen, L. D., Piepmeier, J. M., Huttner, A. J., Becker, K. P., Fulbright, R. K., and Baehring, J. M. (2020). "Leptomeningeal dissemination of low-grade neuroepithelial CNS tumors in adults: a 15-year experience." *Neuro-oncology Practice*, 7, 1, 118-126.

[102] Parambil, J. G., Gossage, J. R., McCrae, K. R., Woodard, T. D., Menon, K. V., Timmerman, K. L., Pederson, D. P., Sprecher, D. L., and Al-Samkari, H. (2022). "Pazopanib for severe bleeding and transfusion-dependent anemia in hereditary hemorrhagic telangiectasia." *Angiogenesis*, 25, 1, 87-97.

[103] Paludetto, M., Puisset, F., Le Louedec, F., Allal, B., Lafont, T., Chatelut, E., and Arella, C. (2018). "Simultaneous monitoring of pazopanib and its metabolites by UPLC–MS/MS." *Journal of Pharmaceutical and Biomedical Analysis*, 154, 373-383.

[104] Masui, K., Hama, Y., Sawada, A., Akamatsu, S., Sakaue, J., Okabe, M., and Kobayashi, T. (2022). "A rare case of intestinal lymphangiectasia induced by pazopanib." In *International Cancer Conference Journal*, vol. 11, 1, 87-90.

[105] Xue, X., Zhang, Y., Liao, Y., Sun, D., Li, L., Liu, Y., Wang, Y., Jiang, W., Zhang, J., Luan, Y., and Zhao, X. (2022). s"Design, synthesis and biological evaluation of dual HDAC and VEGFR inhibitors as multitargeted anticancer agents." *Investigational New Drugs*, 40, 1, 10-20.

[106] Frampton, J. E. (2021). "Entrectinib: a review in NTRK+ solid tumours and ROS1+ NSCLC." *Drugs*, 81, 6, 697-708.

[107] Jiang, Q., Li, M., Li, H., and Chen, L. (2022). "Entrectinib, a new multi-target inhibitor for cancer therapy." *Biomedicine & Pharmacotherapy*, 150, 112974.

[108] Meneses-Lorente, G., Bentley, D., Guerini, E., Kowalski, K., how-Maneval, E., Yu, L., Brink, A., Djebli, N., Mercier, F., Buchheit, V., and Phipps, A. (2021). "Characterization of the pharmacokinetics of entrectinib and its active M5 metabolite in healthy volunteers and patients with solid tumors." *Investigational New Drugs*, 39, 3, 803-811.

[109] Liang, X., Yang, Q., Wu, P., He, C., Yin, L., Xu, F., Yin, Z., Yue, G., Zou, Y., Li, L., Song, X., Lv, C., Zhang, W., and Jing, B. (2021). "The synthesis review of the approved tyrosine kinase inhibitors for anticancer therapy in 2015–2020." *Bioorganic Chemistry*, 113, 105011.

[110] Heinrich, M. C., Jones, R. L., Mehren, M. v., Schöffski, P., Serrano, C., Kang, Y., Cassier, P. A., Mir, O., Eskens, F., Tap, W. D., Rutkowski, P., Chawla, S. P., Trent, J., Tugnait, M., Evans, E. K., Lauz, T., Zhou, T., Roche, M., Wolf, B. B., Bauer, S., and George, S. (2020). "Avapritinib in advanced PDGFRA D842V-mutant gastrointestinal stromal tumour (NAVIGATOR): a multicentre, open-label, phase 1 trial." *The Lancet Oncology*, 21, 7, 935-946.

[111] Wu, C., Lusvarghi, S., Wang, J., Hsiao, S., Huang, Y., Hung, T., and Ambudkar, S. V. (2019). "Avapritinib: a selective inhibitor of KIT and PDGFRα that reverses ABCB1 and ABCG2-mediated multidrug resistance in cancer cell lines." *Molecular Pharmaceutics*, 16, 7, 3040-3052.

[112] https://www.fda.gov/drugs/resources-information-approved-drugs/fda-approves-avapritinib-gastrointestinal-stromal-tumor-rare-mutation. Updated on 01/09/2020, visited on September 04, 2022.

[113] https://patents.google.com/patent/CN110938077A/en. Accessed on 14 Sept. 2022.

[114] Blair, L. M., and Sperry, J. (2013). "Natural products containing a nitrogen–nitrogen bond." *Journal of Natural Products*, 76, 4, 794-812.

[115] Elgemeie, G. H., Zaghary, W. A., Amin, K. M., and Nasr, T. M. (2005). "New trends in synthesis of pyrazole nucleosides as new antimetabolites." *Nucleosides, Nucleotides, and Nucleic Acids*, 24, 8, 1227-1247.

[116] De Bernardo, S., and Weigele, M. (1976). "Synthesis of the pyrazomycins." *The Journal of Organic Chemistry*, 41, 2, 287-290.

[117] Yuan, S., Luo, Y., Zuo, J., Liu, H., Lin, F., and Yu, B. (2021). "New drug approvals for 2020: Synthesis and clinical applications." *European Journal of Medicinal Chemistry*, 215, 113284.

[118] Shaw, A. T., Bauer, T. M., Marinis, F., Felip, E., Goto, Y., Liu, G., Mazieres, J., Kim, D., Mok, T., Polli, A., Thurm, H., Calella, A. M., Peltz, G., and Solomon, B. J. (2020). 'First-Line Lorlatinib or Crizotinib in Advanced Alk-Positive Lung Cancer', *New England Journal of Medicine*, 383, 2018-2029.

[119] Yun, K. M., and Bazhenova, L. A. (2022). 'Update on Lorlatinib: Role in Reducing the Risk of Disease Progression in Alk-Positive Nsclc', *Cancer Manag Res*, 14, 843-850.

[120] https://en.wikipedia.org/wiki/Lorlatinib#cite_note-EMA-public-8. Accessed on 16 Sept. 2022.

[121] Drilon, A., Tan, S. W., Lassen, U. N., Leyvraz, S., Liu, Y., Patel, J. D., Rosen, L., Solomon, B., Norenberg, R., Dima, L., Brega, N., Shen, L., Moreno, V., Kummar, S., and Lin, J. J. (2022). 'Efficacy and Safety of Larotrectinib in Patients with Tropomyosin Receptor Kinase Fusion–Positive Lung Cancers', *JCO Precision Oncology*, e2100418.

[122] Filippi, R., and Depetris, I. (2021). '*Evaluating Larotrectinib for the Treatment of Advanced Solid Tumors Harboring an Ntrk Gene Fusion*', 22, 677-684.

[123] Hong, S., DuBois, S. G., Kummar, S., Farago, A. F., Albert, C. M., Rohrberg, K. S., van Tilburg, C. M., Nagasubramanian, R., Berlin, J. D., Federman, N., Mascarenhas, L., Geoerger, B., Dowlati, A., Pappo, A. S., Bielack, S., Doz, F., McDermott, R., Patel, J. D., Schilder, R. J., Tahara, M., Pfister, S. M., Witt, O., Ladanyi, M., Rudzinski, E. R., Nanda, S., Childs, B. H., Laetsch, T. W., Hyman, D. M., and Drilon, A. (2020). 'Larotrectinib in Patients with Trk Fusion-Positive Solid Tumours: A Pooled Analysis of Three Phase 1/2 Clinical Trials', *Lancet Oncol*, 21, 531-540.

[124] Lunning, M., Vose, J., Nastoupil, L., Fowler, N., Burger, J. A., Wierda, W. G., Schreeder, M. T., Siddiqi, T., Flowers, C. R., Cohen, J. B., Sportelli, P., Miskin, H. P., Weiss, M. S., and O'Brien, S. (2019). 'Ublituximab and Umbralisib in Relapsed/Refractory B-Cell Non-Hodgkin Lymphoma and Chronic Lymphocytic Leukemia', *Blood*, 134, 1811-1820.

[125] Fowler, N. H., Samaniego, F., Jurczak, W., Ghosh, N., Derenzini, E., Reeves, J. A., Knopińska-Posłuszny, W., Cheah, C. Y., Phillips, T., Lech-Maranda, E., Cheson, B. D., Caimi, P. F., Grosicki, S., Leslie, L. A., Chavez, J. C., Fonseca, G., Babu, S., Hodson, D. J., Shao, S. H., Burke, J. M., Sharman, J. P., Law, J. Y., Pagel, J. M., Miskin, H. P., Sportelli, P., O'Connor, O. A., Weiss, M. S., and Zinzani, P. L.

(2021). 'Umbralisib, a Dual Pi3kδ/Ck1ε Inhibitor in Patients with Relapsed or Refractory Indolent Lymphoma', *Journal of Clinical Oncology*, 39, 1609-1618.

[126] Aguilera A., Cecilia, C., Valero, T., Lorente-Macías, Á., Baillache, D. J., Croke, S., and Unciti-Broceta, A. (2021). "Small molecule kinase inhibitor drugs (1995–2021): medical indication, pharmacology, and synthesis." *Journal of Medicinal Chemistry*, 65, 2, 1047-1131.

[127] Hughes, T. P., Mauro, M. J., Cortes, J. E., Minami, H., Rea, D., DeAngelo, D. J., Breccia, M., Goh, Y. T., Talpaz, M., Hochhaus, A., le Coutre, P., Ottmann, O., Heinrich, M. C., Steegmann, J. L., Deininger, M. W. N., Jjwm Janssen, Mahon, F. X., Minami, Y., Yeung, D., Ross, D. M., Tallman, M. S., Park, J. H., Druker, B. J., Hynds, D., Duan, Y., Meille, C., Hourcade-Potelleret, F., Vanasse, K. G., Lang, F., and Kim, D. W. (2019). 'Asciminib in Chronic Myeloid Leukemia after Abl Kinase Inhibitor Failure', *N Engl J Med*, 381, 2315-2326.

[128] Schoepfer, J., Jahnke, W., Berellini, G., Buonamici, S., Cotesta, S., Cowan-Jacob, S. W., Dodd, S., Drueckes, P., Fabbro, D., Gabriel, T., Groell, J., Grotzfeld, R. M., Hassan, A. Q., Henry, C., Iyer, V., Jones, D., Lombardo, F., Loo, A., Manley, P. W., Pellé, X., Rummel, G., Salem, B., Warmuth, M., Wylie, A. A., Zoller, T., Marzinzik, A. L., and Furet, P. (2018). apixaban', *Journal of Medicinal Chemistry*, 61, 8120-8135.

[129] Tran, P., Hanna, I., Eggimann, F. K., Schoepfer, J., Ray, T., Zhu, B., and Wang, L. (2020). "Disposition of asciminib, a potent BCR-ABL1 tyrosine kinase inhibitor, in healthy male subjects." *Xenobiotica*, 50, 2, 160-179.

[130] Tiz, D. B., Bagnoli L., Rosati, O., Marini, F., Sancineto, L., and Santi, C. (2022) "New Halogen-Containing Drugs Approved by FDA in 2021: An Overview on Their Syntheses and Pharmaceutical Use." *Molecules*, 27, 5, 1643.

[131] Blair, H. A. (2021). "Belumosudil: first approval." *Drugs*, 81, 14, 1677-1682

[132] Przepiorka, D., Le, R. Q., Ionan, A., Li, R., Wang, Y., Gudi, R., Mitra, S., Vallejo, J., Okusanya, O. O., Ma, L., Yang, Y., Paresma Patel, Djelila Mezaache, Rakhi Shah, Anamitro Banerjee, Sherita McLamore, Adam N. Maung, Kirsten B. Goldberg, Richard Pazdur, Marc R. Theoret, Angelo De Claro, R. (2022) "FDA Approval Summary: Belumosudil for Adult and Pediatric Patients 12 Years and Older with Chronic GvHD after Two or More Prior Lines of Systemic Therapy." *Clinical Cancer Research*, OF1-OF5.

[133] Schueller, O., Skucas, E., Regev, G., Shaw, I., Singh, N., Sanghvi, M., Croft, M., Lohmer, L., Alabanza, A., and Patel, J. (2022). "Absolute Bioavailability, Mass Balance, and Metabolic Profiling Assessment of [14C]-Belumosudil in Healthy Men: A Phase 1, Open-Label, 2-Part Study." *Clinical Pharmacology in Drug Development*.

[134] Lentjes, E., Tan, Y., and Van Ginneken, C. A. M. (1985). "Determination of sulfinpyrazone and four metabolites in plasma and urine by high pressure liquid chromatography." *Pharmaceutisch Weekblad*, 7, 6, 252-259.

[135] https://m.chemicalbook.com/ProductChemicalPropertiesCB2122351_EN.htm. Accessed on 16 Sept. 2022.

[136] He, M., Rettie, A. E., Neal, J., and Trager, W. F. (2001). "Metabolism of sulfinpyrazone sulfide and sulfinpyrazone by human liver microsomes and cDNA-expressed cytochrome P450s." *Drug Metabolism and Disposition*, 29, 5, 701-711.

[137] Margulies, E. H., White, A. M., and Sherry, S. (1980). "Sulfinpyrazone: a review of its pharmacological properties and therapeutic use." *Drugs*, 20, 3, 179-197.

[138] Zhu, S., Long, L., Hu, Y., Tuo, Y., Li, Y., and Yu, Z. (2021). "GnRHa/Stanozolol combined therapy maintains normal bone growth in central precocious puberty." *Frontiers in Endocrinology*, 12, 678797.

[139] Lemiński, A., Kubis, M., Kaczmarek, K., Gołąb, A., Kazimierczak, A., Kotfis, K., Słojewski, M. (2022). "When Bodybuilding Goes Wrong—Bilateral Renal Artery Thrombosis in a Long-Term Misuser of Anabolic Steroids Treated with AngioJet Rheolytic Thrombectomy." *International Journal of Environmental Research and Public Health*, 19, 4, 2122-2218.

[140] Sharma, R., Balasinor, N. H., and Inamdar, L. S. (2021). "High, not low-dose of stanozolol (Anabolic–androgenic steroid) impedes embryo implantation by attenuating endometrial receptivity in the mouse, Mus musculus." *Steroids*, 165, 108752.

[141] Alves, J. C., Santos, A., Jorge, P., Lavrador, C., and Carreira, L. M. (2022). "Effect of a single intra-articular administration of stanozolol in a naturally occurring canine osteoarthritis model: a randomised trial." *Scientific Reports*, 12, 1, 5887-5898.

[142] Liu, Y., Yang, C., Xue, H., Ye, F., Sun, W., Wang, J., Qi, W., Niu, H., Hua, L., Wang, H., Han, B., and Shao, Z. (2021). "Stanozolol improves the progression-free survival of patients with high-risk myelodysplastic syndrome after decitabine treatment." *International Journal of Hematology*, 113, 6, 807-814.

[143] Ozcagli, E., Kara, M., Kotil, T., Fragkiadaki, P., Tzatzarakis, M. N., Tsitsimpikou, C., Stivaktakis, P. D., Tsoukalas, D., Spandidos, D. A., Tsatsakis, A. M., Alpertunga, B. (2018). "Stanozolol administration combined with exercise leads to decreased telomerase activity possibly associated with liver aging." *International Journal of Molecular Medicine*, 42, 1, 405-413.

[144] Kintz, P., Gheddar, L., and Raul, J. (2021). "Simultaneous testing for anabolic steroids in human hair specimens collected from various anatomic locations has several advantages when compared with the standard head hair analysis." *Drug Testing and Analysis*, 13, 7, 1445-1451.

[145] Deshmukh, N. I., Zachar, G., Petroczi, A., Székely, A. D., Barker, J., and Naughton, D. P. (2012). "Determination of stanozolol and 3′-hydroxystanozolol in rat hair, urine and serum using liquid chromatography tandem mass spectrometry." *Chemistry Central Journal*, 6, 1, 1-12.

[146] Matos, R. R., Anselmo, C. D., Sardela, V. F., H. M., and Pereira, G. (2021). "Phase II stanozolol metabolism study using the zebrafish water tank (ZWT) model." *Journal of Pharmaceutical and Biomedical Analysis*, 195, 113886.

[147] Mégarbane, B. (2010). "Treatment of patients with ethylene glycol or methanol poisoning: focus on fomepizole." *Open Access Emergency Medicine*, *OAEM* 2, 67-75.

[148] Sande, M., Thompson, D., and Monte, A. A. (2012). "Fomepizole for severe disulfiram-ethanol reactions." *The American Journal of Emergency Medicine*, 30, 1, 262.e3-5.

[149] Lepik, K. J., Brubacher, J. R., DeWitt, C. R., Lam, G. S. S., Lawson, E. J., Erhardt, G. D., Purssell, R. A., Kennedy, J. R., and Brignall, J. L. (2008). "Bradycardia and hypotension associated with fomepizole infusion during hemodialysis." *Clinical Toxicology*, 46, 6, 570-573.

[150] Bestic, M., Blackford, M., and Reed, M. (2009). "Fomepizole: a critical assessment of current dosing recommendations." *The Journal of Clinical Pharmacology*, 49, 2, 130-137.

[151] Kenneth, M. E., Hedström, K., Tolf, B., Östling-Wintzell, H., and Blomstrand, R. (1980). "Studies on the metabolic interactions between 4-methylpyrazole and methanol using the monkey as an animal model." *Archives of Biochemistry and Biophysics*, 199, 2, 606-614.

[152] "Synthesis of Fomepizole." Orphan Medical Inc. (WO2006/115626 A2).

[153] Leal, M. A., Aires, R., Pandolfi, T., Marques, V. B., Campagnaro, B. P., Pereira, T. M. C., Meyrelles, S. S., Campos-Toimil, M., and Vasquez, E. C. (2020). "Sildenafil reduces aortic endothelial dysfunction and structural damage in spontaneously hypertensive rats: Role of NO, NADPH and COX-1 pathways." *Vascular Pharmacology*, 124, 106601.

[154] Assemat, G., Balayssac, S., Gilard, V., Martins-Froment, N., Fabing, I., Rodriguez, F., Génisson, Y., Martino, R., and Malet-Marti, M. (2020). "Isolation and identification of ten new sildenafil derivatives in an alleged herbal supplement for sexual enhancement." *Journal of Pharmaceutical and Biomedical Analysis*, 191, 113482.

[155] Lee, S. K., Kim, D., and Yoo, H. H. (2011). "Comparative metabolism of sildenafil in liver microsomes of different species by using LC/MS-based multivariate analysis." *Journal of Chromatography B*, 879, 28, 3005-3011.

[156] Dunn, P. J., Galvin, S., and Hettenbach, K. (2004). "The development of an environmentally benign synthesis of sildenafil citrate (Viagra™) and its assessment by Green Chemistry metrics." *Green Chemistry*, 6, 1, 43-48.

[157] Spitz, M., Novis, B. H., Ebert, R., Trestian, S., LeRoith, D., and Creutzfeldt, W. (1982). "Betazole-induced GIP secretion is not mediated by gastric HCl." *Metabolism*, 31, 4, 380-382.

[158] Jones, R. G., and Mann, M. J. (1953). "New methods of synthesis of β-aminoethylpyrazoles." *Journal of the American Chemical Society*, 75, 16, 4048-4052.

[159] Romero-Fernandez, M., and Paradisi, F. (2021). "Biocatalytic access to betazole using a one-pot multienzymatic system in continuous flow." *Green Chemistry*, 23, 12, 4594-4603.

[160] Meneses, A., Perez-Garcia, G., Liy-Salmeron, G., Flores-Galvez, D., Castillo, C., and Castillo, E. (2008). "The effects of the 5-HT6 receptor agonist EMD and the 5-HT7 receptor agonist AS19 on memory formation." *Behavioural Brain Research*, 195, 1, 112-119.

[161] Pinto, D. J. P., Orwat, M. J., Koch, S., Rossi, K. A., Alexander, R. S., Smallwood, A., Wong, P. C., Rendina, A. R., and Luettgen, J. M. "Knab, br M." *He, K*, 5339-5356.

[162] Byon, W., Garonzik, S., Boyd, R. A., and Frost, C. E. (2019). "Apixaban: a clinical pharmacokinetic and pharmacodynamic review." *Clinical Pharmacokinetics*, 58, 10, 1265-1279.

[163] Raghavan, N., Frost, C. E., Yu, Z., He, K., Zhang, H., Humphreys, W. G., Pinto, D., Chen, S., Bonacorsi, S., Wong, P. C., and Zhang, D. (2009). "Apixaban metabolism and pharmacokinetics after oral administration to humans." *Drug Metabolism and Disposition*, 37, 1, 74-81.

[164] Wang, L., Zhang, D., Raghavan, N., Yao, M., Ma, L., Frost, C. A., and Maxwell, B. D. (2010) "*In vitro* assessment of metabolic drug-drug interaction potential of apixaban through cytochrome P450 phenotyping, inhibition, and induction studies." *Drug Metabolism and Disposition*, 38, 3, 448-458.

[165] Alsumali, D. L., Liu, R., Patel, D., Nanji, S., and Djatche, L. M. (2021). 'Budget Impact Analysis of Vericiguat for the Treatment of Chronic Heart Failure with Reduced Ejection Fraction Following a Worsening Event', *Adv Ther*, 38, 2631-2643.

[166] Armstrong, P. W., Pieske, B., Anstrom, K. J., Ezekowitz, J., Hernandez, A. F., Butler, J., Lam, C. S. P., Ponikowski, P., Voors, A. A., Jia, G., McNulty, S. E., Patel, M. J., Roessig, L., Koglin, J., and O'Connor, C. M. (2020). 'Vericiguat in Patients with Heart Failure and Reduced Ejection Fraction', *New England Journal of Medicine*, 382 x, 1883-1893.

[167] Coats, J. S., and Tolppanen, H. (2021). 'Drug Treatment of Heart Failure with Reduced Ejection Fraction: Defining the Role of Vericiguat', *Drugs*, 81, 1599-604.

[168] Kang, and Lamb, Y. N. (2022). 'Vericiguat: A Review in Chronic Heart Failure with Reduced Ejection Fraction', *Am J Cardiovasc Drugs*, 22, 451-459.

[169] Janssen, W., Schwarz, T., Bütehorn, U., Steinke, W., Sandmann, S., Lang, D., Kern, A., Hucke, F., and Gerisch, M. (2022). "Pharmacokinetics and mass balance of vericiguat in rats and dogs and distribution in rats." *Xenobiotica,* just-accepted, 1-30.

[170] Follmann, M., Ackerstaff, J., Redlich, G., Wunder, F., Lang, D., Kern, A., and Fey, P. (2017). "Discovery of the soluble guanylate cyclase stimulator vericiguat (BAY 1021189) for the treatment of chronic heart failure." *Journal of Medicinal Chemistry,* 60, 12, 5146-5161.

[171] Gandolfi, O., Roncada, P., Dall'Olio, R., and Montanaro, N. (1988). "Molecular mechanisms in the action of minaprine." *Progress in Neuro-Psychopharmacology & Biological Psychiatry*, 12, 5, 629-637.

[172] Fung, M., Thornton, A., Mybeck, K., Wu, J. H., Hornbuckle, K., and Muniz, E. (2001). "Evaluation of the characteristics of safety withdrawal of prescription drugs from worldwide pharmaceutical markets-1960 to 1999." *Drug Information Journal* 35, 1, 293-317.

[173] Lacarelle, E., Marre, F., Durand, A., Davi, H., and Rahmani, R. (1991). "Metabolism of minaprine in human and animal hepatocytes and liver microsomes—prediction of metabolism *in vivo*." *Xenobiotica*, 21, 3, 317-329.

[174] Wermuth, C. G., Schlewer, G., Bourguignon, J. J., Maghioros, G., Bouchet, M. J., Moire, C., Kan, J. P., Worms, P., and Biziere, K. (1989). "3-Aminopyridazine derivatives with atypical antidepressant, serotonergic and dopaminergic activities." *Journal of Medicinal Chemistry*, 32, 3, 528-537.

[175] Dirany, A., Sires, I., Oturan, N., Özcan, A., and Oturan, M. A. (2012). "Electrochemical treatment of the antibiotic sulfachloropyridazine: kinetics, reaction pathways, and toxicity evolution." *Environmental Science & Technology*, 46, 7, 4074-4082.

[176] Druey, J., Meier, K., and Eichenberger, K. (1954). "Heilmittelchemische Studien in der heterocyclischen Reihe. 4. Mitteilung. Pyridazine I. Derivate des cyclischen Maleinsäure-und Citraconsäure-hydrazids." *Helvetica Chimica Acta*, 37, 1, 121-133.

[177] Xing, L., Liu, Y., Yao, H., Wang, T., Xie, F., Luo, S., Luo, P., and Tang, S. (2022). "An Efficient UPLC-MS/MS Method Established to Detect Relugolix Concentration in Rat Plasma." *Frontiers in Pharmacology*, 13, 874973.

[178] Flick, C., Leverett, C. A., Ding, H. X., McInturff, E., Fink, S. J., Mahapatra, S., and Carney, D. W. (2021). "Synthetic approaches to the new drugs approved during 2019." *Journal of Medicinal Chemistry*, 64, 7, 3604-3657.

[179] Wu, Z., Teng, Q., Yang, Y., Acharekar, N., Wang, J., He, M., Yoganathan, S., Lin, J., Wang, J., and Chen, Z. (2022). "MET inhibitor tepotinib antagonizes multidrug resistance mediated by ABCG2 transporter: *In vitro* and *in vivo* study." *Acta Pharmaceutica Sinica*, B, 12, 5, 2609-2618.

[180] Friese-Hamim, M., Clark, A., Perrin, D., Crowley, L., Reusch, C., Bogatyrova, O., Zhang, H., Crandall, T., Lin, J., Ma, J., and Bachner, D. (2022). "Brain penetration and efficacy of tepotinib in orthotopic patient-derived xenograft models of MET-driven non-small cell lung cancer brain metastases." *Lung Cancer*, *163*, 77-86.

[181] Jing, G., Yu, F., and Xue, H. (2022). "Tepotinib suppresses proliferation, invasion, migration, and promotes apoptosis of melanoma cells via inhibiting MET and PI3K/AKT signaling pathways." *Oncology Letters*, 23, 170.

[182] Johne, A., Scheible, H., Becker, A., van Lier, J. J., Wolna, P., and Meyring, M. (2020). "Open-label, single-center, phase I trial to investigate the mass balance and absolute bioavailability of the highly selective oral MET inhibitor tepotinib in healthy volunteers." *Investigational New Drugs*, 38, 5, 1507-1519.

[183] Zhang, N., An, B., Zhou, Y., Li, X., and Yan, M. (2019). "Synthesis, evaluation, and mechanism study of new tepotinib derivatives as antiproliferative agents." *Molecules*, 24, 6, 1173.

[184] Deeks, E. D. (2015). 'Olaparib: First Global Approval', *Drugs*, 75, 231-40.

[185] Paik, J. (2021). 'Olaparib: A Review as First-Line Maintenance Therapy in Advanced Ovarian Cancer', *Targeted Oncology*, 16, 847-856.

[186] Wang, L., and Wang, M. (2020). "*In vitro* metabolism of olaparib in liver microsomes by liquid chromatography/electrospray ionization high-resolution mass spectrometry." *Rapid Communications in Mass Spectrometry*, 34, 3, e8575.

[187] Chen, Z., Wang, S., Liu, K., Zhang, R., Li, Q., Bian, W., Qiao, R., and Li, C. (2022). 'Practical and Scalable Manufacturing Process for the Key Intermediate of Poly (Adp-Ribose) Polymerase Inhibitor Olaparib', *ACS Omega*, 7, 6313-6321.

[188] Murai, J., Huang, S. Y., Das, B. B., Renaud, A., Zhang, Y., Doroshow, J. H., Ji, J., Takeda, S., and Pommier, Y. (2012). 'Trapping of Parp1 and Parp2 by Clinical Parp Inhibitors', *Cancer Res*, 72, 5588-5599.

[189] Rouleau, M., Patel, A., Hendzel, M. J., Kaufmann, S. H., and Poirier, G. G. (2010). 'Parp Inhibition: Parp1 and Beyond', *Nat Rev Cancer*, 10, 293-301.

[190] Gao, C., Dong, W., Cui, Z., Yuan, Q., Hu, X., Wu, Q., Han, X., Xu, Y., and Min, Z. (2019). 'Synthesis, Preliminarily Biological Evaluation and Molecular Docking Study of New Olaparib Analogues as Multifunctional Parp-1 and Cholinesterase Inhibitors', *Journal of Enzyme Inhibition and Medicinal Chemistry*, 34, 150-162.

[191] Lou, X., Yang, X., Ding, Y., Wang, J., Yan, Q., Huang, X., Guo, Y., Wang, X., and Xiang, W. (2013). 'Synthesis of Olaparib Derivatives and Their Antitumor Activities', *Chemical Research in Chinese Universities*, 29, 231-235.

[192] Essel, K. G., Behbakht, K., Lai, T., Hand, L., Evans, E., Dvorak, J., Ding, K., Konecny, G., and Moore, K. N. (2021). 'Parpi after Parpi in Epithelial Ovarian Cancer', *Gynecologic Oncology Reports*, 35, 100699.

[193] Strickland, K. C., Howitt B. E., Shukla, S. A., Rodig, S., Ritterhouse, L. L., Liu, J. F., Garber, J. E., Chowdhury, D., Wu, C. J., D'Andrea, A. D., Matulonis, U. A., and Konstantinopoulos, P. A. (2016). 'Association and Prognostic Significance of Brca1/2-Mutation Status with Neoantigen Load, Number of Tumor-Infiltrating Lymphocytes and Expression of Pd-1/Pd-L1 in High Grade Serous Ovarian Cancer', *Oncotarget*, 7, 13587-13598.

[194] Helleday, T. (2011). 'The Underlying Mechanism for the Parp and Brca Synthetic Lethality: Clearing up the Misunderstandings', *Mol Oncol*, 5, 387-393.

[195] Javle, M., and Curtin, N. J. (2011). 'The Potential for Poly (Adp-Ribose) Polymerase Inhibitors in Cancer Therapy', *Ther Adv Med Oncol*, 3, 257-267.

[196] Hoy, S. M. (2018). 'Talazoparib: First Global Approval', *Drugs*, 78, 1939-46.

[197] Turk, A. A., and Wisinski, K. B. (2018). 'Parp Inhibitors in Breast Cancer: Bringing Synthetic Lethality to the Bedside', *Cancer*, 124, 2498-2506.

[198] Boussios, S., Abson, C., Moschetta, M., Rassy, E., Karathanasi, A., Bhat, T., Ghumman, F., Sheriff, M., and Pavlidis, N. (2020). 'Poly (Adp-Ribose) Polymerase Inhibitors: Talazoparib in Ovarian Cancer and Beyond', *Drugs in R&D,* 20, 55-73.

[199] Kim, H., Kim, S., Kim, H. J., Kim, Y. K., Kwak, J. Y., Yhim, H. Y., Kim, S. H., Do, Y. R., Oh, S., Lee, S. E., Jootar, S., Cui, J. W., and Kim, D. W. (2018). 'Comparison of Frequency and Sensitivity of Bcr-Abl1 Kinase Domain Mutations in Asian and White Patients with Imatinib-Resistant Chronic-Phase Chronic Myeloid Leukemia', *Clin Lymphoma Myeloma Leuk*, 18, e391-e399.

[200] Baldwin, P., Ohman, A. W., Medina, J. E., McCarthy, E. T., Dinulescu, D. M., and Sridhar, S. (2019). 'Nanoformulation of Talazoparib Delays Tumor Progression and Ascites Formation in a Late Stage Cancer Model', *Frontiers in Oncology*, 9.

[201] Li, J., Xiao, D., Liu, L., Xie, F., Li, W., Sun, W., Yang, X., and Zhou, X. (2020). 'Design, Synthesis, and *in Vitro* Evaluation of the Photoactivatable Prodrug of the Parp Inhibitor Talazoparib', *Molecules*, 25, 407.

[202] Zhou, D., Chen, H., Mpoy, C., Afrin, S., Rogers, B. E., Garbow, J. R., Katzenellenbogen, J. A., and Xu, J. (2021). 'Radiosynthesis and Evaluation of Talazoparib and Its Derivatives as Parp-1-Targeting Agents', *Biomedicines*, 9.

[203] Yu, Y., Chung, C., Plotka, A., Quinn, K., Shi, H., Pápai, Z., Nguyen, L., and Wang, D. (2019). "A Phase 1 Mass Balance Study of 14C-Labeled Talazoparib in Patients With Advanced Solid Tumors." *The Journal of Clinical Pharmacology*, 59, 9, 1195-1203.

[204] Bower, H., Björkholm, M., Dickman, P. W., Höglund, M., Lambert, P. C., and Andersson, T. M. (2016). 'Life Expectancy of Patients with Chronic Myeloid Leukemia Approaches the Life Expectancy of the General Population', *J Clin Oncol*, 34, 2851-2857.

[205] Kim, H., Kim, S., Kim, H. J., Kim, Y. K., Kwak, J. Y., Yhim, H. Y., Kim, S. H., Do, Y. R., Oh, S., Lee, S. E., Jootar, S., Cui, J. W., and Kim, D. W. (2018). 'Comparison of Frequency and Sensitivity of Bcr-Abl1 Kinase Domain Mutations in Asian and White Patients with Imatinib-Resistant Chronic-Phase Chronic Myeloid Leukemia', *Clin Lymphoma Myeloma Leuk*, 18, e391-e399.

[206] Pulte, E. D., Chen, H., Price, L. S. L., Gudi, R., Li, H., Okusanya, O. O., Ma, L., Rodriguez, L., Vallejo, J., Norsworthy, K. J., de Claro, R. A., Theoret, M. R., and Pazdur, R. (2022). 'Fda Approval Summary: Revised Indication and Dosing Regimen for Ponatinib Based on the Results of the Optic Trial', *Oncologist*, 27, 149-157.

[207] O'Hare, T., Shakespeare, W. C., Zhu, X., Eide, C. A., Rivera, V. M., Wang, F., Adrian, L. T., Zhou, T., Huang, W. S., Xu, Q., Metcalf, C. A., 3rd, Tyner, J. W., Loriaux, M. M., Corbin, A. S., Wardwell, S., Ning, Y., Keats, J. A., Wang, Y., Sundaramoorthi, R., Thomas, M., Zhou, D., Snodgrass, J., Commodore, L., Sawyer, T. K., Dalgarno, D. C., Deininger, M. W., Druker, B. J., and Clackson, T. (2009). 'Ap24534, a Pan-Bcr-Abl Inhibitor for Chronic Myeloid Leukemia, Potently Inhibits the T315i Mutant and Overcomes Mutation-Based Resistance', *Cancer Cell*, 16, 401-412.

[208] Lu, L., Kok, C. H., Saunders, V. A., Wang, J., McLean, J. A., Hughes, T. P., and White, D. L. (2018). 'Modelling Ponatinib Resistance in Tyrosine Kinase Inhibitor-Naïve and Dasatinib Resistant Bcr-Abl1+ Cell Lines', *Oncotarget*, 9, 34735-47.

[209] Zabriskie, M. S., Eide, C. A., Tantravahi, S. K., Vellore, N. A., Estrada, J., Nicolini, F. E., Khoury, H. J., Larson, R. A., Konopleva, M., Cortes, J. E., Kantarjian, H., Jabbour, E. J., Kornblau, S. M., Lipton, J. H., Rea, D., Stenke, L., Barbany, G., Lange, T., Hernández-Boluda, J. C., Ossenkoppele, G. J., Press, R. D., Chuah, C., Goldberg, S. L., Wetzler, M., Mahon, F. X., Etienne, G., Baccarani, M., Soverini, S., Rosti, G., Rousselot, P., Friedman, R., Deininger, M., Reynolds, K. R., Heaton, W. L., Eiring, A. M., Pomicter, A. D., Khorashad, J. S., Kelley, T. W., Baron, R., Druker, B. J., Deininger, M. W., and O'Hare, T. (2014). 'Bcr-Abl1 Compound Mutations Combining Key Kinase Domain Positions Confer Clinical Resistance to Ponatinib in Ph Chromosome-Positive Leukemia', *Cancer Cell*, 26, 428-442.

[210] Ye, Y. E., Woodward, C. N., and Narasimhan, N. I. (2017). "Absorption, metabolism, and excretion of [14C] ponatinib after a single oral dose in humans." *Cancer Chemotherapy and Pharmacology*, 79, 3. 507-518.

[211] Moulder, S. L., Borges, V. F., Baetz, T., McSpadden, T., Fernetich, G., Murthy, R. K., Chavira, R., Guthrie, K., Barrett, E., and Chia, S. K. (2017). 'Phase I Study of Ont-380, a Her2 Inhibitor, in Patients with Her2+-Advanced Solid Tumors, with an

Expansion Cohort in Her2+ Metastatic Breast Cancer (Mbc)', *Clinical Cancer Research*, 23, 3529-3536.
[212] Spector, N. L., and Blackwell, K. L. (2009). 'Understanding the Mechanisms Behind Trastuzumab Therapy for Human Epidermal Growth Factor Receptor 2-Positive Breast Cancer', *J Clin Oncol*, 27, 5838-5847.
[213] Sirhan, Z., Thyagarajan, A., and Sahu, R. P. (2022). 'The Efficacy of Tucatinib-Based Therapeutic Approaches for Her2-Positive Breast Cancer', *Military Medical Research*, 9, 39.
[214] Schlam, S. M. Swain, (2021). 'Her2-Positive Breast Cancer and Tyrosine Kinase Inhibitors: The Time Is Now', 7, 56.
[215] Sun, H., Cardinal, K. A., Wienkers, L., Chin, A., Kumar, V., Neace, C., and Henderson, C. (2022). "Elimination of tucatinib, a small molecule kinase inhibitor of HER2, is primarily governed by CYP2C8 enantioselective oxidation of gem-dimethyl." *Cancer Chemotherapy and Pharmacology*, 1-14.
[216] Lyu, Y., Huang, L., Zhu, X., Lu, S., and Mao, Y. (2021). "A Practical Alternate Synthesis of Tucatinib." *Organic Preparations and Procedures International*, 53, 6, 554-561.
[217] Vansteenkiste, J. F., Kerkhove, C. V. D., Wauters, E., and Mol, P. V. (2019). 'Capmatinib for the Treatment of Non-Small Cell Lung Cancer', *Expert Review of Anticancer Therapy*, 19, 659-671.
[218] Liu, X., Wang, Q., Yang, G., Marando, C., Koblish, H. K., Hall, L. M., Fridman, J. S., Behshad, E., Wynn, R., Li, Y., Boer, J., Diamond, S., He, C., Xu, M., Zhuo, J., Yao, W., Newton, R. C., and Scherle, P. A. (2011). 'A Novel Kinase Inhibitor, Incb28060, Blocks C-Met–Dependent Signaling, Neoplastic Activities, and Cross-Talk with Egfr and Her-3', *Clinical Cancer Research*, 17, 7127-7138.
[219] Glaenzel, U., Jin, Y., Hansen, R., Schroer, K., Rahmanzadeh, G., Pfaar, U., and van Lier, J. J. (2020). "Absorption, distribution, metabolism, and excretion of capmatinib (INC280) in healthy male volunteers and *in vitro* aldehyde oxidase phenotyping of the major metabolite." *Drug Metabolism and Disposition*, 48, 10, 873-885.
[220] Wang, L., Li, R., Song, C., Chen, Y., Long, H., and Yang, L. (2021). "Small-Molecule Anti-Cancer Drugs From 2016 to 2020: Synthesis and Clinical Application." *Natural Product Communications* 16, 9, 1934578X211040326.
[221] Brooks, B. R., Miller, R. G., Swash, M., and Munsat, T. L. (2000). 'El Escorial Revisited: Revised Criteria for the Diagnosis of Amyotrophic Lateral Sclerosis', *Amyotroph Lateral Scler Other Motor Neuron Disord*, 1, 293-299.
[222] Cruz, M. P. (2018). 'Edaravone (Radicava): A Novel Neuroprotective Agent for the Treatment of Amyotrophic Lateral Sclerosis'. *P t*, 43, 25-28.
[223] Cho, H., and Shukla, S. (2020). '*Role of Edaravone as a Treatment Option for Patients with Amyotrophic Lateral Sclerosis*', 14, 29.
[224] Jaiswal, M. K. (2019). "Riluzole and Edaravone: A Tale of Two Amyotrophic Lateral Sclerosis Drugs", *Medicinal Research Reviews*, 39, 733-748.
[225] Pérez-González, A. (2011). Galano, 'Oh Radical Scavenging Activity of Edaravo: Mechanism and Kinetics', *J Phys Chem B*, 115, 1306-1314.

[226] Vu, M., Tortorice, K., Zacher, J., Dong, D., Hur, K., Zhang, R., Good, C. B., Glassman, P. A., and Cunningham, F. E. (2020). 'Assessment of Use and Safety of Edaravone for Amyotrophic Lateral Sclerosis in the Veterans Affairs Health Care System', *JAMA Network Open*, 3, e2014645-e45.
[227] Watanabe, T., Yuki, S., Egawa, M., and Nishi, H. (1994). 'Protective Effects of Mci-186 on Cerebral Ischemia: Possible Involvement of Free Radical Scavenging and Antioxidant Actions', *J Pharmacol Exp Ther*, 268, 1597-1604.
[228] Watanabe, T., Morinaka, Y., Iseki, K., Watanabe, T., Yuki, S., and Nishi, H. (2003). 'Structure-Activity Relationship of 3-Methyl-1-phenyl-2-pyrazolin-5-one (Edaravone)', *Redox Rep*, 8, 151-5.
[229] Murota, S., Morita, I., and Suda, N. (1990). 'The Control of Vascular Endothelial Cell Injury', *Ann N Y Acad Sci*, 598, 182-187.
[230] Maples, R., Ma, F., and Zhang, Y. K. (2001). 'Comparison of the Radical Trapping Ability of Pbn, S-Ppbn and Nxy-059', *Free Radic Res*, 34, 417-426.
[231] Cruz, M. P. (2018). "Edaravone (Radicava): a novel neuroprotective agent for the treatment of amyotrophic lateral sclerosis." *Pharmacy and Therapeutics*, 43, 1, 25.
[232] Zhou, S., Hong, Q., Mei, W., He, Y., Wu, C., and Sun, T. (2021). "Scale-Up of a Continuous Manufacturing Process of Edaravone." *Organic Process Research & Development*, 25, 9, 2146-2153.
[233] Michelson, D., Snyder, E., Paradis, E., Chengan-Liu, M., Snavely, D. B., Hutzelmann, J., Walsh, J. K., Krystal, A. D., Benca, R. M., Cohn, M., Lines, C., Roth, T., and Herring, W. J. (2014). 'Safety and Efficacy of Suvorexant During 1-Year Treatment of Insomnia with Subsequent Abrupt Treatment Discontinuation: A Phase 3 Randomised, Double-Blind, Placebo-Controlled Trial', *Lancet Neurol*, 13, 461-471.
[234] Herring, W. J., Connor, K. M., Ivgy-May, N., Snyder, E., Liu, K., Snavely, D. B., Krystal, A. D., Walsh, J. K., Benca, R. M., Rosenberg, R., Sangal, R. B., Budd, K., Hutzelmann, J., Leibensperger, H., Froman, S., Lines, C., Roth, T., and Michelson, D. (2016). 'Suvorexant in Patients with Insomnia: Results from Two 3-Month Randomized Controlled Clinical Trials', *Biol Psychiatry*, 79, 136-148.
[235] Kuriyama, A., and Tabata, H. (2017). 'Suvorexant for the Treatment of Primary Insomnia: A Systematic Review and Meta-Analysis', *Sleep Medicine Reviews*, 35, e0136910.
[236] Cui, D., Cabalu, T., Yee, K. L., Small, J., Li, X., Liu, B., and Maciolek, C. (2016). "*In vitro* and *in vivo* characterisation of the metabolism and disposition of suvorexant in humans." *Xenobiotica*, 46, 10, 882-895.
[237] Baxter, C. A., Cleator, E., Brands, K. M. J., Edwards, J. S., Reamer, R. A., Sheen, F. J., Stewart, G. W., Strotman, N. A., and Wallace, D. J. (2011). "The first large-scale synthesis of MK-4305: a dual orexin receptor antagonist for the treatment of sleep disorder." *Organic Process Research & Development*, 15, 2, 367-375.
[238] Erdos, J., and Wild, C. (2022). 'Mid- and Long-Term (at Least 12 Months) Follow-up of Patients with Spinal Muscular Atrophy (Sma) Treated with Nusinersen, Onasemnogene Abeparvovec, Risdiplam or Combination Therapies: A Systematic Review of Real-World Study Data', *European Journal of Paediatric Neurology*, 39, 1-10.

[239] Markati, T., Fisher, G., Ramdas, S., and Servais, L. (2022). 'Risdiplam: An Investigational Survival Motor Neuron 2 (Smn2) Splicing Modifier for Spinal Muscular Atrophy (Sma)', *Expert Opinion on Investigational Drugs*, 31, 451-461.

[240] Sturm, S., Günther, A., Jaber, B., Jordan, P., Al Kotbi, N., Parkar, N., Cleary, Y., Frances, N., Bergauer, T., Heinig, K., Kletzl, H., Marquet, A., Ratni, H., Poirier, A., Müller, L., Czech, C., and Khwaja, O. (2019). 'A Phase 1 Healthy Male Volunteer Single Escalating Dose Study of the Pharmacokinetics and Pharmacodynamics of Risdiplam (Rg7916, Ro7034067), a Smn2 Splicing Modifier', *Br J Clin Pharmacol*, 85, 181-193.

[241] Baranello, G., Darras, B. T., Day, J. W., Deconinck, N., Klein, A., Masson, R., Mercuri, E., Rose, K., El-Khairi, M., Gerber, M., Gorni, K., Khwaja, O., Kletzl, H., Scalco, R. S., Seabrook, T., Fontoura, P., and Servais, L. (2021). 'Risdiplam in Type 1 Spinal Muscular Atrophy', *N Engl J Med*, 384, 915-923.

[242] Mercuri, E., Deconinck, N., Mazzone, E. S., Nascimento, A., Oskoui, M., Saito, K., Vuillerot, C., Baranello, G., Boespflug-Tanguy, O., Goemans, N., Kirschner, J., Kostera-Pruszczyk, A., Servais, L., Gerber, M., Gorni, K., Khwaja, O., Kletzl, H., Scalco, R. S., Staunton, H., Yeung, W. Y., Martin, C., Fontoura, P., and Day, J. W. (2022). 'Safety and Efficacy of Once-Daily Risdiplam in Type 2 and Non-Ambulant Type 3 Spinal Muscular Atrophy (Sunfish Part 2): A Phase 3, Double-Blind, Randomised, Placebo-Controlled Trial', *Lancet Neurol*, 21, 42-52.

[243] Zuppa, C. De Michelis, Meo, G., Prada, V., Gemelli, C., Infantino, M., Manfredi, M., Pesce, G., Tagliafico, A. S., Benedetti, L., Fiorillo, C., Schenone, A., Quartuccio, L., and Grandis, M. (2021). 'Maintenance Treatment with Subcutaneous Immunoglobulins in the Long-Term Management of Anti-Hmcgr Myopathy', *Neuromuscul Disord*, 31, 134-138.

[244] Finkel, R. S., Farrar, M. A., Vlodavets, D., Servais, L., Zanoteli, E., Al-Muhaizea, M., Nelson, L., Prufer, A., Wang, Y., Fisher, C., Gerber, M., Gorni, K., Kletzl, H., Palfreeman, L., Scalco, R., and Bertini, E. (2022). 'Rainbowfish: Preliminary Efficacy and Safety Data in Risdiplam-Treated Infants with Presymptomatic Sma (P17-5.003)', *Neurology*, 98, 1636.

[245] Ratni, H., Ebeling, M., Baird, J., Bendels, S., Bylund, J., Chen, K. S., and Denk, N. (2018). "Discovery of risdiplam, a selective survival of motor neuron-2 (SMN2) gene splicing modifier for the treatment of spinal muscular atrophy (SMA)." 6501-6517.

[246] Muccini, C., Canetti, D., Castagna, A., and Spagnuolo, V. (2022). '*Efficacy and Safety Profile of Fostemsavir for the Treatment of People with Human Immunodeficiency Virus-1 (Hiv-1): Current Evidence and Place in Therapy*', 16, 297-304.

[247] Nettles, R. E., Schürmann, D., Zhu, L., Stonier, M., Huang, S. P., Chang, I., Chien, C., Krystal, M., Wind-Rotolo, M., Ray, N., Hanna, G. J., Bertz, R., and Grasela, D. (2012). 'Pharmacodynamics, Safety, and Pharmacokinetics of Bms-663068, an Oral Hiv-1 Attachment Inhibitor in Hiv-1-Infected Subjects', *Journal of Infect Dis*, 206, 1002-1011.

[248] Lalezari, J. P., Latiff, G. H., Brinson, C., Echevarría, J., Treviño-Pérez, S., Bogner, J. R., Thompson, M., Fourie, J., Sussmann Pena, O. A., Mendo Urbina, F. C.,

Martins, M., Diaconescu, I. G., Stock, D. A., Joshi, S. R., Hanna, G. J., and Lataillade, M. (2015). 'Safety and Efficacy of the Hiv-1 Attachment Inhibitor Prodrug Bms-663068 in Treatment-Experienced Individuals: 24 Week Results of Ai438011, a Phase 2b, Randomised Controlled Trial', *Lancet HIV*, 2, e427-e437.

[249] Kozal, M., Aberg, J., Pialoux, G., Cahn, P., Thompson, M., Molina, J., Grinsztejn, B., Diaz, R., Castagna, A., Kumar, P., Latiff, G., DeJesus, E., Gummel, M., Gartland, M., Pierce, A., Ackerman, P., Llamoso, C., and Lataillade, M. (2020). 'Fostemsavir in Adults with Multidrug-Resistant Hiv-1 Infection', *New England Journal of Medicine*, 382, 1232-1243.

[250] Lataillade, M., Lalezari, J. P., Kozal, M., Aberg, J. A., Pialoux, G., Cahn, P., Thompson, M., Molina, J., Moreno, S., Grinsztejn, B., Diaz, R. S., Castagna, A., Kumar, P. N., Latiff, G. H., De Jesus, E., Wang, M., Chabria, S., Gartland, M., Pierce, A., Ackerman, P., and Llamoso, C. (2020). 'Safety and Efficacy of the Hiv-1 Attachment Inhibitor Prodrug Fostemsavir in Heavily Treatment-Experienced Individuals: Week 96 Results of the Phase 3 Brighte Study', *The Lancet HIV*, 7, e740-e51.

[251] Yuan, S., Luo, Y., Zuo, J., Liu, H., Li, F., and Yu, B. (2021). "New drug approvals for 2020: Synthesis and clinical applications." *European Journal of Medicinal Chemistry*, 215, 113284.

[252] Eastman, R. T., Roth, J. S., Brimacombe, K. R., Simeonov, A., Shen, M., Patnaik, S., and Hall, M. D. (2020). 'Remdesivir: A Review of Its Discovery and Development Leading to Emergency Use Authorization for Treatment of Covid-19', *ACS Central Science*, 6, 672-683.

[253] Andersen, G., and Rambaut, A. (2020). '*The Proximal Origin of Sars-Cov-2*', 26, 450-452.

[254] Bansal, V., Mahapure, K. S., Bhurwal, A., Gupta, I., Hassanain, S., Makadia, J., Madas, N., Armaly, P., Singh, R., Mehra, I., O'Horo, J. C., and Kashyap, R. (2020). 'Mortality Benefit of Remdesivir in Covid-19: A Systematic Review and Meta-Analysis', *Front Med (Lausanne)*, 7, 606429.

[255] Lee, T. C., Murthy, S., Del Corpo, O., Senécal, J., Butler-Laporte, G., Sohani, Z. N., Brophy, J. M., and McDonald, E. G. (2022). 'Remdesivir for the Treatment of Covid-19: A Systematic Review and Meta-Analysis', *Clin Microbiol Infect*.

[256] Ansems, K., Grundeis, F., Dahms, K., Mikolajewska, A., Thieme, V., Piechotta, V., Metzendorf, M. I., Stegemann, M., Benstoem, C., and Fichtner, F. (2021). "Remdesivir for the Treatment of Covid-19", *Cochrane Database Syst Rev*, 8, CD014962.

[257] Wang, Y., Zhou, F., Zhang, D., Zhao, J., Du, R., Hu, Y., Cheng, Z., Gao, L., Jin, Y., Luo, G., Fu, S., Lu, Q., Du, G., Wang, K., Lu, Y., Fan, G., Zhang, Y., Liu, Y., Ruan, S., Liu, W., Jaki, T., Hayden, F. G., and Horby, P. W. (2020). "*Evaluation of the Efficacy and Safety of Intravenous Remdesivir in Adult Patients with Severe Covid-19: Study Protocol for a Phase 3 Randomized, Double-Blind, Placebo-Controlled, Multicentre Trial*", 21, 422.

[258] Gupte, V., Hegde, R., Sawant, S., Kalathingal, K., Jadhav, S., Malabade, R., and Gogtay, J. (2022). "Safety and Clinical Outcomes of Remdesivir in Hospitalised

Covid-19 Patients: A Retrospective Analysis of Active Surveillance Database", *BMC Infectious Diseases*, 22, Article 1.
[259] Gannedi, V., Villuri, B. K., Reddy, S. N., Ku, C., Wong, C., and Hung, S. (2021). "Practical remdesivir synthesis through one-pot organocatalyzed asymmetric (S)-P-Phosphoramidation." *The Journal of Organic Chemistry*, 86, 7, 4977-4985.
[260] Jallow, S., and Govender, N. P. (2021) "Ibrexafungerp: A First-in-Class Oral Triterpenoid Glucan Synthase Inhibitor", *Journal of Fungi*, 7, 3, 163.

Biographical Sketches

Dr. Neetu Talreja

Affiliation: Alliance University, Department of Science, Faculty of Science and Technology, Anekal, Bengaluru-562 106, Karnataka, India

Education: PhD

Business Address: Alliance University, Department of Science, Faculty of Science and Technology, Anekal, Bengaluru, Karnataka, India

Research and Professional Experience:

- June 2020-Feb 2022: Researcher in Advanced Ceramics and Nanotechnology Laboratory, Department of Materials Engineering, Faculty of Engineering, University of Concepción, Concepción, Chile.
- May 2019-May 2020: Researcher in Multidisciplinary Research Institute for Science and Technology, IIMCT, University of La Serena, Benavente 980, La Serena, Chile.
- September 2017-April 2019: Postdoctoral-Fellow in Department of Bio-Nanotechnology, Gachon University, South Korea.
- December 2016-September 2017: Post-Doctoral Researcher in the Department of Chemical Engineering, Beijing University of Chemical Technology, Beijing, P.R. Chi.
- April 2010-June 2015: Project Associate in Department of Chemical Engineering & Centre for Environment science & Engineering, Indian Institute Of Technology (I.I.T) Kanpur, (UP), India.

Professional Appointments: Assistant Professor, Alliance University, Department of Science, Faculty of Science and Technology, Anekal, Bengaluru.

Publications from the Last 3 Years:

Patents:

[1] Preparation of three-dimensional (3D) Polymeric-metal-carbon framework and application thereof in environmental remediation, Indian patent (Filed). Application number: 202041041356, 24/09/2020. Inventor: Sasidharan.V, Divya Chauhan, Neetu Talreja, Mohammad Ashfaq.

[2] Method for the Synthesis of Bimetallic Povidon-Iodine Microflower composite, Indian patent (Filed). Application number: 202041056205, dated 24/12/2020. Inventor: Mohammad Ashfaq, Neetu Talreja, Divya Chauhan.

[3] Method for the synthesis of Graphene Oxide Incorporated Copper Bismuth Nanorods, Indian Patent (Filed). Application number: 202141016915 dated 10/04/2021. Inventor: Mohammad Ashfaq, Neetu Talreja, Divya Chauhan.

Publications:

Talreja, Neetu, Mohammad Ashfaq, Divya, Adriana C Mera, CA. Rodríguez, RV Mangalaraja, Facile synthesis of Zn metal-doped BiOI microstructure based photocatalyst material: Efficient photodegradation ability against tetracycline and antibacterial activity. *New Journal of Chemistry*. 2021

Talreja, Neetu, Shagufta Afreen, Mohammad Ashfaq, alreja, Divya, Adriana C Mera, CA. Rodríguez, RV Mangalaraja, Bimetal (Fe/Zn) doped BiOI photocatalyst: An effective photodegradation of tetracycline and bacteria. *Chemosphere*, 2021, 280, 130803

Talreja, Neetu, Mohammad Ashfaq, alreja, Divya, Adriana C Mera, CA. Rodríguez, Strategic doping approach of Fe-BiOI microstructure: An improved photodegradation efficiency of tetracycline. *ACS Omega*, 2021, 6(2), 1575-1583.

Talreja, Neetu, Sung Hoon Jung, Haiyen Lee, TaeYoung Kim, Phenol-formaldehyde-resin based activated carbons for high-performance supercapacitors. *Chemical Engineering Journal*. 2020, 379, 122332.

Mr. Shashank Madhukar Sonkusare

Affiliation: Institute of Chemical Technology, Mumbai, India

Education: B. Phamacy

Business Address: 2VF4+GX4, Nathalal Parekh Marg, near Khalsa College, Matunga, Mumbai, Maharashtra, MS, India

Professional Appointments: Pursuing a Master's in Pharmacy (Pharmaceutical Chemistry)

Honors:

1. Secured AIR 45 in GPAT Entrance Examination 2021.
2. Secured AIR 63 in NIPER JEE Entrance Examination 2021.

Chapter 3

Carbon Based Nanomaterials: Applications towards the Electrochemical Determination of Hydrazine

Narmatha Sivaraman[1,2]
Sakkarapalayam Murugesan Senthil Kumar[1,2]
and Rangasamy Thangamuthu[1,2,*]
[1] Electroorganic and Materials Electrochemistry Division (EMED),
CSIR-Central Electrochemical Research Institute, Karaikudi, India
[2] Academy of Scientific and Innovative Research (AcSIR), Ghaziabad, India

Abstract

Nanomaterials have been widely explored as an amenable sensing platform to further boost the electrochemical kinetics with diminution of overpotential. Besides, carbon-based nanomaterials have gained substantial amenities of being stable, economical, chemically inert which will nurture significant properties like relatively wide potential window with diminished background current made them as an appropriate electrocatalyst for sensing applications. In addition, it has a promising conductivity with higher surface area and easy to functionalize.

Herein, unique configuration of carbonaceous materials and their composites are explored for assaying the detection of hydrazine (HZ) owing to the possibility of quantitative measurement with high precision, sensitivity and reproducibility, which are necessary for accurate detection in water and environmental samples. In this context, we mainly focused on overview of carbon-based nanomaterials for the detection of

* Corresponding Author's Email: thangamuthu@cecri.res.in.

In: A Review of Hydrazine and Its Applications
Editor: Thomas W. Gerdes
ISBN: 979-8-88697-671-7

toxic HZ in a worthy manner. Additionally, carbon-based materials with surface modifications were summarized along with the inclusion of carbon nanotubes, carbon quantum dots, graphene, ordered mesoporous carbon (OMC), metal organic framework (MOF) derived carbon, carbon nanofibers and carbon black and their optimized sensing parameters are outlined.

This chapter mainly discusses the advantages and limitations of existing electrode alterations and also providing a futuristic idea to improve the detection strategy for HZ assay and covers the in-depth mechanism of hydrazine oxidation in water and other environmental samples.

Keywords: hydrazine, electrochemical sensor, carcinogenic pollutant, carbon-based nanomaterials, chemically modified electrodes

Introduction

As the population around globe is progressing along with socioeconomic growth, there is an ever-growing demand over all resources and water quality has been steadily depleting. Accordingly, there is always an increasing need to provide safe drinking water and even more of the world's population is concentrated in urban developments. However, the aging/inadequate water infrastructure as well as treatment of municipal and industrial waste water (effluents) exist within many cities produces severe problems, rising from contaminants (environmental pollutants) entering into the domestic water supply. Further, major environmental issues like uncapped water contamination and depreciated groundwater level owing to less rainfall have further exacerbated water scarcity and pollution levels. The estimation of hazardous pollutant is highly significant and acquired much attention for the detection of risks in healthcare and environment in recent decades (Kumunda et al. 2021).

In the series of hazardous pollutants, hydrazine (HZ) is an imperative with a chemical formula of N_2H_4 which is widely utilized as pharmaceutical intermediates, agricultural chemicals, air bags, textile dyes, spacecraft, fuel cells, photographic chemicals, rocket fuel, blowing agent, insecticides, emulsifier and corrosion inhibitors in numerous industries (Krittayavathananon et al. 2014; Crapnell and Banks 2022). In addition, hydrazine is known as colourless, toxic, flammable substance which is categorised as an irritant with carcinogenic effect on human (group B2) by

U.S. Environmental Protection Agency (US-EPA) with a threshold value of not exceeding 10 ppb. Moreover oral, inhalation and transdermal routes are feasible to absorb the excess of HZ present in the environment. The prolonged exposure of carcinogenic HZ through these routes could leads to eye irritation, inflammation, and intense result of corrosion on both skin and mucous membranes with higher toxicity on humans. The production of HZ is not endogenous in human bodies, instead some specific drugs like isoniazid which will metabolize in human body to form toxic HZ (Liu et al. 2019). Due to inevitable industrial significance of HZ and its toxicological effects, detection of the HZ is extensively explored (Crapnell and Banks 2022). HZ has mutagenic nature which can severely damage liver, kidney, lungs, brain and spinal cord. Moreover, HZ is a severe threat to soil and water pollutions due to its strong permeability and higher solubility, would directly affect the central nervous system and many biological organs even at trace level with longer exposure to hydrazine (Ma et al. 2021). Accordingly, the effective, rapid, accurate determination of trace level existence of HZ is crucial for current environmental crisis.

The detection of carcinogenic HZ is crucial which can be exactly quantified by plenty of conventional techniques such as gas chromatography-mass spectrometry, matrix assisted laser desorption/ionization time-of-flight mass spectrometry, solid-phase spectrophotometry, and high-performance liquid chromatography along with tandem mass spectrometry. In addition, numerous methods like ion chromatography, flow injection analysis and chemiluminescence techniques are available for the detection of HZ (Links 2012). Nonetheless, the above-mentioned techniques had few shortcomings like expensive instruments are needed to carry out the experiment, complicated operation with tedious procedure for sample preparation, linear ranges are relatively narrow and have low precision and trained operator is needed, etc., thus limits their application towards the on-site detection of carcinogenic HZ.

Advantages of Electrochemical Sensors for Pollutant Detection

Currently existing conventional techniques like colorimetric and spectrometric methods for the detection of pollutants require tedious sample preparation with lower sensitivity (it may interfere with the common interferents which may lead to erroneous response on detection). To sort out this issue, electrochemical sensing platform can be utilized to monitor the healthcare and environmental sector.

The promising electrochemical sensing procedure is superior than all other existing conventional techniques with many advantages such as low cost, simple preparation, fast response, easy operation, rapid on-site detection and miniaturisation also possible with appealing electrochemical performances (Ma et al. 2021). The electrochemical sensing method has significant application in hazardous pollutant detection, which is widely explored across the globe.

Principle of Electrochemical Sensors

An electrochemical sensor is a device which is used to recognize, detect and transmits the information regarding the physiological change or a process in a chemical reaction with excellent sensitivity and selectivity. An electrochemical sensor (EC sensors) is a self-contained integrated device that detects certain chemical species using transducer element as an electrode by utilizing oxidation or reduction current and providing the quantitative or semi-quantitative information about the analyte.

Generally, sensor devices are made up of an analyte-selective interface along with transducer. A transducer is a tool which transform distinct changes (chemical or physical) into a conceivable signal. The term 'transducer' is borrowed from Latin *traduce,* it signifies the transfer of energy from one system to another and with the same or another version (Shalini Devi, Sasya, and Krishnan 2020). They are primarily categorized into optical, mechanical, electrochemical and thermal. By minimizing the distance within recognition element and transducer, small quantity of sample tends to enhance the accuracy of signal. For speedy detection of analytes, minimal distance is required for swift flow from analyte to transducer. This interaction is mainly depending on the analyte molecule and surface of sensing platform. The above-mentioned interface is highly responsible and capable to recognize, sense and regulate the sensitivity as well as specificity of the analyte. Potential of working electrode depending upon the concentration of analyte which is present over the electrolyte (Vinoth, Shalini Devi, and Pandikumar 2021). Where, reference electrode provides distinct potential. The desirable output of sensor is accurate, selective, sensitive and rapid detection of various (bio) molecules, which is essential for basic biomedical research applications.

Over the last few decades, electrochemical sensor field is grown steeply because it furnishes fast, simple, reliable, cost-effective, shorter response time and its higher performance and also, they are highly selective, sensitive and specific along with lower detection limit (Simões and Xavier 2017). Electrochemical techniques such as cyclic voltammetry (CV), chrono-amperometry (CA), electrochemical impedance spectroscopy (EIS), square wave voltammetry (SWV) and differential pulse voltammetry (DPV) has been employed for the detection of pollutants. Electrochemical sensor can generate an electrical signal into a digital signal for the analysis through a sequence of steps as shown in Figure 1. To conclude, electrochemical sensors have acquired slight edge because of their cost-effectiveness, high sensitivity and rapid response.

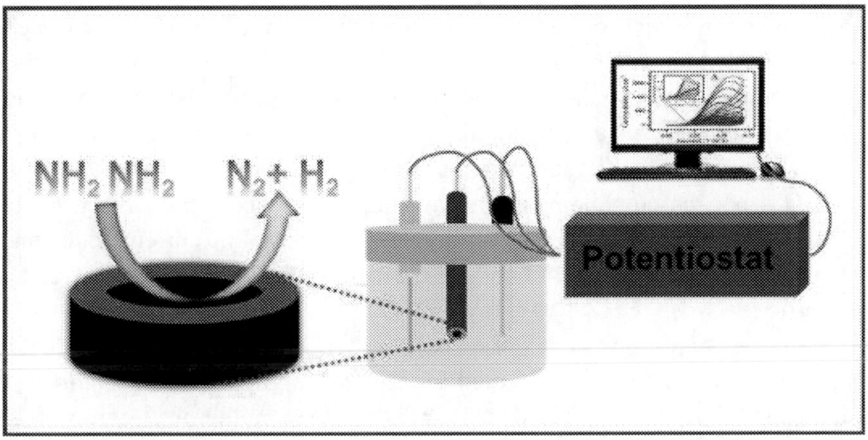

Figure 1. Schematic representation of HZ electro-catalytic oxidation at the modified surface and the electrode setup with potentiostat.

Mechanism of Hydrazine Detection

Electrochemical sensing of hydrazine generally functions through monitoring the concentration of species by analysing the electrochemical signal, additionally the total electrons were consumed during the electrochemical oxidation of hydrazine with three-electrode setup.

$$N_2H_5^+ \rightleftharpoons H^+ + N_2H_4 \quad (1)$$

$$N_2H_4 \rightarrow N_2H_4^{\cdot+} + e \quad (2)$$

$$N_2H_4^{\cdot+} \rightarrow N_2 + 4H^+ + 3e \quad (3)$$

Compton et al. (Miao and Compton 2021; Miao, Chen, and Compton 2021) meticulously proposed that HZ is highly electroactive in unprotonated form (N_2H_4), however, protonated species of HZ ($N_2H_5^+$) is electro-inactive. The electrochemical oxidation of HZ occurring by 4 e^- process along with rate determining second step shown in above equations.

Material Choice

The on-site detection of hazardous pollutant HZ is well explored with plenty of nanomaterials with higher selectivity and precision. On top of that, carbon-based nanomaterials gain much interest into the sensing platform on account of low cost, high surface area and high surface to volume ratio with greater catalytic activity due to the existence of porous structure with graphitic active sites. Carbon-based nanomaterials have multiple benefits such as reasonable, sustainable, wide potential window and lesser background current which made them as appropriate material for the electrochemical sensing application, in specific towards the HZ.

In this chapter, we focus especially on furnishing a thorough outlook of latest research endeavours devoted for electroanalytical sensing platform for detecting HZ with recent advances using numerous carbon-based nanomaterial systems.

Electrochemical Sensing of Hydrazine with Different Carbon-Based Nanomaterials

Up to now, plenty of electrochemical sensing platforms are available for the detection of environmental pollutants. Recently, lots of activities are engaged to improvise the linear detection range, response time, reusability and sensitivity by electrode modifications. To improve the sensing parameters, nanostructured materials are introduced as an electrode modifier that can serve as a matrix to facilitate rapid electron transfer at electrode interface with the target analyte (Shalini Devi, Sasya, and Krishnan 2020). For precise and potentially selective target drug detection, adequate number of literatures are existing based on metal-based nanomaterials, biomaterials and carbonaceous materials as modifiers. However, lesser stability and higher cost of bio-based materials such as enzymes, antibiotics, proteins and DNA aptamer, etc. in chemical environments like O_2 and pH restrict their scope. Whereas carbonaceous materials hold immense potential towards commercialization, because they are cost effective and have wide potential window with outstanding catalytic activity. Due to the unique structural and excellent mechanical, electrical, optical and thermal properties, carbon-based nanomaterials gained a significant interest in diverse research areas such as electrochemical sensor, biomedical applications, supercapacitors, drug delivery, photocatalysts, batteries, solar cells, energy storage and conversion (Patel and Kim 2019).

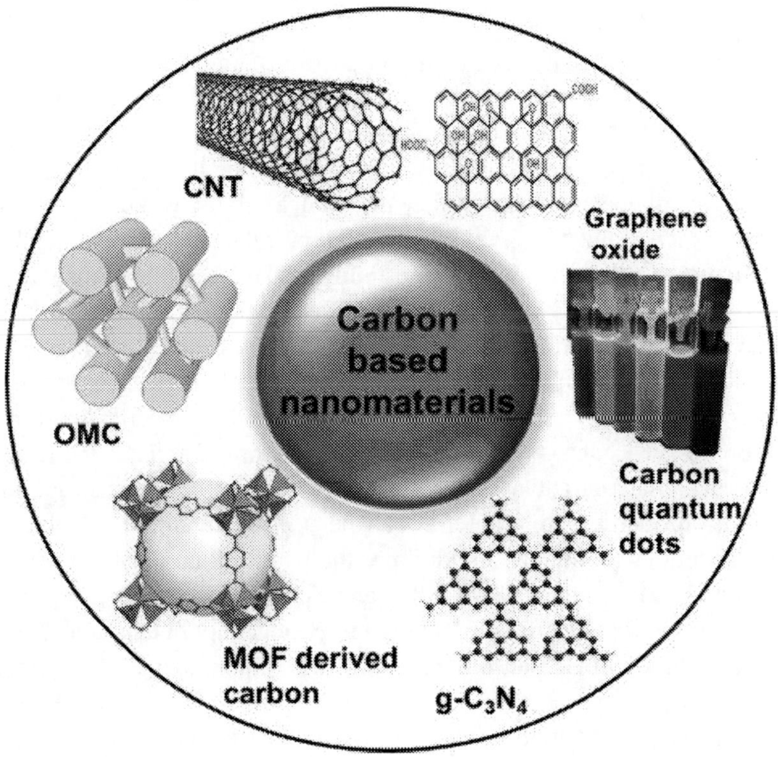

Figure 2. Schematic representation of different types of carbon-based nanomaterials (CBNs).

Continuous efforts are made to upsurge the sensing performance of environmental hazardous pollutants with the usage of carbonaceous materials like graphene, carbon nanotubes, porous carbons, bio-derived carbon, carbon quantum dots, carbon nano-fibre, metal organic framework (MOF) derived carbon, graphitic carbon nitride and carbon black etc (Figure 2), which will provide a novel route to enhance the sensing activity (Hui 2019). Additionally, modification of these carbonaceous materials with functional groups, doping, induced defect sites, creating different nanostructure, porosity and surface area allows the opportunity to augment their properties.

Carbon Nanotubes

Starts from the discovery of carbon nanotubes (CNTs), this superior material gained popularity and widely utilized so far owing to its enhanced properties. Commonly, CNTs are produced through arc discharge method or by chemical vapor deposition (CVD) technique from graphite. Further, the cylindrical construction of CNT holds both optical and electrical characteristics but not limited to the expanded sp^2 carbon, also can be tuned by physical properties like length, diameter, single walled as well as multi walled, chirality and surface functionalization. Owing to the pluralistic features, CNTs had extensively investigated towards enormous industrial applications (C. Cha et al. 2013). Additionally, CNTs are classified into two main groups, single walled carbon nanotubes (SWCNT) and multi walled carbon nanotubes (MWCNTs) are commonly have a nanoscale tube of single or otherwise with multilayer graphite, respectively which are curled at specific helical angle position around centre axis (Crevillen, Escarpa, and García 2019; Congcong Zhang and Du 2020). On account of its extensive thermal, structural and electronic properties, this material can be utilized as improved platform to sense hazardous pollutants. In addition, the extraordinary stability of tubular structure as well as side wall of CNTs can be employed as superlative template for nanomaterials, specifically nanowires (Congcong Zhang and Du 2020). The performance of HZ sensors with CNT and its composites are summarized in Table 1.

Electrodeposited gold nanoparticles attached over ZnO-MWCNT films were employed for the exact electrochemical quantification of HZ. Here, the electrodeposited nano-Au was immobilized by utilizing ZnO-MWCNT as a biomimic membrane with the detection limit of 0.15 μM and has wide linear range from 0.5 to 1800 μM with higher sensitivity (Cuihong Zhang et al.

2010). Sung Phil Kim et al. proclaimed that chemically reduced bimetallic PdPt nanoparticles gradually deposited over thiolated CNTs was employed as electron mediator to fabricate amperometric hydrazine chemical sensor with an at % of Pt and Pd around 3.2 and 3.0, respectively. This sensor provides a wide linear response of 0.55 to 1200 μM with lower detection limit of 0.28 μM (S. P. Kim et al. 2015). The introduction of metallic nanoparticles on CNTs greatly influences the electrocatalytic activity.

In 2017, Tuantuan Zhou et al. reported that cobalt oxide nanostructures with multi-walled carbon nanotubes (Co_3O_4/MWCNTs) composites were prepared by hydrothermal synthesis and HZ sensing performance was examined with a low detection limit of 0.449 μM (T. Zhou et al. 2017). In an interesting strategy, the electrochemical properties of CNT were improved by compositing with cobalt pyrite (CoS_2) by a simple wet chemical method to quantify HZ in aqueous solution by i-t technique in the concentration range of 0.1 nM to 1.0 mM (M. M. Rahman et al. 2016). Recently, carbon paste electrode (CPE) modified by ferrocene- derivative (EFTA- ethyl 2- (4-ferrocenyl [1,2,3] triazol-1-yl) acetate) mixed with ionic liquid and composite made of CoS_2- carbon nanotube (EFTA/IL/CoS_2-CNT/CPE) was explored for the detection of HZ. Here, synergistic impact of IL with ferrocene derivative and CNT nanocomposite played a vital role for the detection of HZ with a LOD of 0.015 μM (Tajik et al. 2021). Bin Fang et al. employed ammonia-evaporation-induced method for synthesizing ZnO nanoflowers which was immobilized with MWCNT on GCE utilized for electrochemical sensing of HZ. Here, CNT was employed to stabilize and bind the ZnO on electrode surface to augment mass transport with huge surface area (Fang et al. 2009). Further, hydrothermally synthesized bimetallic oxides were employed as an electrode modifier for the quantification of HZ. Nickel ferrite ($NiFe_2O_4$) nanoparticles over MWCNT were utilized as an electrode alterer for the detection of HZ. Due to the large surface area, pseudocapacitive behaviour, electrochemical stability and various oxidation states of $NiFe_2O_4$ are highly responsible for the electrocatalytic activity (Fang et al. 2011).

Another work employed curcumin multi-walled carbon nanotubes modified over GCE was fabricated using electrodeposition. Here, curcumin (CM) electrodeposited on MWCNT for quantifying HZ with LOD of 1.4 μM and has a considerable sensitivity of 22.9 nA μM^{-1}. Additionally catalytic rate constant, k_{cat} of 6.26×10^3 M^{-1} s^{-1} was obtained by CA (Zheng and Song 2009).

Table 1. Various carbon-based nanomaterials and their composites are described for the detection of HZ with analytical parameters

S. No	Electrodes	Method of detection	LOD	Linear range	Sensitivity	Ref.
1	CNT-PdPt	CA	0.28 μM	0.55–1200 μM	NR	(S. P. Kim et al. 2015)
2	nano-Au/ZnO-MWCNTs/GCE	CA	0.15 μM	0.5 - 1800 μM	0.0428 μA mM^{-1}	(Cuihong Zhang et al. 2010)
3	Co_3O_4/MWCNTs	CA	0.449 μM	1.0–187.4 mM	120.26 mA mM^{-1}	(T. Zhou et al. 2017)
4	CoS_2-CNT	DPV	0.015 μM	0.03–500 μM	0.073 μA μM^{-1}	(Tajik et al. 2021)
5	CoS_2/CNT NCs/Nafion/GCE	CA	0.1 nM	0.1 nM - 1.0 mM	4.430 μA $nM^{-1}m^{-2}$	(Online et al. 2016)
6	PEDOP/MWCNTs–Pd/GCE	CA	4 × 10−8 M	1.0 × 10−7 - 5.0 × 10^{-3} M	NR	(S. K. Kim et al. 2011)
7	CdO/CNT NCs	CA	4 pM	0.01 nM - 0.1 mM	25.7911 μA μM cm^{-2}	(Crapnell and Banks 2022)
8	AuNPs/CNTs-ErGO/GCE	CA	0.065 μM	0.3 - 319 μM	9.73 μA μM^{-1} cm^{-2}	(Y. Zhang et al. 2019)
10	EFTA/IL/CoS_2-CNT/CPE	DPV	0.015 μM	0.03 - 500.0 μM	0.073 μA μM^{-1}	(Tajik et al. 2021)
11	ZnO/MWCNTs/GCE	CA	0.18 μM	0.6 - 250 μM	0.2469 μA μM^{-1}	(Fang et al. 2009)
12	$NiFe_2O_4$/MWCNTs nanocomposites	CA	1.5 μM	5.0 μM - 2.5 mM	0.1 μA μM^{-1}	(Fang et al. 2011)
13	CoHCF modified MWCNTs/graphite composite	CA	0.91 μM	2.0 - 1200 μM	0.34 nA μM^{-1}	(Abbas et al. 2020)
14	GNF	CA	0.3 μM	0.5 - 7.5 μM	0.028 μA μM^{-1}	(Mutyala and Mathiyarasu 2015)
15	Au@Pt-nFs/GO	CA	0.43 μM	0.8 μM - 0.429 mM	1695.3 μA mM^{-1} cm^{-2}	(Z. Yang, Zheng, and Zheng 2017)
16	PdNPs/rGO	CA	0.007 μM	0.1 - 1000 μM	NR	(Krittayavathananon et al. 2014)
17	Pt-Pd nanoparticles on ERGO	CA	1.7 μM	0.007 - 5.5 mM	NR	(Ghasemi et al. 2019)
18	rGO/Au-Pd composite	CA	0.016 μM	0.1 - 200 μM	NR	(H. Zhou et al. 2020)
19	Au@Ni-BCS/rGO	CA	0.9 μM	3 μM -20 mM	NR	(X. Zhang and Zheng 2019)
20	Pd-PEDOT@CM/GNP/GCE	CA	0.28 μM	1 - 5000 μM	0.200 μA μM 1 cm 2	(Promsuwan, Thongtawat, and Limbut 2020)

S. No	Electrodes	Method of detection	LOD	Linear range	Sensitivity	Ref
21	HAP-rGO nanocomposite	CA	0.43 μM	2.5 μM - 1.16 mM	4.21 μA μM^{-1} cm^{-2}	(Gao et al. 2017)
22	MnO_2/3D-N-rGO	CA	085.0 μM	20 - 100 μM	NR	(Wuamprakhon et al. 2017)
23	CoS_2/IL-graphene/GCE	CA	0.39 μM	5 - 400 μM	NR	(Luan et al. 2018)
24	Co–GE/GCE	CA	0.1 μM	0.25 μM - 2200 μM	562.5 mA mM^{-1} cm^{-2}	(Links 2012)
25	GCE/r-GO/*poly*(CoOBImPc)	CA	40 nM	100 - 900 nM	0.3696 μA nM^{-1} cm^{-2}	(Nemakal et al. 2019)
26	HRGO–Co@SnO_2	CV	NR	NR	NR	(Ramanathan et al. 2019)
27	CM-AgNPs decorated rGO-$FeCo_2O_4$	DPV	23.7 nM	2 μM - 1200 μM	0.113 μA μM^{-1}	(Mejri and Mars 2019)
28	RGO/Pt-TPP nanocomposite	CA	5 nM	13 nM - 232 μM	1.44 μA μM^{-1} cm^{-2}	(Sakthinathan et al. 2017)
29	$ZnFe_2O_4$/RGO/SPE	DPV	0.01 μM	0.03 - 610.0 μM	NR	(Tajik et al. 2022)
30	PANI/Gr	CA	15.38 mM	0.01 μM - 0.1 mM	32.54×10^{-5} A cm^{-2} mM^{-1}	(Ameen, Akhtar, and Shik 2012)

NR- not reported; CoS_2-CNT NCs- Cobalt pyrite (CoS_2) decorated carbon nanotube nanocomposites; HAP- hydroxyapatite; CoS_2/IL-graphene- Cobalt disulfide/ ionic liquid; Co–GE/GCE- cobalt (Co)-nanoflowers on petalage-like graphene; r-GO/*poly*(CoOBImPc)- cobalt (II) octabenzimidazolephthalocyanine; CM-AgNPs . Curcumin stabilized silver nanoparticles; Au@Pt-nFs/GO- Au@Pt core-shell nanoflowers supported on GO; Au@Ni-BCS/rGO-Au@Ni bimetallic decorated graphene oxide; PANI/Gr- polyaniline/graphene.

Seul Ki Kim et al. reported poly-ethylene dioxy pyrrole (PEDOP)-coated MWCNTs–palladium nanoparticles modified electrode for electrochemical detection of HZ with improved oxidation peak current and lower overpotential of -0.25 V. PEDOP/MWCNTs–Pd/GCE was utilized for sensing of HZ in analytically significant pH range with lower LOD of 4×10^{-8} M (S. K. Kim et al. 2011).

Furthermore, Xinjin Zhang et al. reported a highly porous gold nanoparticles anchored polydopamine-modified MWCNT composite with rGO with high surface area of 290 m^2 g^{-1} along with large pore volume. The catalytic activity of the prepared materials is enhanced with high surface area for the detection of HZ with a low LOD of 0.31 μM (X. Zhang and Zheng 2020).

Xinchun Li et al. enunciated CoHCF nanoparticles on the MWCNTs/ graphite composite employed for the sensing of HZ by microchip electrophoresis. Further, synergetic effect of cobalt hexacyanoferrate and MWCNTs played a major role and enhanced the detection range to two order magnitude (LOD- 0.91 μM) (Li et al. 2012).

Graphene and Their Derivatives

Graphene, is a pioneering nanomaterial, has attained a tremendous attention since its discovery in 2004. Graphene has a 2D structure with single layer sp^2 hybridized carbon attached through covalent bonds for producing hexagonal lattice with flat structure. Besides, it has significant advantages like high electron mobility, flexibility, thermal stability, conductivity, high strength and can be employed as an electrode modifier in various electroanalytical fields (Congcong Zhang and Du 2020). Moreover, it has excellent mechanical property and biocompatibility which could avoid the problems associated with metal nanoparticles like Fe, Ni, Co as well as CNTs and leads to the potential applicability towards the electrochemical sensing applications. The performance of HZ sensors with graphene and its composites are summarized in Table 1.

Mutyala et al. synthesized graphene nanoflakes (GNF) by cationic surfactant facilitated with exfoliated graphite which produced stable suspension in organic solvent. Additionally, edge sites of GNF promoted the electron transfer kinetics in this nanostructure. Due to the inimitable structural and electronic properties, GNF showed higher current response for HZ with enhanced stability (Mutyala and Mathiyarasu 2015).

Zhenting Zhao stated that electrochemical sensor of hydrazine has been facilitated via Au nanoparticles (AuNPs) coated over carbon nanotubes by mixing with electrochemically r-GO film (CNTs-ErGO). The optimization procedure has been carried out with the variation of CNTs to ErGO ratio with the amount of Au nanoparticles and the calculated LOD was 0.065 μM (Zhao et al. 2016). In situ polyol synthesized different particle sizes of PdNPs over rGO, which contains same structure like graphene connects with carboxylic acid groups in the edge of graphene sheets. The electrochemical detection of HZ was executed by rotating disk electrode (RDE) in wastewater with a LOD of 7 nM (Krittayavathananon et al. 2014). In an interesting way, bimetallic nanoparticles such as Pt and Pd electrodeposited over ErGO was utilized to enhance the sensing of HZ. Here, bimetallic nanoparticles had been employed to improve current density and lower potentials with huge sensitivity than a single metal deposited ERGO/GCE. Here, Pd noble metal was employed for the enhanced dispersion of Pd/Pt over ERGO surface (Ghasemi et al. 2019). Yang et al. synthesized three-dimensional porous core-shell Pt, Au supported GO via a greener route. Here, Au core and Pt dendritic shells were spread over GO provides a large surface area, open structure with abundant active sites as well as accelerates the kinetics for the detection of HZ with wider linear response range of 0.8 μM to 0.429 mM and higher sensitivity (Z. Yang, Zheng, and Zheng 2017). Further, bimetallic catalyst of Au@Ni was decorated over GO by differing the mass ratio of Au and Ni to enhance the catalytic activity by forming bimetallic heterodimer or core-shell structure. It achieved a LOD of 0.06 μM with high selectivity and good reproducibility (X. Zhang and Zheng 2019). Huanfang Zhou and co-workers reported a facile preparation of gold/palladium (Au/Pd) bimetallic nanoparticles intervened rGO synergistically fashioned by multi-layered and well-structured nanohybrid were explored as an electrocatalyst for the determination of HZ. Further, CV technique is employed to synthesize rGO/Au-Pd nanohybrid devoid of any co-reactants with purification and drying process (H. Zhou et al. 2020).

Another work employed a novel electrochemical assembly of CoS_2-decorated ionic liquid-functionalized graphene for the electrochemical detection of HZ. The combination of both CoS_2 and graphene components had advantages to overcome the limitations of each individual components with a relatively low detection limit of 0.39 μM (Luan et al. 2018). Yaping He and co-workers stated hierarchical cobalt (Co)-nanoflowers prepared by one-pot ultrasonic electrochemical technique on petalage-like graphene (GE). Here, GO was seized and reduced to GE by providing active sites for nucleus formation of Co^{2+} with attachment. Further, electrochemically combining the deposited material will forms Co-GE with increased dimensionality (Links 2012). Interestingly, organometallic framework was utilized as a template to synthesize cobalt disulphide rhombic dodecahedral nanocomposites coated with nitrogen coordinated graphene has huge advantages of high porosity. The presence of N coordination boosted the intrinsic properties of active sites by altering the electronic structure of cobalt centre in N-CoS_2-graphene composite which has a good activity towards the detection of HZ with an LOD of 30 nM and linear range of 1.0 μM – 1.0 mM (J. Yang et al. 2022). Phatsawit Wuamprakhon et al. reported layered K-birnessite MnO_2 nanosheets layered over nitrogen-doped rGO aerogel (3D-N-rGO) for the detection of HZ using rotating disk electrode (RDE). Here, the Mn oxidation number is reduced at MnO_2/3D-N-rGO surface and monitored by X-ray absorption spectroscopy (XAS) (Wuamprakhon et al. 2018).

Recently, $ZnFe_2O_4$/RGO nanocomposite was fabricated by a hydrothermal protocol to modify screen-printed electrode (SPE) surface for the detection of micromolar HZ concentration. Moreover, the synergistic impact of $ZnFe_2O_4$ with physicochemical features of rGO provides a platform with enhanced rGO edge-plane structures, large surface area with higher conductivity (Tajik et al. 2022). Additionally, Alma Mejri et al. (Mejri and Mars 2019) revealed curcumin stabilized Ag nanoparticles were modified by $FeCo_2O_4$ – rGO composite which acted as electrode materials for the sensing of HZ at +0.15 V working potential (LOD- 23.7 nM).

Recently, Sushma Rani et al. proclaimed the solvothermal synthesis of Zn -metal organic framework - graphene oxide composite (Zn-MOF@rGO) for the quantification of HZ by amperometric technique. Herein, both Zn-MOF and rGO helps to significantly improving the performance due to huge surface area, excellent adsorption affinity, chemical stability and electrical conductivity (Rani et al. 2020).

In another study, flow-injection amperometric hydrazine detection was established by working elelctrode altered with Pd-poly(3,4-ethylene dioxythiophene) covered over carbon microspheres with graphene nanoplatelets (Pd-PEDOT@CM/GNP/GCE). In flow-based detection procedures, optimization parameters are crucial for better performance such as quantity of Pd-PEDOT@CM/GNP in prepared slurry, flow rate, sample volume and working potential (Promsuwan, Thongtawat, and Limbut 2020). In addition, Feng Gao et al. have studied in-situ grown hydroxyapatite (HAP) over GO under ambient temperature of 85°C, which could be efficiently induce the reduction of GO without any additional reducing agents or high-temperature thermal treatment. Electrochemical sensing assays of toxic chemical hydrazine over HAP/rGO deals with an LOD of 0.43 μM (Gao et al. 2017).

Manjunatha Nemakal et al. synthesized an electropolymerized cobalt (II) octabenzimidazolephthalocyanine (CoOBImPc) with 100 times enhanced electrical conductivity than its precursor. In addition, to amplify conductivity, electron transfer kinetics and surface area, CoOBImPc was combined with r-GO. The modified GCE/*poly*(CoOBImPc) showed good electrochemical oxidation response with a LOD of 33 nM. Further, the designed sensor was selective towards HZ in the presence of various coexisting interferents (Nemakal et al. 2019).

Subramanian Sakthinathan et al. described the noncovalent π–π stacking interaction between the rGO/platinum (II) tetraphenylporphyrin (RGO/Pt-TPP). Hither, the RGO/Pt-TPP composite exhibited an enhanced catalytic activity with low overpotential and higher oxidation current towards the detection of HZ (Sakthinathan et al. 2017).

Sadia Ameen et al. reported in situ electrochemically synthesized aniline composite with GO in HCl. The polyaniline/graphene oxide (PANI/GO) composite exhibited a significant activity towards the electrochemical detection of HZ by higher electrical conductivity of PANI conducting polymer with higher sensitivity (Ameen, Akhtar, and Shik 2012).

Graphitic Carbon Nitride-Based Materials

Graphitic carbon nitride (g-C_3N_4), metal free, graphite like polymeric carbon nitride is a fascinated class of material in recent years due to the inimitable electronic structure, higher stability, optical properties and low cost. It owns an outstanding catalytic property with unique structure due to the presence of covalent bond of carbon and nitrogen by π- conjugation. These materials have a wider application in sensor, degradation and biological applications etc. The bulk g-C_3N_4 can be easily prepared and possess sluggish charge transfer kinetics, lesser electrical conductivity with poor water dispersibility and hence cannot directly be used for the electrochemical sensing applications. Instead of bulk, g-C_3N_4 nanosheets were produced to further enhance the surface area by decreasing the interaction of g-C_3N_4 layers. In addition, the conductivity is increased by post functionalisation techniques like doping with metal oxide, coupled with graphene or by CNT etc (Dai et al. 2017). The performance of HZ sensors with g-C_3N_4 and its composites are summarized in Table 2.

A low-cost S doped graphitic-carbon nitride (S-gC_3N_4) was coated on conductive fluorine doped tin oxide (FTO) plate by doctor blade technique can be utilized for the electrochemical oxidation of HZ. LSV can be applied with an LOD of 0.06 μM and a linear range of 60 μM - 475 μM (Mohammad, Khan, and Cho 2020).

Mohaddeseh Afshari et al. reported silver (Ag) nanoparticle deposited over FTO polyaniline/g-C_3N_4 (PANI/g-C_3N_4/AgNPs) film for the electrochemical detection of HZ with detection limit of 300 μM (S/N = 3). In addition, amperometric current shown here was extremely stable with an existence of interferents (Afshari, Dinari, and Momeni 2019). Mohammad et al. (Mohammad et al. 2021) revealed that biogenic or greener approach is utilized for synthesizing Ag particles decorated tin oxide (SnO_2)-modified graphitic carbon nitride. Herein, fluorine-doped tin oxide (FTO) substrate is used as a working electrode and modified by doctor blade technique and employed for electrochemical assay of HZ with an LOD of 0.16 μM. Herein, the reported works are related to the incorporation of silver nanoparticles with g-C_3N_4 which greatly influences the catalytic activity.

Table 2. Various carbon-based nanomaterials and their composites are described for the detection of HZ with analytical parameters

S. No	Electrode	Method of detection	LOD	Linear range	Sensitivity	Ref
1	Co_3O_4/g-C_3N_4	CA	1 μM	5 - 1000 mM	874.3 mA mM^{-1}	(Dai et al. 2017)
2	S-g-C_3N_4/FTO	LSV	0.06 μM	60 - 475μM	NR	(Mohammad, Khan, and Cho 2020)
3	PANI/g-C_3N_4/AgNPs nanocomposite	CV	300 μM	5 - 300 mM	NR	(Afshari, Dinari, and Momeni 2019)
4	ZIF/g-C_3N_4/RGO nanohybrid	CA	32 nM	0.0001 - 1.0386 mM	93.71 μA mM^{-1} cm^{-2}	(Liang et al. 2021)
5	g-C_3N_4-KBC	CA	0.7 μM	2 -14 μM	2.5 μA mM^{-1}	(Ramanujam and Thirupathi 2018)
6	**Ag@SO-gCN/FTO**	LSV	0.164 μM	2 - 30 μM	2.01 μA μM^{-1} cm^{-2}	(Mohammad et al. 2021)
7	AuNPs@NPC-rGO/GCE)	CA	9.6 nM	0.05 - 1.00 μM	NR	(Y. Zhang et al. 2019)
8	CM - OMC	CA	0.39 μM	12.5 - 225 μM	10.54 nA μM^{-1}	(Wei et al. 2015)
9	CuS-OMC	CA	0.10 μM	0.25 - 40 μM	0.915 μA cm^{-2} μM^{-1}	(Srinidhi et al. 2020)
10	CQD	CA	39.7 mM	125 - 1125 mM	151.5 mA mM^{-1} cm^{-2}	(Sha et al. 2018)

NR- not reported; S-gC_3N_4- sulfur doped graphitic-carbon nitride; PANI/g-C_3N_4/AgNPs- graphitic carbon nitride/polyaniline/silver nanocomposites; ZIF- zeolitic imidazolate framework; CM – OMC- curcumin - ordered mesoporous carbon.

Recently Gaopeng Dai et al. demonstrated flower-like Co_3O_4/g-C_3N_4 nanocomposite prepared by simple hydrothermal technique. Mass fraction of 2% g-C_3N_4 was opted for better catalytic activity with Co_3O_4 which showed excellent performance towards the oxidation of HZ by both CV and CA technique. Importantly, the Co_3O_4/g-C_3N_4 modified electrode provides a LOD of 1 μM at S/N ratio of 3 with large sensitivity of 874.3 mA mM^{-1} (Dai et al. 2017).

In an interesting way, 2D/2D/2D heterojunctions were fabricated by a dense zeolitic imidazolate framework (ZIF) which is moulded with rGO and g-C_3N_4 and directly coated over carbon cloth (CCE) to employ as a modified electrode for the electrooxidation of HZ. The synergy and coupling effects of these nanohybrid endows marvellous electrocatalytic activity towards amperometric detection of HZ with LOD of 32 nM in the monitoring rage of 0.0001 to 1.0386 mM (Liang et al. 2021). Thippani Thirupathi et al. (Ramanujam and Thirupathi 2018) proclaimed that g-C_3N_4 was prepared by pyrolysis of melamine along with Ketjenblack 600JD carbon (KBC) at 500 °C. The synthesized g-C_3N_4-KBC will oxidize hydrazine close to onset potential of 0.145 V. Compared with all other carbon-based materials, g-C_3N_4 type of material is superior in terms of selectivity, sensitivity, detection range and stability.

Metal Organic Framework (MOF) Derived Porous Carbon Materials

Metal-organic framework (MOF) derived carbon nanomaterials has emerged as an innovative platform for preparing porous carbon with huge surface area due the inherent properties like hierarchical structure and tunable pores (Y. Zhang et al. 2019). Confinement preparation process can be used for preparing the zeolitic imidazolate framework-67 (ZIF-67) as reported by Yong Zhang et al. (Y. Zhang et al. 2019). Further, Au embedded over N-doped porous carbon which was affixed on r-GO nanosheets for assaying electrooxidation of HZ. In liquid phase, LOD of HZ electrooxidation was measured to be 9.6 nM with a detection range 0.05 to 1.00 μM. Secondly, in gaseous phase the prepared sensor showed LOD of 1.8 ppm.

Ordered Porous Carbon Materials

Generally, carbon-based materials play a crucial role in earth's civilization due to the strong bond formation ability. On which ordered mesoporous type carbon (OMC) materials have tunable porous structure with high specific surface area which make them as a good candidate for electrochemical sensor applications. OMC is synthesized by a template technique and a variety of shapes and morphologies are available such as nanospheres, monoliths, single crystals and fibres. OMCs are a long range of structurally ordered, made of mesoporous channels which are arranged in hexagonal, worm, cubic, lamellar like structures with huge specific surface area. The performance of HZ sensors with OMC and its composites are summarized in Table 2.

Recently, Srinidhi et al. (Srinidhi et al. 2020) proclaimed that solvothermally synthesized copper sulphide over OMC (CuS-OMC) and nano casting route is utilized for OMC preparation. Herein, CuS-OMC was immobilized over GCE and used for electrochemical detection of HZ at an applied potential of 270 mV with an LOD of 0.10 μM and a linear calibration range of 0.25 to 40 μM.

Qinghe Wei et al. (Wei et al. 2015) developed an ultra-sensitive voltametric sensor using curcumin - ordered mesoporous carbon composite (CM - OMC) for the detection of hydrous hydrazine. CM was electrodeposited over OMC which enhanced the electroactivity of CM-OMC compared to pristine OMC and CM with a LOD of 0.39 μM. Linear relationship was obtained within the range of 12.5 to 225 μM with appreciable sensitivity of 10.54 nA μM^{-1}.

Carbon Quantum Dots

Commonly, carbon quantum dots (CQD) and graphene quantum dots have been an interesting class of materials which are used for the electrochemical sensing of hydrazine. Moreover, CQDs are small carbon nanoparticles with sizes less than 10 nm (Patel and Kim 2019). These materials have an excellent physio-chemical property such as high electrical conductivity, bio-compatibility, excellent redox properties and size tunability as well as higher stability. CQDs usually synthesized by two steps, top-down approach and bottom-up approach. In top-down approach, the CQDs can be prepared by a complicated steps by using higher energy consuming devices (by chemical/electrochemical oxidation etc.,). For synthesizing CQDs, extensive research has been going on for economic growth by using carbon sources as precursors. Amongst the available synthetic techniques of CQDs, pyrolysis was highly recommended with great advantages like lower cost, mild synthesis condition and versality of various carbon precursors (Sha et al. 2018).

Interestingly, Rinky Sha et al. reported a single step pyrolysis of biomass derived (from chia seeds) carbon quantum dots that can be applied for the electrochemical determination of hydrazine. The sensing capability was boosted by the existence of oxygen rich functional groups over the CQD surface with an LOD of 39.7 μM (Sha et al. 2018).

Conclusion and Future Scope

In the past few decades, carbon-based nanomaterials (CBNs) witnessed tremendous growth in their synthesis techniques and can be applied on various applications which elevated the position of carbon-based nanomaterials as a widely utilized class of nanomaterials. Such CBNs are classified as nanotubes, carbon nitrides, activated carbons, MOF derived carbons and graphene oxide are some of the most exemplary forms of carbon with incredible properties. Further, CBNs have brought rapid attention owing to their unique optical, mechanical, electrical properties with large surface area, chemical stability, structural dimensions, facile synthesis techniques from wide carbon sources, electrocatalytic properties, surface tunability with high surface to volume ratio and fluorescence properties etc., (Maduraiveeran and Jin 2021). Currently, fascinating properties of CBNs made them as a potential candidate towards versatile applications such as catalysis, electronic devices, drug delivery, bio-sensing and bio-imaging. The present chapter nurturing a comprehensive overview of electrochemical sensing application by various carbon-based nanomaterials modified electrodes for the electrooxidation of hazardous hydrazine. Moreover, CBNs expose a new horizon with different nanomaterials towards the fabrication of novel sensing platform with improved activities.

To summarize, the studies have been made to develop a sensing platform for the electrochemical detection of HZ with effective carbon-based nanomaterials by tuning their design, doping with heteroatoms, controlling their morphologies and preparing a composite with conductive metal oxide or with noble metals as well as enhancing their physical and chemical properties.

This chapter signifies the comprehensive attempt to rationalize the existing carbon-based nanomaterials, the superstructures formed by combining with different functional groups, and the various predicted carbon nanostructures. To this end, we have classified sensing of hydrazine by different carbon-based materials in terms of their different structures. We have also described the obtained sensing parameters like linear detection range and lower detection limit with different carbon-based nanomaterials and the significant changes in their sensing activity.

It is crucial to mention that carbon-based nanomaterials have been exhaustively employed to develop an electrochemical sensor owing to their higher conductivity. On top of that, these materials have a significant advantage for the electrochemical detection of HZ with higher sensitivity and selectivity (with decreased overpotential). However, pristine graphene and

CNT exhibit high conductivity, still the existence of defects and functional groups could be enhancing the sensing applications. Further, strong acid treated CNT and graphene produces better sensing activity than pristine compounds.

Besides, substantial quantity of research dilemmas and challenges are still need to be addressed for the carbon-based nanomaterials for enhancing the catalytic activity. Also, the on-site detection of hydrazine with higher selectivity and sensitivity is also a major concern while developing a sensing platform to make them commercially scalable. This represents the propitious direction for developing carbon-based nanomaterials to fascinate interest in diverse research fields.

Author Credit

- *Narmatha Sivaraman*: Conceptualization, Writing (Original Draft), Visualization.
- *Sakkarapalayam Murugesan Senthil Kumar*: Writing (Review & Editing).
- *Rangasamy Thangamuthu*: Writing (Review & Editing), Validation, Supervision.

Acknowledgments

The authors would like to acknowledge the Department of Science and Technology, India, for providing DST-INSPIRE fellowship (DST/INSPIRE Fellowship/ 2019/IF190394).

References

Abbas, Yasir, Shafqat Ali, Majid Basharat, Wenqi Zou, Fan Yang, Wei Liu, Shuangkun Zhang, Zhanpeng Wu, Naeem Akhtar, and Dezhen Wu. 2020. "Heteroatom-Doped Carbon Nanoparticle-Ionic Liquid Composites as Electrochemical Sensors for Uric Acid." *ACS Applied Nano Materials* 3 (11): 11383–90. https://doi.org/10.1021/acsanm.0c02466.

Afshari, Mohaddeseh, Mohammad Dinari, and Mohamad Mohsen Momeni. 2019. "The Graphitic Carbon Nitride/Polyaniline/Silver Nanocomposites as a Potential

Electrocatalyst for Hydrazine Detection." *Journal of Electroanalytical Chemistry* 833 (June 2018): 9–16. https://doi.org/10.1016/j.jelechem.2018.11.022.

Ameen, Sadia, M Shaheer Akhtar, and Hyung Shik. 2012. "Sensors and Actuators B : Chemical Hydrazine Chemical Sensing by Modified Electrode Based on in Situ Electrochemically Synthesized Polyaniline / Graphene Composite Thin Film." *Sensors & Actuators: B. Chemical* 173: 177–83. https://doi.org/10.1016/j.snb.2012.06.065.

Cha, C., Shin, S.R., Annabi, N., Dokmeci, M.R., Khademhosseini, A. 2013. Carbon-Based Nanomaterials : Multifunctional Materials for. ACS Nano. 7 : 2891-2897. https://doi.org/10.1021/nn401196a.

Crapnell, Robert D, and Craig E Banks. 2022. "*Diagnostics Electroanalytical Overview : The Electroanalytical Sensing of Hydrazine Introduction : Hydrazine*," 71–86. https://doi.org/10.1039/d1sd00006c.

Crevillen, Agustín G., Alberto Escarpa, and Carlos D. García. 2019. *Chapter 1: Carbon-Based Nanomaterials in Analytical Chemistry. RSC Detection Science*. Vol. 2019-Janua. https://doi.org/10.1039/9781788012751-00001.

Dai, Gaopeng, Jingwen Xie, Cheng Li, and Suqin Liu. 2017. "Flower-like Co3O4/Graphitic Carbon Nitride Nanocomposite Based Electrochemical Sensor and Its Highly Sensitive Electrocatalysis of Hydrazine." *Journal of Alloys and Compounds* 727: 43–51. https://doi.org/10.1016/j.jallcom.2017.08.100.

Fang, Bin, Cuihong Zhang, Wei Zhang, and Guangfeng Wang. 2009. " A Novel Hydrazine Electrochemical Sensor Based on a Carbon Nanotube-Wired ZnO Nanoflower-Modified Electrode" Electrochimica Acta 55: 178–82. https://doi.org/10.1016/j.electacta.2009.08.036.

Fang, Bin, Yuehua Feng, Min Liu, Guangfeng Wang, Xiaojun Zhang, and Meifang Wang. 2011. "Electrocatalytic Oxidation of Hydrazine at a Glassy Carbon Electrode Modified with Nickel Ferrite and Multi-Walled Carbon Nanotubes." *Microchimica Acta* 175 (1–2): 145–50. https://doi.org/10.1007/s00604-011-0662-8.

Gao, Feng, Qingxiang Wang, Ningning Gao, Yizhen Yang, Fuxian Cai, Mayoka Yamane, Fei Gao, and Hidekazu Tanaka. 2017. "Hydroxyapatite/Chemically Reduced Graphene Oxide Composite: Environment-Friendly Synthesis and High-Performance Electrochemical Sensing for Hydrazine." *Biosensors and Bioelectronics* 97 (June): 238–45. https://doi.org/10.1016/j.bios.2017.06.005.

Ghasemi, Shahram, Sayed Reza Hosseini, Faeze Hasanpoor, and Shima Nabipour. 2019. "Amperometric Hydrazine Sensor Based on the Use of Pt-Pd Nanoparticles Placed on Reduced Graphene Oxide Nanosheets." *Microchimica Acta* 186 (9): 601. https://doi.org/10.1007/s00604-019-3704-2.

Kim, Seul Ki, Yu Na Jeong, Mohammad Shamsuddin Ahmed, Jung Min You, Hyun Chul Choi, and Seungwon Jeon. 2011. "Electrocatalytic Determination of Hydrazine by a Glassy Carbon Electrode Modified with PEDOP/MWCNTs-Pd Nanoparticles." *Sensors and Actuators, B: Chemical* 153 (1): 246–51. https://doi.org/10.1016/j.snb.2010.10.039.

Kim, Sung Phil, Seul Gi Lee, Myong Yong Choi, and Hyun Chul Choi. 2015. "Highly Sensitive Hydrazine Chemical Sensor Based on CNT-PdPt Nanocomposites." *Journal of Nanomaterials* 2015. https://doi.org/10.1155/2015/120485.

Krittayavathananon, Atiweena, Pattarachai Srimuk, Santamon Luanwuthi, and Montree Sawangphruk. 2014. "Palladium Nanoparticles Decorated on Reduced Graphene Oxide Rotating Disk Electrodes toward Ultrasensitive Hydrazine Detection: Effects of Particle Size and Hydrodynamic Diffusion." *Analytical Chemistry* 86 (24): 12272–78. https://doi.org/10.1021/ac503446q.

Kumunda, Coster, Abolanle S. Adekunle, Bhekie B. Mamba, Ntuthuko W. Hlongwa, and Thabo T. I. Nkambule. 2021. "Electrochemical Detection of Environmental Pollutants Based on Graphene Derivatives: A Review." *Frontiers in Materials* 7 (February). https://doi.org/10.3389/fmats.2020.616787.

Li, Xinchun, Zuanguang Chen, Yuwen Zhong, Fan Yang, Jianbin Pan, and Yajing Liang. 2012. "Cobalt Hexacyanoferrate Modified Multi-Walled Carbon Nanotubes/Graphite Composite Electrode as Electrochemical Sensor on Microfluidic Chip." *Analytica Chimica Acta* 710: 118–24. https://doi.org/10.1016/j.aca.2011.10.035.

Liang, Cuiyuan, Huiming Lin, Wei Guo, Xing Lu, Dexin Yu, Songjie Fan, Feng Zhang, and Fengyu Qu. 2021. "Amperometric Sensor Based on ZIF/g-C3N4/RGO Heterojunction Nanocomposite for Hydrazine Detection." *Microchimica Acta* 188 (2). https://doi.org/10.1007/s00604-021-04711-6.

Links. 2012. "*Petalage-like Graphene Hybrid Microstructures for Hydrazine Sensing*," Analyst, 137: 4841–48. https://doi.org/10.1039/c2an35672d.

Liu, Jiang, Jianze Jiang, Yandong Dou, Fangfang Zhang, Xin Liu, Jiaojiao Qu, and Qing Zhu. 2019, A novel chemiluminescent probe for hydrazine detection in water and HeLa cells, *Organic & Biomolecular Chemistry,* 17: 6975-6979, doi: 10.1039/c9ob01 407a.

Luan, Feng, Shuang Zhang, Dandan Chen, Kun Zheng, and Xuming Zhuang. 2018. "CoS2-Decorated Ionic Liquid-Functionalized Graphene as a Novel Hydrazine Electrochemical Sensor." *Talanta* 182: 529–35. https://doi.org/10.1016/j.talanta.2018.02.031.

Ma, Xiaoqing, Kang-lai Tang, Kang Lu, Chenke Zhang, Wenbing Shi, and Wenxi Zhao. 2021. "*Structural Engineering of Hollow Micro Fl Ower-like CuS @ C Hybrids as Versatile Electrochemical Sensing Platform for Highly Sensitive Hydrogen Peroxide and Hydrazine Detection.*" https://doi.org/10.1021/acsami.1c11747.

Maduraiveeran, Govindhan, and Wei Jin. 2021. "Carbon Nanomaterials: Synthesis, Properties and Applications in Electrochemical Sensors and Energy Conversion Systems." *Materials Science and Engineering B: Solid-State Materials for Advanced Technology* 272 (January): 115341. https://doi.org/10.1016/j.mseb.2021.115341.

Mejri, Alma, and Abdelmoneim Mars. 2019. "*Voltammetric Simultaneous Quantification of P-Nitrophenol and Hydrazine by Using Magnetic Spinel FeCo 2 O 4 Nanosheets on Reduced Graphene Oxide Layers Modified with Curcumin-Stabilized Silver Nanoparticles.*"

Miao, Ruiyang, and Richard G. Compton. 2021. "The Electro-Oxidation of Hydrazine: A Self-Inhibiting Reaction." *Journal of Physical Chemistry Letters* 12 (6): 1601–5. https://doi.org/10.1021/acs.jpclett.1c00070.

Miao, Ruiyang, Lifu Chen, and Richard G. Compton. 2021. "Electro-Oxidation of Hydrazine Shows Marcusian Electron Transfer Kinetics." *Science China Chemistry* 64 (2): 322–29. https://doi.org/10.1007/s11426-020-9889-1.

Mohammad, Akbar, Mohammad Ehtisham Khan, and Moo Hwan Cho. 2020. "Sulfur-Doped-Graphitic-Carbon Nitride (S-g-C3N4) for Low Cost Electrochemical Sensing of Hydrazine." *Journal of Alloys and Compounds* 816: 152522. https://doi.org/10.1016/j.jallcom.2019.152522.

Mohammad, Akbar, Mohammad Ehtisham Khan, Md Rezaul Karim, Moo Hwan Cho, and Taeho Yoon. 2021. "Ag-Modified SnO2-Graphitic-Carbon Nitride Nanostructures for Electrochemical Sensor Applications." *Ceramics International* 47 (16): 23578–89. https://doi.org/10.1016/j.ceramint.2021.05.076.

Mutyala, Sankararao, and Jayaraman Mathiyarasu. 2015. "Preparation of Graphene Nanoflakes and Its Application for Detection of Hydrazine." *Sensors and Actuators, B: Chemical* 210: 692–99. https://doi.org/10.1016/j.snb.2015.01.033.

Nemakal, Manjunatha, Shambhulinga Aralekallu, Imadadulla Mohammed, and Sreenivasa Swamy. 2019. "Electropolymerized Octabenzimidazole Phthalocyanine as an Amperometric Sensor for Hydrazine." *Journal of Electroanalytical Chemistry* 839 (March): 238–46. https://doi.org/10.1016/j.jelechem.2019.03.050.

Patel, Kapil D, and Hae-won Kim. 2019. "*Materials Horizons Platform for Theranostics.*" https://doi.org/10.1039/c8mh00966j.

Phatsawit Wuamprakhon, Atiweena Krittayavathananon, Nattapol Ma, Nutthaphon Phattharasupakun, Thana Maihom, Jumras Limtrakul, Montree Sawangphruk. 2018. "Layered Manganese Oxide Nanosheets Coated on N-Doped Graphene Aerogel for Hydrazine Detection: Reaction Mechanism Investigated by in Situ Electrochemical X-Ray Absorption Spectroscopy." Journal of Electroanalytical Chemistry 808 (November 2017): 124–32. https://doi.org/10.1016/j.jelechem.2017. 12.003.

Promsuwan, Kiattisak, Jariya Thongtawat, and Warakorn Limbut. 2020. "*Microspheres / Graphene Nanoplatelet – Modified Electrode for Flow-Based-Amperometric Hydrazine Sensor.*"

Rahman, M. M., J. Ahmed, A. M. Asiri, I. A. Siddiquey and M. A. Hasnat. 2016, Development of ultra-sensitive hydrazine sensor based on facile CoS2-CNT nanocomposites, *RSC Advances*, 6: 90470-90479, doi: 10.1039/C6RA08772H.

Ramanathan, Subramanian, Elaiyappillai Elanthamilan, Asir Obadiah, Arulappan Durairaj, Palanisamy SanthoshKumar, Johnson Princy Merlin, Subramanian Ramasundaram, and Samuel Vasanthkumar. 2019. "HRGO–Co@SnO2 Nanocomposite for Electrochemical Detection of Hydrazine." *Journal of Electronic Materials* 48 (1): 542–50. https://doi.org/10.1007/s11664-018-6738-z.

Ramanujam, Kothandaraman, and Thippani Thirupathi. 2018. "Carbon Supported G-C3N4 for Electrochemical Sensing of Hydrazine." *Electrochemical Energy Technology* 4 (1): 21–31. https://doi.org/10.1515/eetech-2018-0003.

Rani, Sushma, Shivani Kapoor, Bharti Sharma, Sandeep Kumar, Rajesh Malhotra, and Neeraj Dilbaghi. 2020. *Fabrication of Zn-MOF@rGO Based Sensitive Nanosensor for the Real Time Monitoring of Hydrazine*. *Journal of Alloys and Compounds*. Vol. 816 : 152509. https://doi.org/10.1016/j.jallcom.2019.152509.

Sakthinathan, Subramanian, Subbiramaniyan Kubendhiran, Shen Ming Chen, Mani Govindasamy, Fahad M.A. Al-Hemaid, M. Ajmal Ali, P. Tamizhdurai, and S. Sivasanker. 2017. "Metallated Porphyrin Noncovalent Interaction with Reduced Graphene Oxide-Modified Electrode for Amperometric Detection of Environmental

Pollutant Hydrazine." *Applied Organometallic Chemistry* 31 (9): 1–10. https://doi.org/10.1002/aoc.3703.

Sha, Rinky, S. Solomon Jones, Nandimalla Vishnu, Bhuvaneswari Soundiraraju, and Sushmee Badhulika. 2018. "A Novel Biomass Derived Carbon Quantum Dots for Highly Sensitive and Selective Detection of Hydrazine." *Electroanalysis* 30 (10): 2228–32. https://doi.org/10.1002/elan.201800255.

Shalini Devi, K. S., Madhurantakam Sasya, and Uma Maheswari Krishnan. 2020. "Emerging Vistas on Electrochemical Detection of Diabetic Retinopathy Biomarkers." *TrAC - Trends in Analytical Chemistry* 125: 115838. https://doi.org/10.1016/j.trac.2020.115838.

Simões, F R, and M G Xavier. 2017. *6 - Electrochemical Sensors. Nanoscience and Its Applications*. Elsevier Inc. https://doi.org/10.1016/B978-0-323-49780-0/00006-5.

Srinidhi, G., S. Sudalaimani, K. Giribabu, S. J.Sardhar Basha, and C. Suresh. 2020. "Amperometric Determination of Hydrazine Using a CuS-Ordered Mesoporous Carbon Electrode." *Microchimica Acta* 187 (6) : 359. https://doi.org/10.1007/s00604-020-04325-4.

Tajik, Somayeh, Hadi Beitollahi, Rahman Hosseinzadeh, Abbas Aghaei Afshar, Rajender S. Varma, Ho Won Jang, and Mohammadreza Shokouhimehr. 2021. "Electrochemical Detection of Hydrazine by Carbon Paste Electrode Modified with Ferrocene Derivatives, Ionic Liquid, and CoS2-Carbon Nanotube Nanocomposite." *ACS Omega* 6 (7): 4641–48. https://doi.org/10.1021/acsomega.0c05306.

Tajik, Somayeh, Mohammad B Askari, Sayed A Ahmadi, Fraiba G Nejad, Zahra Dourandish, Razieh Razavi, Hadi Beitollahi, and Antonio Di Bartolomeo. 2022. "Electrochemical Sensor Based on ZnFe2O4/RGO Nanocomposite for Ultrasensitive Detection of Hydrazine in Real Samples." *Nanomaterials,* 12(3):491 . https://doi.org/10.3390/ nano12030491.

Tran T. Tung, Kumud M. Tripathi. 2019. "*Carbon Nanomaterial Sensors for Cancer and Disease Diagnosis*." https://doi.org/10.1002/9781119373476.ch8.

Vinoth, S., K. S. Shalini Devi, and A. Pandikumar. 2021. "A Comprehensive Review on Graphitic Carbon Nitride Based Electrochemical and Biosensors for Environmental and Healthcare Applications." *TrAC - Trends in Analytical Chemistry* 140: 116274. https://doi.org/10.1016/j.trac.2021.116274.

Wei, Qinghe, Qingqing Zhang, Hongwei Gu, Xiang Gao, and Bin Qi. 2015. "Ultra Sensitive Voltammetric Determination of Hydrazine with Curcumin - Ordered Mesoporous Carbon Composite as Sensing Material." *International Journal of Electrochemical Science* 10 (9): 7083–90.

Wuamprakhon, Phatsawit, Atiweena Krittayavathananon, Nattapol Ma, Nutthaphon Phattharasupakun, Thana Maihom, Jumras Limtrakul, and Montree Sawangphruk. 2017. "Layered Manganese Oxide Nanosheets Coated on N-Doped Graphene Aerogel for Hydrazine Detection: Reaction Mechanism Investigated by in Situ Electrochemical X-Ray Absorption Spectroscopy." *Journal of Electroanalytical Chemistry* 808 (December). https://doi.org/10.1016/j.jelechem.2017.12.003.

Xinjin Zhang and Jianbin Zheng. 2020. "Amperometric Hydrazine Sensor Based on the Use of a Gold Nanoparticle-Modified Nanocomposite Consisting of Porous

Polydopamine, Multiwalled Carbon Nanotubes and Reduced Graphene Oxide." *Microchimica Acta* 187 (1) : 89. https://doi.org/10.1007/s00604-019-4014-4.

Yang, Jie, Lu Yang, Xin Tang, Yu Zhang, Qiaoyan Dong, Zhiyuan He, Nihong Li, Ke Huang, Hong Luo, and Xiaoli Xiong. 2022. "ZIF Derived N-CoS2@graphene Rhombic Dodecahedral Nanocomposites: As a High Sensitivity Sensor for Hydrazine." *Sensors and Actuators B: Chemical* 351: 130967. https://doi.org/https://doi.org/10.1016/j.snb.2021.130967.

Yang, Ziyin, Xiaohui Zheng, and Jianbin Zheng. 2017. "Facile Synthesis of Three-Dimensional Porous Au@Pt Core-Shell Nanoflowers Supported on Graphene Oxide for Highly Sensitive and Selective Detection of Hydrazine." *Chemical Engineering Journal* 327: 431–40. https://doi.org/10.1016/j.cej.2017.06.120.

Zhang, Congcong, and Xin Du. 2020. "*Electrochemical Sensors Based on Carbon Nanomaterial Used in Diagnosing Metabolic Disease*" 8 (August): 1–8. https://doi.org/10.3389/fchem.2020.00651.

Zhang, Cuihong, Guangfeng Wang, Yulan Ji, Min Liu, Yuehua Feng, Zhidan Zhang, and Bin Fang. 2010. "Enhancement in Analytical Hydrazine Based on Gold Nanoparticles Deposited on ZnO-MWCNTs Films." *Sensors and Actuators, B: Chemical* 150 (1): 247–53. https://doi.org/10.1016/j.snb.2010.07.007.

Zhang, Xinjin, and Jianbin Zheng. 2019. "Controllable Synthesis of Highly Active Au@Ni Nanocatalyst Supported on Graphene Oxide for Electrochemical Sensing of Hydrazine." *Applied Surface Science* 493 (July): 1159–66. https://doi.org/10.1016/j.apsusc.2019.07.136.

Zhang, Yong, Ying Zhang, Ding Zhang, Sida Li, Chao Jiang, and Yi Su. 2019. "Confinement Preparation of Au Nanoparticles Embedded in ZIF-67-Derived N-Doped Porous Carbon for High-Performance Detection of Hydrazine in Liquid/Gas Phase." *Sensors and Actuators, B: Chemical* 285 (January): 607–16. https://doi.org/10.1016/j.snb.2019.01.105.

Zhao, Zhenting, Yongjiao Sun, Pengwei Li, Wendong Zhang, Kun Lian, Jie Hu, and Yong Chen. 2016. "Preparation and Characterization of AuNPs/CNTs-ErGO Electrochemical Sensors for Highly Sensitive Detection of Hydrazine." *Talanta* 158: 283–91. https://doi.org/10.1016/j.talanta.2016.05.065.

Zheng, Li, and Jun-feng Song. 2009. "*Sensors and Actuators B : Chemical Curcumin Multi-Wall Carbon Nanotubes Modified Glassy Carbon Electrode and Its Electrocatalytic Activity towards Oxidation of Hydrazine*" 135: 650–55. https://doi.org/10.1016/j.snb.2008.09.035.

Zhou, Huanfang, Limin Chen, Shuting Li, Songqing Huang, Yingying Sun, Yanhong Chen, Zhao Wang, Wei Liu, and Xinchun Li. 2020. "One-Step Electroreduction Preparation of Multilayered Reduced Graphene Oxide/Gold-Palladium Nanohybrid as a Proficient Electrocatalyst for Development of Sensitive Hydrazine Sensor." *Journal of Colloid and Interface Science* 566: 473–84. https://doi.org/10.1016/j.jcis. 2020.01.105.

Zhou, Tuantuan, Wanlin Gao, Yanshan Gao, Qiang Wang, and Ahmad Umar. 2017. Co3O4 nanoparticles/MWCNTs composites: a potential scaffold for hydrazine and glucose electrochemical detection, *RSC Advances*, 7: 50087–96. https://doi.org/10.1039/C7RA10892C.

Bibliography

An introduction to bioreactor hydrodynamics and gas-liquid mass transfer

LCCN	2013048397
Type of material	Book
Personal name	Kadic, Enes, 1973-
Main title	An introduction to bioreactor hydrodynamics and gas-liquid mass transfer / Enes Kadic, Theodore J. Heindel.
Published/Produced	Hoboken, New Jersey: Wiley, 2014.
Description	1 online resource.
Links	Cover image http://catalogimages.wiley.com/images/db/jimages/9781118104019.jpg
ISBN	9781118869895 (pdf) 9781118869833 (epub)
LC classification	TP248.25.B55
Related names	Heindel, Theodore J., 1969-
Summary	"This book reviews and compares the major types of bioreactors used to produce renewable fuels, chemicals, medicines, and proteins, by providing an overview of the hydrodynamics and gas-liquid mass transfer operations in this equipment. These operations are important because they influence the quality and quantity of the desired material produced in the reactor. The text also discusses advantages and disadvantages of each bioreactor and provides a procedure for optimal bioreactor selection based on current process needs, giving chemical and mechanical engineers a practical, working reference"-- Provided by publisher. "This book provides an overview of hydrodyanmics and gas-liqiud mass transfer operations for various bioreactor configurations. It provides a summary of the available information for various bioreactors, reviews the relevant theories and experimental

procedures, and provide a procedure for bioreactor selection"-- Provided by publisher.

Contents 1 Introduction 1 2 Modes of Operation 3 2.1 Batch Bioreactors 3 2.2 Continuous Bioreactors 9 2.3 Summary 15 3 Gas-Liquid Mass Transfer Models 17 4 Experimental Measurement Techniques 28 4.1 Measuring Bioreactor Hydrodynamic Characteristics 28 4.1.1 Flow regime measurements 29 4.1.2 Local pressure drop 30 4.1.3 Mixing or residence time 32 4.1.4 Axial diffusion coefficient 33 4.1.5 Gas-liquid interfacial area 34 4.1.6 Bubble size and velocity 35 4.1.7 Global and local liquid velocity 37 4.1.8 Gas holdup 40 4.1.8.1 Bed expansion 41 4.1.8.2 Pressure drop measurements 41 4.1.8.3 Dynamic gas disengagement (DGD) 46 4.1.8.4 Tomographic techniques 47 4.1.9 Liquid holdup 50 4.1.10 Power measurements 51 4.2 Gas-Liquid Mass Transfer 53 4.2.1 Dissolved oxygen measurement techniques 54 4.2.1.1 Chemical method 54 4.2.1.2 Volumetric method 56 4.2.1.3 Tubing method 56 4.2.1.4 Optode method 57 4.2.1.5 Electrochemical electrode method 58 4.2.1.5.1 Polarographic electrodes 59 4.2.1.5.2 Galvanic probes 61 4.2.1.5.3 Electrochemical electrode time constant 61 4.2.1.5.4 Electrochemical electrode response time (τe) 64 4.2.1.5.5 Electrochemical electrode response models 66 4.2.1.5.6 Summary of electrochemical electrode response models 72 4.2.2 Dissolved carbon monoxide measurements 72 4.2.2.1 Bioassay overview 74 4.2.2.2 Needed materials 75 4.2.2.3 Liquid sample collection 76 4.2.2.4 Identifying the concentrated myoglobin solution concentration 77 4.2.2.5 Sample preparation for analysis 78 4.2.2.6 Determining the dissolved CO concentration 79 4.2.3 Determining volumetric gas-liquid mass transfer coefficient, kLa 80 4.2.3.1 Gas balance method 81 4.2.3.2 Dynamic method 82 4.2.3.2.1 Biological dynamic method 82 4.2.3.2.2 Non-biological dynamic method 85 4.2.3.2.3 Variations of the inlet

	Sparger Design 272 8.5 Correlations 277 8.6 Needed Research 280 8.7 Summary 284 9 Fixed Bed Bioreactors 295 9.1 Introduction 295 9.2 Column Geometry and Components 299 9.3 Flow Regime 307 9.4 Liquid Properties 314 9.5 Packing Material 316 9.5.1 Random Packing 319 9.5.2 Structured Packing 321 9.6 Biological Considerations 324 9.7 Correlations 325 9.8 Needed Research 327 9.9 Summary 328 10 Novel Bioreactors 333 10.1 Introduction 333 10.2 Novel Bubble-Induced Flow Designs 333 10.3 Miniaturized Bioreactors 341 10.3.1 Microreactors 343 10.3.2 Nanoreactors 348 10.4 Membrane Reactor 349 10.5 Summary 353 11 Figures of Merit 355 12 Concluding Remarks 363 13 Nomenclature 367 Abbreviations 375 Greek Symbols 377 Dimensionless numbers 379 14 Bibliography 382 .
LC Subjects	Bioreactors. Bioreactors--Fluid dynamics.
Other Subjects	Science / Mechanics / Dynamics / Fluid Dynamics.
Notes	Includes bibliographical references and index.
Additional formats	Print version: Kadic, Enes, 1973- Introduction to bioreactor hydrodynamics and gas-liquid mass transfer Hoboken, New Jersey: Wiley, 2014 9781118104019 (DLC) 2013042733

An introduction to bioreactor hydrodynamics and gas-liquid mass transfer

LCCN	2013042733
Type of material	Book
Personal name	Kadic, Enes, 1973-
Main title	An introduction to bioreactor hydrodynamics and gas-liquid mass transfer / Enes Kadic and Theodore J. Heindel, Department of Mechanical Engineering, Iowa State University.
Published/Produced	Hoboken, New Jersey: Wiley, [2014] ©2014
Description	x, 314 pages: illustrations; 25 cm

Links	Cover image http://catalogimages.wiley.com/images/db/jimages/9781118104019.jpg
ISBN	9781118104019 (hardback) 1118104013 (hardback)
LC classification	TP248.25.B55 K33 2014
Related names	Heindel, Theodore J., 1969-
Summary	"This book reviews and compares the major types of bioreactors used to produce renewable fuels, chemicals, medicines, and proteins, by providing an overview of the hydrodynamics and gas-liquid mass transfer operations in this equipment. These operations are important because they influence the quality and quantity of the desired material produced in the reactor. The text also discusses advantages and disadvantages of each bioreactor and provides a procedure for optimal bioreactor selection based on current process needs, giving chemical and mechanical engineers a practical, working reference"-- Provided by publisher. "This book provides an overview of hydrodyanmics and gas-liqiud mass transfer operations for various bioreactor configurations. It provides a summary of the available information for various bioreactors, reviews the relevant theories and experimental procedures, and provide a procedure for bioreactor selection"-- Provided by publisher.
Contents	1 Introduction 1 2 Modes of Operation 3 2.1 Batch Bioreactors 3 2.2 Continuous Bioreactors 9 2.3 Summary 15 3 Gas-Liquid Mass Transfer Models 17 4 Experimental Measurement Techniques 28 4.1 Measuring Bioreactor Hydrodynamic Characteristics 28 4.1.1 Flow regime measurements 29 4.1.2 Local pressure drop 30 4.1.3 Mixing or residence time 32 4.1.4 Axial diffusion coefficient 33 4.1.5 Gas-liquid interfacial area 34 4.1.6 Bubble size and velocity 35 4.1.7 Global and local liquid velocity 37 4.1.8 Gas holdup 40 4.1.8.1 Bed expansion 41 4.1.8.2 Pressure drop measurements 41 4.1.8.3 Dynamic gas

	Nomenclature 367 Abbreviations 375 Greek Symbols 377 Dimensionless numbers 379 14 Bibliography 382 .
LC Subjects	Bioreactors. Bioreactors--Fluid dynamics.
Other Subjects	Science / Mechanics / Dynamics / Fluid Dynamics.
Notes	Includes bibliographical references (pages 275-310) and index.
Additional formats	Online version: Kadic, Enes, 1973- Introduction to bioreactor hydrodynamics and gas-liquid mass transfer Hoboken, New Jersey: Wiley, 2014 9781118869895 (DLC) 2013048397

Fluorophores: characterization, synthesis and applications

LCCN	2019716006
Type of material	Book
Personal name	Watanabe, Shota, author.
Main title	Fluorophores: characterization, synthesis and applications / Shota Watanabe.
Published/Produced	Hauppauge, New York: Nova Science Publishers, Incorporated, [2013]
Description	1 online resource.
ISBN	9781628082692 soft cover
LC classification	TP245.F6
Contents	Fluorescent probes based on upconverting lanthanide nanostructures / Valerio Voliani, Julia Pérez-Prieto - Applications of fluorescent molecular probes in cell biology / Xin Zhang, Jing Liu, Wenchao Wang, Yundan Jia, Junfeng Wang, Gingsong Liu - Aromatic hydrazine-based fluorophores: synthesis, spectroscopy and bioapplications / Ozlem Dilek, Susan Bane - Concentration dependence of humic acid fluorescence intensity / N.I Lavrik, N.U. Mulloev.
LC Subjects	Fluorine compounds. Fluorescent probes. Fluorescence spectroscopy.
Notes	Includes bibliographical references and index.

Additional formats	Print version: Fluorophores Hauppauge, New York: Nova Science Publishers, Inc., [2013] 9781628082685 (DLC) 2013021115
Series	Chemical engineering methods and technology

Fluorophores: characterization, synthesis and applications

LCCN	2013021115
Type of material	Book
Personal name	Watanabe, Shota, author.
Main title	Fluorophores: characterization, synthesis and applications / Shota Watanabe.
Published/Produced	Hauppauge, New York: Nova Science Publishers, Inc., [2013]
Description	ix, 102 pages: illustrations; 23 cm.
ISBN	9781628082685 (soft cover)
LC classification	TP245.F6 W38 2013
Contents	Fluorescent probes based on upconverting lanthanide nanostructures / Valerio Voliani, Julia Pérez-Prieto - Applications of fluorescent molecular probes in cell biology / Xin Zhang, Jing Liu, Wenchao Wang, Yundan Jia, Junfeng Wang, Gingsong Liu - Aromatic hydrazine-based fluorophores: synthesis, spectroscopy and bioapplications / Ozlem Dilek, Susan Bane - Concentration dependence of humic acid fluorescence intensity / N.I Lavrik, N.U. Mulloev.
LC Subjects	Fluorine compounds. Fluorescent probes. Fluorescence spectroscopy.
Notes	Includes bibliographical references and index.
Series	Chemical engineering methods and technology

Formation of hydrazine as a chloramine by-product

LCCN	2007299335
Type of material	Book
Main title	Formation of hydrazine as a chloramine by-product / prepared by Issam Najm ... [et al.]; sponsored by Awwa Research Foundation.

Published/Created	Denver, CO: AWWA Research Foundation, American Water Works Association, IWA Pub., c2006.
Description	xviii, 51 p.: ill.; 28 cm.
Links	Publisher description http://www.loc.gov/catdir/enhancements/fy1305/2007299335-d.html
	Table of contents only http://www.loc.gov/catdir/enhancements/fy1305/2007299335-t.html
ISBN	1583214461
	9781583214466
LC classification	TD462 .F676 2006
Related names	Najm, Issam N.
	AWWA Research Foundation.
	American Water Works Association.
	IWA Publishing.
LC Subjects	Water--Purification--Chloramination--By-products.
	Chloramines.
	Hydrazine.
Notes	Includes bibliographical references (p. 49-50).

Functional Polymer Blends and Nanocomposites: A Practical Engineering Approach

LCCN	2014935853
Type of material	Book
Main title	Functional Polymer Blends and Nanocomposites: A Practical Engineering Approach / edied by Gennady E. Zaikov, DSc. Liliya I. Bazylak, PhD, and A.K. Haghi, PhD.
Published/Produced	Tornoto: Apple Academic Press Inc., 2014.
Description	xviii, 349 pages; 24 cm
ISBN	9781926895895
	1926895894
LC classification	TA418.9.N35 .F85 2014
Related names	Zaikov, Gennady E., editor.
	Bazylak, Liliya I., editor.
	Haghi, A. K., editor.
	Zaikov, Gennady E.
	Bazylyak, Liliya I.
	Haghi, A. K.

Contents Distribution of internal links of the polymer chain in the self-avoiding random walks statistics / Yu. G. Medvedevskikh - Polymeric stars in self-avoiding random walks statistics / Yu. G. Medvedevskikh, L. I. Bazylyak, and G. E. Zaikov - The number of configurations of polymeric chain in the self-avoiding random walks statistics / Yu. G. Medvedevskikh - Conformation of linear polymeric chains at the interface layer liquid / solid and adsorption isotherm / Yu. G. Medvedevskikh, and G. I. Khovanets - Phenomenological coefficients of the viscosity for low-molecular elementary liquids and solutions / Yu. G. Medvedevskikh, and O. Yu. Khavunko - Viscoelastic properties of the polystyrene in concentrated solutions and melts / Yu. G. Medvedevskikh, O. Yu. Khavunko, L. I. Bazylyak, and G. E. Zaikov - Polyaniline / Nano-TiO2-S hybride composites / M. M. Yatsyshyn, A. S. Kun'ko, and O.V. Reshetnyak - Simulation of corrosive dissolution of PtBinary nano-cluster in acid environment of polymer electrolyte membrane (PEM) fuel cells / S. A. Korniy, V. I. Kopylets', and V. I. Pokhmurskiy - Effect of amino naphthalene sulfonic acid nature on the structure and physical properties of their copolymers with aniline / O. I. Aksimentyeva, and V. P. Dyakonov - UV / Vis-spectra of silver nanoparticles as characteristics of their sizes and sizes distribution / A. R. Kytsya, O. V. Reshetnyak, L. I. Bazylak, and Yu. M. Hrynda - SynthesisandAnticorrosionActivity of the zinc phosphate nanoparticles / L. I. Bazylyak, A. R. Kytsya, I. M. Zin', and S. A. Korniy - An influence of kinetic parameters of reaction on the size of obtained nanoparticles under reduction of silver ions by hydrazine / A. R. Kytsya, L. I. Bazylyak, Yu. M. Hrynda, and Yu. G. Medvedevskikh - Case studies on nanopolymrs and their chemicals complexity -new approaches, limitations and control / A. K. Haghi, and G. E. Zaikov.

LC Subjects	Nanocomposites (Materials)--Research. Conducting polymers--Congresses.
Notes	Includes bibliographical references and index.
Series	AAP Research Notes on Polymer Engineering Science & Technology AAP Research Notes on Polymer Engineering Science & Technology.

Hydrazine and its derivatives: preparation, properties, applications

LCCN	2001026195
Type of material	Book
Personal name	Schmidt, Eckart W. (Eckart Walter), 1935-
Main title	Hydrazine and its derivatives: preparation, properties, applications / Eckart W. Schmidt.
Edition	2nd ed.
Published/Created	New York: Wiley-Interscience, 2001.
Description	2 v. (liii, 2121 p.): ill.; 25 cm.
Links	Contributor biographical information http://www. loc. gov/catdir/bios/wiley043/2001026195.html Publisher description http://www.loc.gov/catdir/ description/wiley035/2001026195.html Table of Contents http://www.loc.gov/catdir/ toc/onix07/2001026195.html
ISBN	0471415537 (set: cloth)
LC classification	TP248.H5 S36 2001
LC Subjects	Hydrazine.
Notes	Includes bibliographical references (p. 1639-2080) and index.

Hydrazines and cancer: a guidebook on the carcinogenic activities of hydrazines, related chemicals, and hydrazine-containing natural products

LCCN	2002421271
Type of material	Book
Personal name	Tóth, Béla.
Main title	Hydrazines and cancer: a guidebook on the carcinogenic activities of hydrazines, related chemicals, and hydrazine-containing natural products / Bela Toth.

Published/Created	Amsterdam: Harwood Academic Publishers, c2000.
Description	x, 243 p.: ill.; 26 cm.
Links	Table of contents http://www.loc.gov/catdir/toc/fy031/2002421271.html Publisher description http://www.loc.gov/catdir/enhancements/fy0745/2002421271-d.html
ISBN	9057026317
LC classification	RC268.7.H88 T68 2000
LC Subjects	Hydrazines--Carcinogenicity.
Other Subjects	Hydrazines--toxicity. Neoplasms--chemically induced. Hydrazines--Carcinogenicity.
Notes	Includes bibliographical references.

Inorganic hydrazine derivatives: synthesis, properties, and applications

LCCN	2013038588
Type of material	Book
Main title	Inorganic hydrazine derivatives: synthesis, properties, and applications / edited by K.C. Patil and Tanu Mimani Rattan.
Published/Produced	Chichester, West Sussex, United Kingdom: John Wiley & Sons Inc., 2014.
Description	1 online resource.
ISBN	9781118693568 (ePub) 9781118693582 (Adobe PDF)
LC classification	QD181.N1
Related names	Patil, K. C., editor of compilation. Rattan, Tanu Mimani, editor of compilation.
Summary	"Inorganic Hydrazine Derivatives: Synthesis, Properties and Applications presents a comprehensive review of the research carried out in this field during the last four decades"-- Provided by publisher.
LC Subjects	Hydrazines. Hydrazines--Industrial applications.
Other Subjects	Science / Chemistry / Inorganic.
Notes	Includes index. Includes bibliographical references.

Additional formats	Print version: Inorganic hydrazine derivatives Chichester, West Sussex, United Kingdom: John Wiley & Sons Inc., 2014 9781118715130 (DLC) 2013037822

Inorganic hydrazine derivatives: synthesis, properties, and applications

LCCN	2013037822
Type of material	Book
Main title	Inorganic hydrazine derivatives: synthesis, properties, and applications / edited by K.C. Patil, Department of Inorganic and Physical Chemistry, Indian Institute of Science, Bangalore, India and Tanu Mimani Rattan, Department of Physics, Sri Sathya Sai Institute of Higher Learning, Prasanthi Nilayam, India.
Published/Produced	Chichester, West Sussex, United Kingdom: Wiley, 2014.
Description	xxii, 260 pages: illustrations; 24 cm
ISBN	9781118715130 (hardback) 1118715136 (hardback)
LC classification	QD181.N1 I57 2014
Related names	Patil, K. C., editor of compilation. Rattan, Tanu Mimani, editor of compilation.
Summary	"Inorganic Hydrazine Derivatives: Synthesis, Properties and Applications presents a comprehensive review of the research carried out in this field during the last four decades"-- Provided by publisher.
LC Subjects	Hydrazines. Hydrazines--Industrial applications.
Other Subjects	Science / Chemistry / Inorganic.
Notes	Includes bibliographical references and index.
Additional formats	Online version: Inorganic hydrazine derivatives Chichester, West Sussex, United Kingdom: John Wiley & Sons Inc., 2014 9781118693568 (DLC) 2013038588

Nuclear Radiation Nanosensors and Nanosensory Systems

LCCN	2019754566
Type of material	Book
Main title	Nuclear Radiation Nanosensors and Nanosensory Systems / edited by Paata J. Kervalishvili, Panayotis H. Yannakopoulos.
Edition	1st ed. 2016.
Published/Produced	Dordrecht: Springer Netherlands: Imprint: Springer, 2016.
Description	1 online resource (XVI, 200 pages 54 illustrations, 23 illustrations in color.) PDF
ISBN	9789401774680
Related names	Kervalishvili, Paata J. editor. Yannakopoulos, Panayotis H. editor.
Summary	This collection of selected review papers focuses on topics such as digital radiation sensors and nanosensory systems for nanotechnology applications and integrated X-ray/PET/CT detectors; nanophosphors and nanocrystal quantum dots as X-ray radiation sensors; the luminescence efficiency of CdSe/ZnS QD and UV-induced luminescence efficiency distribution; investigations devoted to the quantum and multi-parametrical nature of disasters and the modeling thereof using quantum search and quantum query algorithms; sum-frequency-generation, IR fourier and raman spectroscopy methods; as well as investigations into the vibrational modes of viruses and other pathogenic microorganisms aimed at creating optical biosensory systems. This is followed by a review of radiation resistant semiconductor sensors and magnetic measurement instrumentation for magnetic diagnostics of high-tech fission and fusion set-ups and accelerators; the evaluation of the use of neutron-radiation, 10B-enriched semiconducting materials as thin-film, highly reliable, highly sensitive and fast-acting robust solid-state electronic neutron-detectors; and the irradiation of n-Si crystals with protons,

which converts the "metallic" inclusions to "dielectric" ones in isochronous annealing, therefore leading to opto/micro/nanoelectronic devices, including nuclear radiation nanosensors. The book concludes with a comparative study of the nitride and sulfide chemisorbed layers; a chemical model that describes the formation of such layers in hydrazine-sulfide and water sodium sulfide solution; and recent developments in the microwave-enhanced processing and microwave-assisted synthesis of nanoparticles and nanomaterials using Mn(OH)2.

Contents Preface - Acknowledgments - Quantum Approach to Sensory Information Processing for Modeling of Disasters - Digital radiation sensors and nanosensory systems - Sensors for magnetic fields measurement in high level of penetrating radiation in accelerators and new generation fusion reactors - Formation and annealing of nano-sized atomic clusters in n-Si crystals irradiated with high-energy protons - Luminescence Efficiency of Cadmium Selenide/Zinc Sulfide (CdSe/ZnS) Quantum Dot Nanoparticle sensors under X-ray excitation - Nitride and sulfide chemisored layers as the surface passivants for A3B5 semiconductors - Efficiency of luminescence of (Lu,Gd)2SiO5:Ce (LGSO:Ce) crystal sensory material in the X-ray imaging range - Microwave in environmental technologies and synthesis of nano-materials: The Georgian experience - Microwave enhanced producing of high-purity metallic manganese and composite manganese based alloys - From Chernobyl to Fukushima. The Environment is acting as a friend or an enemy? - Oscillation and optical properties of viruses and other pathogenic microorganisms: A review article - Subject Index - Author Index.

LC Subjects Materials--Surfaces.
Nanoscale science.
Nanoscience.
Nanostructures.

	Radiation protection. Radiation-Safety measures. Thin films.
Other Subjects	Nanoscale Science and Technology. Effects of Radiation/Radiation Protection. Surfaces and Interfaces, Thin Films.
Additional formats	Print version: Nuclear radiation nanosensors and nanosensory systems 9789401774666 (DLC) 2016937709 Printed edition: 9789401774666 Printed edition: 9789401774673 Printed edition: 9789401774727
Series	NATO Science for Peace and Security Series B: Physics and Biophysics, 1874-6500 NATO Science for Peace and Security Series B: Physics and Biophysics, 1874-6500

Quantifying hydrazine in chloraminated water

LCCN	2011277523
Type of material	Book
Personal name	Najm, Issam N., author.
Main title	Quantifying hydrazine in chloraminated water / prepared by: Issam Najm, Nancy Patania Brown, and Karl Gramith, Water Quality & Treatment Solutions, Inc.; sponsored by Water Research Foundation.
Published/Produced	Denver, CO: Water Research Foundation, [2011]
Description	xviii, 64 pages: illustrations; 28 cm
ISBN	9781605731513 160573151X
LC classification	MLCM 2017/44080 (T)
Contents	1. Introduction - 2. Materials and methods - 3. Results and discussion - 4. Quality assurance/quality control - 5. Summary and conclusions - 6. Recommendations to utilities - References.
LC Subjects	Hydrazine. Water--Purification--Chloramination.
Notes	Minimal Level Cataloging Plus. Includes bibliographical references (page 61).

Uranous nitrate production for purex process applications using PtO00 Z 8200 Z00 catalyst and hydrazine nitrate as reductant

LCCN	2004327071
Type of material	Book
Main title	Uranous nitrate production for purex process applications using PtO00 Z 8200 Z00 catalyst and hydrazine nitrate as reductant [microform] / K. Sreenivasa Rao ... [et. al.].
Published/Created	Mumbai: Bhabha Atomic Research Centre, 2003.
Description	ii, 48 p.: ill.; 29 cm.
LC classification	Microfiche 2004/60087 (T)
Related names	Sreenivasa Rao, K. Bhabha Atomic Research Centre.
LC Subjects	Reactor fuel reprocessing--India.
Notes	At head of title: Government of India, Atomic Energy Commission. "BARC report"--Cover. Includes bibliographical references (p. 15.). Abstract in English and Hindi.
Reproduction no./Source	Library of Congress - New Delhi Overseas Office
Additional formats	Microfiche. New Delhi: Library of Congress Office; Washington, D.C.: Library of Congress Photoduplication Service, 2004. 1 microfiche. Master microform held by: DLC.
Series	BARC external; BARC/2003/E/009

Wood Polymer Nanocomposites: Chemical Modifications, Properties and Sustainable Applications

LCCN	2019769678
Type of material	Book
Personal name	Rahman, Md Rezaur, author.
Main title	Wood Polymer Nanocomposites: Chemical Modifications, Properties and Sustainable Applications / by Md Rezaur Rahman.
Edition	1st ed. 2018.
Published/Produced	Cham: Springer International Publishing: Imprint: Springer, 2018.

Description	1 online resource (XV, 314 pages 153 illustrations, 49 illustrations in color.) PDF
ISBN	9783319657356
Summary	This book shows how chemical modifications influence some properties of wood nanocomposites. It describes suitable and effective chemical modifications that strengthen the physico-mechanical, thermal and morphological properties of wood. The authors provide intuitive explanation of the various types of chemical modifications applied to polymer cell walls in wood. They emphasize the reaction changes in wood cell walls due to the chemical modifications. Increased mechanical strength, improved thermal stability as well as the efficient retardancy against fungi attack are described. This book concludes summarizing the potential applications of wood-based nanocomposites taking into account sustainability and economic aspects.
Contents	Introduction and reinforcing potential various clay and monomers dispersed wood nanocomposites - Preparation and characterizations of various clay and monomers dispersed wood nanocomposites - Combined styrene/MMA/nanoclay cross-linker effect on wood-polymer nanocomposites - Oxidation of wood species by sodium meta periodate and impregnation with phenyl hydrazine - Characterization of N, N dimethylacetamid impregnated wood polymer nanocomposites - Mechanical and thermal characterization of urea formaldehyde impregnated wood polymer nanocomposites - Characterization of epoxy/nanoclay/wood polymer nanocomposites - Influence of nanoclay/phenol formaldehyde resin on wood polymer nanocomposites - Clay dispersed styrene-co-glycidyl methacrylate impregnated wood polymer nanocomposites: Impact on mechanical and morphological properties - Physico-mechanical,

morphological and thermal properties of clay dispersed styrene-co-maleic acid impregnated wood polymer nanocomposites - Preparation and characterizations of clay dispersed styrene-co-ethylene glycol dimethyl acrylate impregnated wood polymer nanocomposites - Evaluation of clay dispersed styrene-co-3-trimethyl silyl propyl methacrylate impregnated wood polymer nanocomposites on physical, mechanical, thermal and morphological properties - Investigations on the physico-mechanical, thermal and morphological properties of nanoclay dispersed acrylonitrile-co-butyl methacrylate wood polymer nanocomposites - Studies on the physical, mechanical, thermal and morphological properties of impregnated furfuryl alcohol-co-glycidyl methacrylate/nanoclay wood polymer nanocomposites - Nanoclay dispersed furfuryl alcohol-co-ethyl methacrylate wood polymer nanocomposites: The enhancement on physico-mechanical and thermal properties - Sustainable application of various monomer/clay dispersed wood polymer nanocomposites.

LC Subjects Ceramics.
Composite materials.
Composites (Materials).
Engineering--Materials.
Forest products.
Glass.
Polymers.

Other Subjects Ceramics, Glass, Composites, Natural Materials.
Materials Engineering.
Polymer Sciences.
Wood Science & Technology.

Additional formats Print version: Wood polymer nanocomposites. 9783319657349 (DLC) 2017949506
Printed edition: 9783319657349
Printed edition: 9783319657363
Printed edition: 9783319880945

Series Engineering Materials, 1612-1317

Index

D